Synthese Library

Studies in Epistemology, Logic, Methodology, and Philosophy of Science

Volume 516

The aim of *Synthese Library* is to provide a forum for the best current work in the methodology and philosophy of science and in epistemology, all broadly understood. A wide variety of different approaches have traditionally been represented in the Library, and every effort is made to maintain this variety, not for its own sake, but because we believe that there are many fruitful and illuminating approaches to the philosophy of science and related disciplines.

Special attention is paid to methodological studies which illustrate the interplay of empirical and philosophical viewpoints and to contributions to the formal (logical, set-theoretical, mathematical, information-theoretical, decision-theoretical, etc.) methodology of empirical sciences. Likewise, the applications of logical methods to epistemology as well as philosophically and methodologically relevant studies in logic are strongly encouraged. The emphasis on logic will be tempered by interest in the psychological, historical, and sociological aspects of science. In addition to monographs *Synthese Library* publishes thematically unified anthologies and edited volumes with a well-defined topical focus inside the aim and scope of the book series. The contributions in the volumes are expected to be focused and structurally organized in accordance with the central theme(s), and should be tied together by an extensive editorial introduction or set of introductions if the volume is divided into parts. An extensive bibliography and index are mandatory.

Xavier de Donato-Rodríguez • José L. Falguera
• Concha Martínez-Vidal
Editors

Deflationist Conceptions of Abstract Objects

Springer

Editors
Xavier de Donato-Rodríguez
Department of Philosophy and Anthropology
University of Santiago de Compostela
Santiago de Compostela, Spain

José L. Falguera
Department of Philosophy and Anthropology
University of Santiago de Compostela
Santiago de Compostela, Spain

Concha Martínez-Vidal
Department of Philosophy and Anthropology
University of Santiago de Compostela
Santiago de Compostela, Spain

ISSN 0166-6991 ISSN 2542-8292 (electronic)
Synthese Library
ISBN 978-3-031-98716-8 ISBN 978-3-031-98717-5 (eBook)
https://doi.org/10.1007/978-3-031-98717-5

This work was supported by [Ministerio de Ciencia, Tecnología e Innovación Productiva] ([PID2020-115482GB-I00]).

This Springer imprint is published by the registered company Springer Nature Switzerland AG
The registered company address is: Gewerbestrasse 11, 6330 Cham, Switzerland

Contents

Chapter 1
Introduction

Xavier de Donato-Rodríguez, José L. Falguera, and Concha Martínez-Vidal

Since 2021, the editors of this volume have been addressing various issues related to ontological and metaontological deflationism. *Deflationist Conceptions of Abstract Objects* is one of the milestones of the research project pursued along these years. The volume provides an opportunity to explore various deflationist accounts of abstract objects and to discuss a number of issues in relation to some of the metaontological approaches in which such conceptions are framed.

The question of the existence of putative abstract objects of various kinds, such as numbers, sets, propositions, possible worlds, concepts, works of fiction, characters, scientific theories, scientific models, scientific laws, and others—perhaps even theoretical and observational entities—has recently received various ontological and metaontological deflationist responses. Some of these answers claim to find their roots in Carnap's seminal "Empiricism, Semantics, and Ontology" (1950[1956]); for example, accounts like those of Putnam (2004), Schiffer (1996), Hirsch (2011), Price (2004, 2007, 2009), Thomasson (2014, 2015), among others. But there are other deflationist proposals besides those rooted in Carnap's account. For example, Azzouni's "pure deflationary metaphysics" (2004), Bueno's "agnostic nominalism" (2008), and his recent defense of the contingent character of mathematical objects (2020), or those who advocate metaontological deflationism mainly in the philosophy of mathematics, such as Linnebo's minimalism of 'thin' objects (2018) and Rayo's trivial Platonism (2013). Obviously, each of these responses has raised several objections and important philosophical debates, including considerations that argue that the path leading from Quine's ontological standpoint (1948) to

X. de Donato-Rodríguez · J. L. Falguera · C. Martínez-Vidal (✉)
Department of Philosophy and Anthropology, University of Santiago de Compostela, Santiago de Compostela, Spain
e-mail: xavier.dedonato@usc.es; joseluis.falguera@usc.es; mconcepcion.martinez@usc.es

X. de Donato-Rodríguez et al. (eds.), *Deflationist Conceptions of Abstract Objects*, Synthese Library 516, https://doi.org/10.1007/978-3-031-98717-5_1

Carnap's accounts of ontology (1950 [1956]) is closer than is usually assumed, as, for example, Eklund (2009) or Alspector-Kelly (2001) suggest.[1]

Deflationism is a thesis that can be maintained at the ontological as well as at the *meta*ontological level. While ontology is concerned with "being qua being" (Aristotle) or "what there is" (Quine) —to cite two well-known examples—, meta-ontology encompasses any thinking about ontology. The latter was practiced long before the term was coined. Berto, Kroon, and Voltolini (2014) credit Peter van Inwagen (1998) with introducing the term "meta-ontology" into the analytic tradition, while Berto and Plebani (2015, p. 2) describe van Inwagen's intentions "by using the prefix 'meta'" as "suggesting a kind of higher-level reflection." Meta-ontology thus accompanies reflections on the methodology of ontology, on how to understand ontological concepts such as existence or object, or how to understand ontological and metaphysical debates, among others. In short, as Turner (2016) depicts it, *meta*ontology is the study of ontology.

Metaontological deflationism is to be understood as related to (although different from, and indeed logically independent from[2]) ontological deflationism. While ontological deflationism is relative to the notions of existence, object, or kind of object, deflationism at the metaontological level supposes a higher level of abstraction than mere ontological deflationism. It is a form of deflationism about the character or nature of ontology itself and introduces alternative ways of understanding traditional ontological (or metaphysical) disputes. Notwithstanding, in the literature, there seem to be some discrepancies as to how to understand the expression 'metaontological deflationism.' For example, in the case of Hirsch (2011) metaontological deflationism becomes something linked with quantifier variantism, while for Thomasson (2015, pp. 65–69) it does not have this implication at all, since, for her, the existential quantifier has just one meaning, that of the Fregean second-order concept. Although Thomasson argues that Carnap's deflationism does not involve the quantifier variance thesis advocated by Hirsch and Putnam, the fact is that (meta)ontological deflationism has been commonly associated with quantifier variance (see, for instance, Sider, 2009, p. 391; Eklund 2009, p. 137). Thomasson's metaontological deflationism is not related to this antirealist thesis. Quite at the contrary, her "easy" ontology involves a form of simple realism at the first-order level in relation to most disputed entities, like mathematical objects, propositions, or properties. In Thomasson's view (2008, 2015), from some uncontested truths we can affirm the existence of these apparently problematic entities via trivial inferences. At the metalevel, the easy approach implies that ontological disputes are deflated because there is nothing substantial about the notion of existence (e.g., causal relevance, mind-independence, or "real metaphysical nature"). According to Marschall and Schindler (2020, pp. 99–100), Thomasson's metaontological

[1] For a recent examination of how Carnap's views on ontology have determined contemporary debates, see Blatti and Lapointe (eds., 2016).

[2] Thomasson (2014) has argued that semantic deflationism implies metaontological deflationism and feed into one another, but this is disputed by Marshall and Schindler (2020), who argue that the relation is much weaker than Thomasson claims.

deflationism consists of two related views: the view that "existence has no deep nature" and the view that the neo-Quinean approach to ontology is misguided. The neo-Quinean approach, that has been described as 'mainstream' in metaphysics (see Berto & Plebani, 2015), involves a naturalist treatment of ontology, maintaining a view according to which we should be committed to those entities over which our best scientific theories quantify. The methodology presupposed here is that of a 'hard ontology', i.e., one that takes ontological questions as being substantial. This is precisely what Thomasson rejects. Marschall and Schindler (ibid.) nevertheless carefully distinguish between deflationism about truth and reference (semantic deflationism), deflationism about the notion of existence (existence deflationism) and deflationism about how to understand ontology and ontological debates (meta-ontological deflationism).

Metaontological deflationism rejects that metaphysical disputes about the real existence and nature of entities of any kind, including abstract entities—numbers, theoretical entities postulated by science, properties, possible theories worlds, or propositions (to give just a few examples)—have any substantive significance. This results in positions according to which neither existence nor being an object are substantial qualities. A deflationist approach to the notion of existence

> … involves the idea that there is no call for a theory aiming to uncover a deep and substantial nature of existence, for there is nothing more to the notion than is captured in the rules of use that enable it to fulfill its function. (Thomasson, 2015, p. 116)

According to Thomasson, the only thing that matters in order to be able to claim that entities of kind K exist is whether the application conditions actually associated with term 'K' are fulfilled, and, since the conditions will vary considerably depending on the kind of entities, the search for a single, substantive criterion for existence is hopeless. Given the rules of use for 'exist', that are univocal and formal, ontological questions become straightforwardly answerable, by conceptual or empirical means.

Thomasson classifies neo-Fregean accounts in the philosophy of mathematics and Stephen Schiffer's pleonastic approach as (meta)ontologically deflationary. All of them share the idea that existence questions can be easily answerable by conceptual or empirical means, so that, according to her, all they can be labeled as "easy ontologies." Rayo's trivial platonism is also quoted among them (Thomasson, 2015, p. 127, n. 1) and we can also mention Linnebo's minimalism. All these approaches are (meta)ontologically deflationary in as much as they take debates about existence as non-substantial. Moreover, they all agree about the univocity of existence. Neo-Fregeans take Frege's idea of transforming the issue of the reference to abstract objects into a question about how whole sentences succeed in having their purported arithmetical meanings (Linnebo, 2018, p. 8). This comes with adopting Frege's familiar context principle. Starting with sentences is crucial to them because they defend one version or another of what Linnebo calls 'the flexible conception' as contrasted with a 'rigid conception' (Linnebo, 2018, p. 31); namely, the idea that there is no unique real structure of reality to be captured by our 'apparatus of predication and identification,' by our language. This is not to deny that there is a world out there or to claim that entities come into being when we define a language to talk about them. It is not a kind of constructivism. Instead, the thought is that reality gives different answers depending on the conceptual apparatus we use to describe it.[3] There is place for choice, though not for *any* choice; for instance, unfortunately, we can't make it rain or that a mare gives birth to a lion. Thus, allowing for sentences with different logical form that express the same proposition makes room for the flexible conception.

Some contributions to this volume explicitly focus on Carnap's views on meta-ontology. Carnap (1950 [1956]) advances a vital contrast to explain how to understand traditional ontological and metaphysical disputes: the internal versus external distinction. Carnap's distinction enables him to sidestep conventional ontological debates in favor of adopting deflationary ontological commitments within a linguistic framework that is deemed useful for a specific purpose. Carnap's adoption of a semantic deflationist approach comes with the tenet that ontological questions make sense only when they are posed as internal to a linguistic framework. Philosophers of mathematics usually disagree about the existence of numbers, because, due to their abstract nature, their existence is problematic. For the same reason, philosophers disagree about the existence of alleged entities such as propositions or properties. These are two well-known disputes between nominalist and realist (in the metaphysical sense) positions about putative abstract entities. According to Carnap, once we have established a linguistic framework for, say, arithmetic or mathematics in general, we can ask whether the instances of the kind(s) of numbers that underlie the accepted linguistic formulation of the corresponding theory exist. Internal questions have two kinds of answers: analytic, those that are answered by considering the semantic rules of the linguistic framework alone, and synthetic, those that are answered both in terms of semantic rules and empirical inquiry. The demand whether numbers exist posed as a question internal to a number-frame is easily answered, it is an analytic question. Considered as an external question, the issue about whether numbers really exist beyond a mathematical framework is completely meaningless (becomes a pseudo-question). Notwithstanding, Carnap transforms external questions from pseudo-questions of a metaphysical character into pragmatic questions about the usefulness of a linguistic framework; we decide to use one linguistic framework or another for pragmatic reasons. Of course, traditional ontologists take their questions about numbers or propositions to be external to a linguistic framework, but Carnap thinks they are wrong. It makes no sense to ask whether numbers, properties or propositions exist without previously adopting a given linguistic framework.[4] 1F Neo-Carnapian proposals have been adopted in the last years attending to different interpretations of Carnap's ideas on semantics and ontology, like those authored by Thomasson, Schiffer, Hirsch, or Rayo, to name a few.

[3] In Linnebo's (2018, 31) words: "As Frege [1953, §43] observed, reality provides different answers to ontological questions about what there is depending on which concepts we bring to bear".

[4] See also Chap. 2, De Donato and Falguera, this volume.

Other metaontological deflationists have continued to build on Carnap's proposal. So, when two metaphysicians disagree about an ontological question, it may be the case that they disagree about the meanings of the terms—they understand the meaning of the terms differently—, hence they talk past each other; or perhaps they disagree about the meaning speakers (or groups of speakers) associate to each word and hence they are not having a disagreement about the existence of certain entities but a verbal disagreement. Authors agreeing to this metaontological deflationist view have maintained that, given variance of existential quantifier meaning (Hirsch, 2011), many ontological disputes are mere verbal disputes or that we cannot know what counts as an appropriate answer to paradigmatic ontological questions such as 'What is this *really*?' 'What is the *essence* of this type of object?', 'What kind of entities *really exist*?' The reason being that there is no "fact of the matter" that can help us establish whether the answer to any of these questions has cognitive value; that is, whether it has a truth value. (cf. Yablo, 2009).

Nonetheless, it is possible to wonder whether adopting a deflationary metaontological viewpoint is consistent with asking whether there is some sense in which, despite acknowledging that metaphysical questions have no objective answers, we can say that ontological disputes go beyond mere linguistic disputes. Maybe by understanding them to have a pragmatic value, maybe by understanding there are no determinate answers.

The volume also includes several discussions about the semantics and ontological commitments of discourses in special domains, and about how to conceive of the entities in these domains. The views at stake usually take as starting point deflationist understandings of existence and the notion of object to argue that it is possible to commit to the existence of awkward entities like numbers, propositions, properties or possible worlds, without endorsing the problems that more traditional platonistic views about them address.

The preceding paragraphs provide a comprehensive framework for situating the subject matter of this volume. But it is the chapters themselves that are intended to acquaint us with some of the various *Deflationist conceptions of abstract objects*. We have divided the chapters into five parts, although some of them deal with issues in one or more of the other parts. Thus, the volume can be seen as dealing with different issues that fall under the following headings:

Part I. Carnapian background
Part II. Themes on Thomassian easy realism
Part III. Themes on Linnebo's thin realism
Part IV. Agnosticism /quietism and fictionalism
Part V. Themes focusing on metasemantics and semantics

Part I includes chapters discussing different issues related to Carnap's view. **De Donato-Rodríguez & Falguera** begin by comparing Carnap's position on the debate between Platonism and Nominalism with his position on the debate between Realism and Anti-Realism (in a metaphysical sense). They then analyze the Carnapian approach to entities posited with scientific terms (theoretical and observational terms) to argue, against views such as Psillos's (1999 and 2000), that

Carnap held an agnostic position about these entities as substantial entities on the grounds that Carnap regarded metaphysical questions (and their answers) as nonsense. In addition, they defend that Carnap alternatively adopted a deflationist account of the references of such terms (also for the Ramsey variables). They then argue for a neo-Carnapian way of considering these references as abstract artifacts.

Alexius takes as a starting point a Carnapian background to provide an argument against the tenet that ontological questions are analytic and hence trivial. His argument goes in two steps. First, he reasons that the appropriate notion of analyticity the easy realist defends has to be characterized in terms of what he calls the 'broad characterization of meaning.' Then he argues that the ensuing characterization of analyticity—"A use S of an expression E in a language L is analytic for E in L iff the broad majority of competent speakers of L are disposed to accept S as a correct use of E"—overgenerates. That is, statements that should not be regarded analytic are analytic according to it. As a result, the linking premises in easy arguments cease to work as needed in this sort of arguments, and the arguments fail.

In his chapter, **Schindler** offers a critique of the Carnapian-inspired deflationary conceptions of abstract entities. These conceptions defend the view that "all questions regarding abstract objects can be answered (insofar as they can be answered at all) by reference to our linguistic and conceptual practices." Schindler takes as his starting points his version of semantic deflationism and the Quinean principle, which states that "we have to look to our best overall theories to answer existence questions." He concurs with Alexius that "all questions regarding abstract objects can be answered (insofar as they can be answered at all) by reference to our linguistic and conceptual practices." However, he does not dismiss the possibility of analytic truths nor does he deny that "most philosophical questions about abstract objects can be answered (largely) by looking at our linguistic and conceptual practice." He develops his deflationist view about abstract entities in parallel with his own deflationary account of truth. In particular, he advances a proposal for properties and another for numbers, both of which begin by identifying the function that those entities serve in our best scientific theory, as determined by our linguistic and conceptual practices.

The next group of papers, Part II, deals, one way or another, with themes related to Amie Thomasson's easy realism. Contrary to orthodox Carnapian and neo-Carnapian views, **Båve,** in his chapter, proposes that the best option for the Carnapian is to see the traditional metaphysician arguing about existence questions—without any linguistic or conceptual preliminaries—as asking ordinary internal questions. His argument questions the link the Carnapian seems to establish between being competent with a language and understanding meaning. Given the way the Carnapian understands the link, in terms of some sort of inferential semantics, there are some questions that have easy answers. But linguistic competence can be explained in a reliabilist way, hence not entailing that *certain* answers are easy and allowing for classical metaphysicians to be seen as answering internal questions. The author claims the view allows for an improved version of easy ontology and for a more plausible interpretation of what metaphysicians do.

María de Ponte argues for a deflationary view of possible worlds. She begins by examining two contrasting views of modality, those put forth by Lewis and Thomasson, with the aim of advocating for what she terms the "commonsense view." According to Thomasson, Lewis' so called *paraphrase argument*—an argument that goes from "Things might be otherwise than they are" to the paraphrasis "that there are many ways things could have been besides the way that they actually are" and from there to the existence of possible worlds—, is an *easy* argument. De Ponte rejects both Lewis's paraphrase argument and Thomasson's version of it on the grounds that, although Thomasson interprets existential quantification in a deflationary way, it nonetheless involves a paraphrase that is similar to Lewis's. Ponte complements her critique of these views about possible worlds by challenging Thomasson's easy arguments and Thomasson's understanding of modal properties and facts. Finally, de Ponte presents her own account of these issues, a version of critical referentialism according to which there are modal facts.

The book contains a section dedicated to thin realism, Part III, which commences with a contribution from Linnebo himself. In his chapter, **Linnebo** develops his thin-objects view to examine what he terms "the algebraic conception of logic." He commences by delineating the distinction between the rigid and the flexible conceptions of reality. While the former contends that reality is articulated into objects "independently of the concepts we bring to bear" (Linnebo, 2018, p. 31), the latter advocates that it is not. The algebraic conception of first-order logic is the conception of logic that aligns with the flexible view. In accordance with this perspective, just as there is no definitive interpretation of any algebraic relation—he cites the partial order relations as an example—or no singular interpretation of a set of axioms (whichever structure in which they are true suffices), there is no true interpretation of first-order logic. Clearly, according to the rigid conception, there is a canonical interpretation of first-order logic, which means that first-order logic quantifies over those objects in which reality is articulated independently of any concept we might consider. Then he continues to give further details about what the algebraic conception is not; interestingly he explicitly denies that the algebraic conception is a form of anti-realism, though "we cannot speak of a determinate totality of all objects that exist."

Martínez-Vidal following Linnebo's lead, introduces a deflationary notion of object Quinean style, a notion of object that allows for thin objects. She presents Linnebo's concept of thin objects in its different modalities and proposes parallel distinctions "Quinean" style. Subsequently, she presents a reinterpretation of the Quine-Putnam indispensability argument. Rather than establishing the existence of abstract objects, this alternative version of the argument reads as committing us to mathematical entities being heavyweight objects, objects that make demand on the world. This commits its proponents to heavy-duty platonism, a metaphysical picture that is unpalatable from a naturalist viewpoint. She asserts that this new interpretation illuminates the nature of the platonism implied by the standard version of the indispensability argument and, consequently, the implausibility of the position that results from its success.

Topey questions that—contrary to the problem Benacerraf-Field Challenge poses for platonist non deflationist views—accounts of abstract objects committed to the idea that their existence does not amount to much while being mind-independent and 'causally inert' have any epistemological advantage, from a naturalist standpoint. He mentions proposals based on Boghossian's epistemic analyticity and pragmatic accounts of how we get to know basic beliefs, though he sustains that his argument does not depend on the details of the different epistemic proposals. His contention is that 'easy knowledge', knowledge of objects with which no cognitive causal relation is available, is not possible since there is no way to explain how we get true beliefs about those entities without being lucky; since luck is incompatible with knowledge, there is no way to justify that knowledge of arithmetical entities is reliable. As a result, one of the assumptions must go: either abstract objects are not mind-independent or they are not causally inert.

A fourth battery of chapters, Part IV, discusses alternative views that include agnostic or fictionalist tenets. **Bueno**, after initially contrasting the advantages and disadvantages of adopting a Platonic versus a deflationary approach to mathematical objects, addresses the question of which conception of an object underlies such views. To clarify the latter, he takes into consideration four conceptions of objects: two classical ones of a substantive nature, semantical and metaphysical, and two deflationary ones, the thin conception and the ontologically neutral conception. The Fregean and the Quinean conceptions are both semantic views since, in the first case reference for singular terms is guaranteed by the introduction of appropriate definitions and principles of abstraction, while according to the second there exist those objects our best scientific theories quantify over. The metaphysical conception considers an object to be whatever has determined and substantial identity conditions. He contends that the deflationary views mentioned above are metaphysical because they pose conditions for objecthood that are not just logical. In particular, he sees the thin conception, Linnebo's view, as a combination of semantic and metaphysical elements; to be an object is to be the referent of a singular term—their existence does not make demand on the world—but objects must also count on identity criteria that go beyond mere formality since their identity conditions have to be provided by means of predicative dynamic abstraction. Ontologically neutral conceptions, Azzouni's and his, consider that quantifying over objects does not commit us to their existence but while Azzouni advocates a nominalist version of the view, Bueno goes for agnostic neutralism. He then continues to argue for his agnostic position by signaling the difficulties the thin conception faces and pointing at the impossibility of establishing that putative abstract objects such as mathematical objects do not exist as Azzouni contends.

Finn in her chapter suggests understanding ontological debates "as being based around the practical advantages of utilizing a way of speaking in a fictionalist manner." A new version of what she calls Carnap's quietism, *neo-Carnapian fictionalism*. She sees her proposal as bringing benefits to both fictionalism and quietism (neo-Carnapian deflationism). To the neo-Carnapian it brings the advantage Yablo (2000) pointed out in his posing fictionalism as a way of escaping Carnap's reliance on the analytic versus synthetic distinction criticized by Quine. To the fictionalist it keeps the pragmatism while abandoning anti-realism.

In her analysis, **Leng** compares fictionalist talk to thin-objects talk. She initially examines, from a fictionalist standpoint, the question of whether thin objects can be classified as fictional entities. She gives several reasons to support that thin objects are fictional objects— for instance, they do not contribute to the truth of the sentences in which they occur since their contribution is grounded on the contributions of the objects that stand in the abstraction relation. Then she considers, from the perspective of the proponent of thin objects, whether fictional objects are thin objects. Subsequently, rather than undertaking an analysis of this possibility, she concedes that it may be a matter of taste to determine the terms in which to speak and proceeds to evaluate whether the proponent of thin objects can perform a more effective job than the fictionalist in responding the difficulties this second position faces, namely the problem of explaining how numbers, understood as fictional entities, contribute to the truth of mixed statements and that of being committed to modal facts. Ultimately, she determines that the proponent of thin objects is not in a superior position.

García-Carpintero addresses the issue of reference for empty (fictional) names in the context of fictional discourse. To this end, he identifies three distinct applications of fictional names in declarative sentences pertaining to fiction: textual uses, which are the ones characteristic of a work of fiction; paratextual uses: those which establish something in relation to the work of fiction—with or without some explicit prefix such as "according to … [title of the work of fiction] …". so that the established information can be classified as either true or false within the context of the work; and metatextual ones which establish something about fictional expressions in relation to the actual world, such as "Sherlock Holmes is a fictional character." The objective of this study is to challenge realist proposals concerning the use of fictional names in textual and paratextual contexts. In particular, it seeks to refute the view that such names should be understood as deflationary referents. Instead, it proposes that metatextual uses of fictional names can be understood as a form of hypostatization.

Finally a last group of chapters, Part V, deals with issues on semantics and metasemantics. **Eklund** questions that the top-down strategy—a version of which consists in adopting the context principle as a metasemantic strategy—actually plays any part in allowing for the ontological view many of those authors advocate. He gives special thought to both Linnebo's (2018) and Warren's (2020) accounts.

On his part, **Collin**—his chapter, this volume—argues for an inferentialist account of abstract objects to be added to the list of deflationist proposals; he sustains it is a deflationist view since it starts with language to provide a non-representational semantics and agrees that ontological disputes about existence are not substantial, among other features. However, it differs in some respects from the ones Thomasson calls 'easy approaches.' One of the differences is that the inferentialist advocates for various existence predicates.

Azzouni in his chapter brings up the problem of how to understand existence to advance an alternative inspired in the Gricean notion of conversational implicature.

His proposal is still a deflationist view, one that rejects Quine's dictum on ontological commitment, a dictum that Carnapians seem to accept.[5] Azzouni's contention is that ontological commitments do not come with the semantics of the sentences in our languages. Rather they result from the implicatures of uttered sentences.

Schwartzkopff puts forth a proposal to understand number words as concept-denoting, as opposed to object-denoting. He begins by presenting the various uses of number words in natural language, including attributive, predicative, and as numerals. He then briefly discusses the work of Snyder, Samuels, and Shapiro, who have proposed a non-uniform, polymorphic account of the semantic function of "number words." He continues to present a uniform account of terms for number as concept words. To this end, he employs standard type-theoretic semantics. He regards himself as following Frege's lead, given that Frege himself was not always clear that number terms denote objects. He presents this alternative view in detail and uses it to reformulate simple arguments. Given the revised reading he proposes, simple arguments establish "the existence of numbers conceived of as higher-order entities, viz. first-level (cardinality) concepts," a result that may align with the deflationist perspective.

In closing, we would like to thank the participants for their contribution to this volume. We hope that the discussions and insights here raised will contribute to a better understanding of ontological and metaontological deflationism, how they relate to each other, the challenges they raise, as well as some of the main ways to address them.

Acknowledgments This research was made possible thanks to the funding provided by FEDER/ Ministerio de Ciencia, Innovación y Universidades—Agencia Estatal de Investigación/ through grant PID2020-115482GB-I00, as well as by the project "Ingeniería conceptual y desacuerdo epistémico: una perspectiva ontosemántica deflacionista (INCODE) /Conceptual Engineering and Epistemic Disagreement. A Deflationist Onto-semantic Perspective" funded by Ayudas FBBVA a Proyectos de Investigación Científica 2021.

Furthermore, we would like to express our gratitude to the reviewer for her very helpful feedback on an earlier version of this preface.

References

Alspector-Kelly, M. (2001). On Quine on Carnap on Ontology. *Philosophical Studies, 102*(1), 93–122. https://doi.org/10.1023/A:1010353825006

Azzouni, J. (2004). *Deflating existential consequence: A case for nominalism*. Oxford University Press.

Berto, F., & Plebani, M. (2015). *Ontology and metaontology: A contemporary guide*. Bloomsbury Academic.

Berto, F., Kroon, F., & Voltolini, A. (2014). Metaontology: Introduction. *The Monist, 97*(4), 423–429. http://www.jstor.org/stable/44012696

[5] Though Carnapians should, more precisely, be seen as adopting an understanding of 'ontological commitment' which is neither realistic in a metaphysical sense nor anti-realistic, but agnostic.

Blatti, S., & Lapointe, S. (Eds.). (2016). *Ontology after Carnap*. Oxford University Press UK.
Bueno, O. (2008). Truth and proof. *Manuscrito, 31*(1), 419–440.
Bueno, O. (2020). Contingent abstract objects. In J. L. Falguera & C. Martínez-Vidal (Eds.), *Abstract objects: For and against. Synthese library* (Vol. 422). Springer. https://doi.org/10.1007/978-3-030-38242-1_5
Carnap, R. (1950 [1956]) Empiricism, semantics and ontology. *Revue Internationale de Philosophie, 4*, 20–40. [used version in: Rudolf Carnap (1956). *Meaning and necessity* (pp. 205–221). The University of Chicago Press.)]
Eklund, M. (2009). Carnap and ontological pluralism. In D. J. Chalmers, D. Manley, & R. Wasserman (Eds.), *Metametaphysics: New essays on the foundations of ontology* (pp. 130–156). Oxford University Press.
Hirsch, E. (2011). *Quantifier variance and realism: Essays in Metaontology*. OUP.
Linnebo, Ø. (2018). *Thin Objects*. Oxford University Press.
Marschall B., & Schindler T. (2020). Does semantic deflationism entail meta-ontological deflationism? *The Philosophical Quarterly, 71*(1, January 2021), 99–119. https://doi.org/10.1093/pq/pqaa019
Price, H. (2004). Naturalism without representationalism. In M. De Caro & D. Macarthur (Eds.), *Naturalism in question* (pp. 71–88). Harvard University Press.
Price, H. (2007). Quining Naturalism. *The Journal of Philosophy, 104*(8), 375–402.
Price. (2009). Metaphysics after Carnap: The ghost who walks? In D. J. Chalmers, D. Manley, & R. Wasserman (Eds.), *Metametaphysics. New essays on the foundations of ontology* (pp. 320–346). OUP.
Psillos, S. (1999). *Scientific realism: How science tracks truth*. Routledge.
Psillos, S. (2000). Carnap, the Ramsey-sentence and realistic empiricism. *Erkenntnis, 52*, 253–279.
Putnam, H. (2004). *Ethics without ontology*. Harvard University Press.
Quine, W. V. (1948). On what there is. *The Review of Metaphysics, 2*(5), 21–38. http://www.jstor.org/stable/20123117
Rayo, A. (2013). *The construction of logical space*. Oxford University Press.
Schiffer, S. (1996). Language-created language-independent entities. *Philosophical Topics, 24*(1), 149–167.
Sider, T. (2009). Ontological realism. In D. Chalmers, D. Manley, & R. Wasserman (Eds.), *Metametaphysics: New essays on the foundations of ontology* (pp. 384–423). Oxford University Press.
Thomasson, A. (2008). Existence Questions. *Philosophical Studies, 141*(1), 63–78. https://doi.org/10.1007/s11098-008-9263-8
Thomasson, A. L. (2014). Deflationism in semantics and metaphysics. In A. Burguess & B. Sherman (Eds.), *Metasemantics: New essays on the foundations of meaning*. Oxford University Press.
Thomasson, A. (2015). *Ontology made easy*. Oxford University Press.
Turner, J. (2016) Metaontology. In *Oxford handbooks* Online, 2016. https://doi.org/10.1093/oxfordhb/9780199935314.013.25
Van Inwagen, P. (1998). Meta-ontology. *Erkenntnis (1975–), 48*(2/3), 233–250. http://www.jstor.org/stable/20012842
Warren, J. (2020). *Shadows of syntax: Revitalizing logical and mathematical conventionalism*. Oxford University Press.
Yablo, S. (2000). A paradox of existence. In T. Hofweber & A. Everett (Eds.), *Empty names, fiction, and the puzzles of non-existence* (pp. 275–312). CSLI Publications.
Yablo, S. (2009). Must existence-questions have answers? In D. Chalmers, D. Manley, & R. Wasserman (Eds.), *Metametaphysics. New essays on the foundations of Ontology* (pp. 507–525). OUP.

Part I
Carnapian Background

Chapter 2
A Deflationary View of Ontology for Scientific Theories: A Neo-Carnapian Approach

Xavier de Donato-Rodríguez and José L. Falguera

Abstract Some of the keys to understanding how Carnap positions himself with respect to theoretical entities after the mid-1950s are contained in his considerations related to abstract entities in "Empiricism, Semantics and Ontology" (Rev Int Philos 4:20–40, 1950[1956]). In this contribution, we defend that Carnap adopted an agnostic position in relation to the real existence (in a metaphysical sense) of both abstract entities and theoretical entities. We also aim to show that the ultimate reason for this agnosticism, in both cases, is Carnap's rejection of external ontological questions as having no cognitive significance. In addition, we defend the idea that he would never have really succumbed to either an instrumentalist position or any sort of weak realism, such as structural realism, in opposition to some interpretations of Carnap's work on the reference of theoretical terms. To demonstrate the validity of this latter stance of ours, we argue that: (i) the appeal to mathematical entities as referents of theoretical terms can only be understood in a deflationary way, and not in an instrumentalist manner (*pace* Lavers and Psillos); and (ii) Carnap's use of the Ramsey sentence does not involve structural realism and the Newman objection does not apply to it (*pace* Psillos, Demopoulos and, perhaps, Friedman). In any case, even adopting a deflationist approach to theoretical entities, it sounds odd to consider them to be mathematical entities, as Carnap posited. As an alternative, we propose considering them as abstract artifacts.

Keywords Carnap · Deflationary ontology · Platonism · Nominalism · Scientific theories · Instrumentalism · Realism · Agnosticism · Ramsey sentence · Newman objection · Neo-Carnapian ontology

---

X. de Donato-Rodríguez · J. L. Falguera (✉)
Department of Philosophy and Anthropology, University of Santiago de Compostela, Santiago de Compostela, Spain
e-mail: xavier.dedonato@usc.es; joseluis.falguera@usc.es

X. de Donato-Rodríguez et al. (eds.), *Deflationist Conceptions of Abstract Objects*, Synthese Library 516, https://doi.org/10.1007/978-3-031-98717-5_2

2.1 Introduction

According to Carnap (1950a [1956]), the admissibility of abstract entities is just a question of the acceptability of an appropriate linguistic framework, where the answers to existence questions regarding these entities are analytic and trivial.[1] But in some other cases, the answers are not analytic. For example, in the case of posited observable entities designated by observational terms, the mere acceptance of a suitable linguistic framework (i.e., the "thing-language"[2]) is not sufficient: empirical enquiries are necessary. It could be thought that something similar happens with the putative entities designated by theoretical terms (henceforth, theoretical entities). After all, theoretical terms are introduced within an empirical theory; and, for Carnap, (the language of) an empirical theory can be considered as adopting a certain linguistic framework. Accepting an empirical theory involves empirical enquires, and so the admissibility of theoretical entities depends on that empirical research. However, Carnap's approach to theoretical entities includes something peculiar as he (unofficially) proposed considering them as mathematical entities.[3] Taking this proposal into account, it is revealing to analyze Carnap's positions in relation to the "realism vs. instrumentalism" debate that has taken place since the mid-1950s. Carnap's view of theoretical entities has been criticized in various ways. Lavers (2016) pointed out that, in fact, Carnap had not overcome his instrumentalist scruples. Before that, Psillos (1999, 2000b) had already argued that Carnap's alleged neutrality and attempt to reconcile realism and instrumentalism was severely compromised given (i) his choice of mathematical entities as the referents of theoretical terms, and (ii) the fact that his use of the Ramsey sentence faces the so-called 'Newman objection'.

In relation to the instrumentalism vs. realism debate, the point of view we defend in this contribution is that Carnap positioned himself against both approaches in the same way (and for the same reason) he had distanced himself equally from Platonism and nominalism, clearly rejecting both of them. We hence maintain that his conception of theoretical entities as mathematical entities cannot be understood as a way of succumbing to instrumentalism and that his use of the Ramsey sentence (*RS*) does not involve the adoption of structural realism; and even less that, given the Newman objection, the only alternative is to assume a (metaphysical) realism with regard to theoretical entities (in our best scientific theories), as Psillos defends.

Both aspects, the consideration that the referents of theoretical terms are mathematical entities and the idea that the existential assertion of posited theoretical entities that the *RS* provides becomes trivial, have been objected to as giving an unacceptable image of the epistemic role of our best scientific theories. In this paper, we analyze these issues and try to find a way out to this problem, which finally goes

[1] See Carnap (1950a [1956, p. 217]).

[2] See Carnap (1950a [1956, p. 206 and ff.]).

[3] See Carnap (1950a [1956, p. 212]; 1956, p. 43; 1958 [1975, pp. 76–81]; 1963b, p. 963; 1966, pp. 253–254).

through maintaining agnosticism[4] with regard to the real existence of theoretical entities that is very close to Carnap's original point of view. However, we avoid the rather odd idea that the theoretical entities involved are mathematical entities.

2.2 Carnap Against Platonism and Nominalism

In "Empiricism, Semantics and Ontology" (1950a [1956]), Carnap maintained a form of agnosticism in relation to the issue of the *real* existence or inexistence of abstract entities (of different kinds, such as propositions, properties, relations, numbers, sets, etc.); i.e., he took a stand against both Platonism and nominalism. Indeed, in that article, Carnap sought to show how to overcome nominalist convictions, especially the scruples of empiricists, by using a language that designates abstract entities without thereby embracing a Platonist position with regard to the *real* existence of such abstract entities (in a metaphysical sense).

There are two main sources of motivation for Carnap's defense of the use of languages that designate abstract entities. The first is related to the difficulty within scientific contexts (for example, within physics) to dispense with languages that designate abstract entities (e.g., real numbers as space-time coordinates or as values of physical quantities, functions, limits, etc.). In this regard, Carnap (1950a [1956]) was skeptical of the proposal that we should conceive of the languages of science (the languages of scientific theories) as mere formal calculuses. Such a position would mean assuming that the part of the language which apparently refers to such abstract entities is uninterpreted and uninterpretable. But Carnap (1950a [1956, p. 205]) points out that in scientific contexts, such as that of physics, "it is more difficult [than in the case of mathematics] to shun the suspected entities because the language of physics serves for the communication of reports and predictions and hence cannot be taken as a mere calculus".[5]

The second source of motivation has to do with semantic analysis, which Carnap incorporated into his scientific philosophy once he had accepted Tarski's elucidation of "truth". Carnap had already incorporated semantic analysis in his (1942) and (1947 [1956]), thereby surpassing the stage of logical syntacticism for the treatment of philosophical problems (which he especially developed in his (1934 [1937])).[6]

---

[4] Here, to be an agnostic means maintaining that we cannot offer any answer (positive or negative) to the issue and thus it implies rejection of both Platonism and nominalism. As we will see later, this position is not to be confused with what others have called neutralism.

[5] In Sect. 2.3 below, we will learn more about the languages of scientific theories and the entities designated by their expressions, as Carnap sees them.

[6] See Lavers (2015) on the changes in Carnap's point of view from his stage of logical syntacticism to the stage when he incorporates semantic treatments for the analysis of languages. Carnap tried to avoid ontological worries in his syntactic stage by maintaining the distinction between the material and the formal mode of speech, and by dispensing with semantic notions (including 'reference') in the latter mode of speech. (See also Lavers (2016, pp. 201–211).)

His inclusion of semantic analysis entailed dealing with (or designating) entities that were considered abstract. However, this was rejected by prominent empiricists, since they understood that it implied surrendering to Platonism and, therefore, to metaphysical positions that were devoid of cognitive content for logical empiricists.[7]

Carnap's response to such criticism is formulated in his (1950a [1956)] with the distinction between two kinds of ontological questions: questions that are external to a linguistic framework and those other that are internal. He indicates in that paper that the former, insofar as they are conceived as questions about the *real* existence of entities, are pseudo-questions and, hence, out of place. As an alternative, Carnap proposes a different way of posing an external question that is no longer as a question about the real existence of entities of a certain kind. Rather, it becomes a question concerning a practical decision about whether to accept (or not) a certain linguistic framework which introduces expressions for those entities, where such a decision depends on whether the framework is fruitful for a certain purpose.[8]

It seems that a linguistic framework is, for Carnap, a way of specifying that certain expressions, together with their corresponding syntactic and semantic rules, allow one to speak of entities of a certain kind in one's language. It is a way of introducing a new kind of entity, such as abstract entities of different kinds (numbers, propositions, properties, or spatiotemporal coordinates). In such cases, a linguistic framework introduces new expressions, with suitable rules, for a new kind of entity.[9] In so doing, a linguistic framework provides an explicit way in which one's language is expanded to deal with a new kind of entity:

> If someone wishes to speak in his language about a new kind of entities, he has to introduce a system of new ways of speaking, subject to new rules; we shall call this procedure the construction of a linguistic *framework* for the new entities in question. (Carnap, 1950a [1956, p. 206])

However, a linguistic framework does not always introduce new expressions for a new kind of entity, since the entities a linguistic framework can speak of are the usual entities that are referred to in one's everyday language, as is the case with observational objects: "the simplest kind of entities dealt with in the everyday language" (Carnap, 1950a [1956, p. 206]). This is a clear example of a linguistic framework which is part of a natural (everyday) language. It has the peculiarity of not requiring the specific inclusion of new expressions in one's natural language, since "we have all accepted the thing language early in our lives as a matter of course"

[7] See Carnap (1963a), where he mentions empiricists such as E. Nagel, W. V. Quine and N. Goodman, among others, as those who were critical of the designation of abstract entities that his semantic analysis favored.

[8] Along with fruitfulness, other virtues that Carnap mentions when deciding on the acceptance of a linguistic framework, as the thing-language, are those of efficiency and simplicity. The thing-language is appropriated for dealing with observable objects and events (see Carnap (1950a [1956, pp. 206–207])).

[9] See Carnap (1950a [1956, p. 208]).

(Carnap, 1950a [1956, p. 207]).[10] (Although, according to Carnap, we could choose to replace it with some other linguistic framework that would talk of phenomenal entities—such as sense-data—instead of talking of ordinary things.)

The expressions *specifically needed* to speak of a (new) kind of entity, which are introduced via a linguistic framework (with their corresponding rules), are a general term for the (new) kind of entity—i.e., a higher-level predicate (such as 'number' or 'property') to apply to expressions for those entities (as in 'five is a number' or 'red is a property')—and variables for the entities of the (new) kind. The variables need quantifiers for universal and existential sentences. Expressions for properties of the (new) entities, for their relations or functions can also be introduced. Names can be incorporated to designate entities of that kind, but they are not necessarily required. (See Carnap, (1950a [1956, pp. 208 and 213–214]).)

Once a linguistic framework has been adopted, internal questions about entities of the kind thus designated are acceptable. These questions are not about the *real* existence of entities (at least, not in a metaphysical sense of 'real'), they are about the entities designated (referred to) by the appropriated expressions assumed within that framework. In this regard, Carnap says:

> From the internal questions we must clearly distinguish external questions, i.e., philosophical questions concerning the existence or reality of the total system of the new entities. Many philosophers regard a question of this kind as an ontological question which must be raised and answered before the introduction of the new language forms. The latter introduction, they believe, is legitimate only if it can be justified by an ontological insight supplying an affirmative answer to the question of reality. In contrast to this view, we take the position that the introduction of the new ways of speaking does not need any theoretical justification because it does not imply any assertion of reality. We may still speak (and have done so) of the "acceptance of the new entities" since this form of speech is customary; but one must keep in mind that this phrase does not mean for us anything more than acceptance of the new framework, i.e., of the new linguistic forms. Above all, it must not be interpreted as referring to an assumption, belief, or assertion of "the reality of the entities. (Carnap, 1950a [1956, p. 214])

From these comments, it seems clear that for Carnap the semantic expression of "designation" (or "denotation" or "reference") can be understood in a deflationary way. There is nothing substantive in saying (in a semantic metalanguage) that some entity is designated by an expression—such as a name (or constant) or a noun—if the language used to make the assertion incorporates a linguistic framework which introduces that expression (e.g.: "'Five' designates five", given the framework for numbers; or "'Red' designates red", given the framework for thing properties). Nor is there anything substantive in asserting (in a semantic metalanguage) that an

[10] It seems to us that if Carnap spoke of linguistic frameworks as expanding *one's language* or as part of the everyday language (as is assumed here), then this is not compatible with interpretations of Carnap (1950a [1956]) along the lines of a "language pluralist" view (with quantifier variance) as established by Hirsch (2002, 2009, 2011). The question here is not whether a "pluralist view" is philosophically interesting, but whether it is an appropriated interpretation of Carnap (1950a [1956]). For an analysis of different recent interpretations of Carnap's paper, among them the framework-relative view of "language pluralism", see Eklund (2013); see also Eklund (2016, pp. 165–171).

expression playing the role of name (or a constant) or a noun designates an entity of the kind corresponding to the appropriate general term, once the framework for that kind of entity has been given (e.g.: "'Five' designates a number"; "'Red' designates a property of things"). The following quotation may serve to illustrate this consideration:

> Generally speaking, any expression of the form "'…' designates …" is an analytic statement provided the term '…' is a constant in an accepted framework. (Carnap, 1950a [1956, p. 217])

Furthermore, Carnap also considers that, given a language *L* that contains sentences such as "Five is a number." Or "Red is a property of things." (given the appropriate linguistic framework within *L*), these sentences are analytic. And with this, the logical consequences that follow from such sentences, for example "There are numbers." or "There are propositions.", are analytic as well.[11]

Not all internal questions have answers that are analytic statements. Some internal questions require empirical investigation, and the answer can be affirmative or negative depending on the confirming or refuting evidence. This is the case of some questions about determined things or about subkinds of things, given the thing-language; e.g., "Is there a laptop in my desk?", or "Are unicorns and centaurs real or merely imaginary?". The latter question may make you suspect that it is in fact an external question which is posed as internal. Carnap (1950a [1956, p. 207]) clarifies that:

> [t]he concept of reality occurring in these internal questions is an empirical scientific non-metaphysical concept. To recognize something as a real thing or event means to succeed in incorporating it into the system of things at a particular space-time position so that it fits together with the other things as real, according to the rules of the framework.

Hence, assertions about the existence of entities that are possible observable things are not substantial either. They only give an affirmative or negative account of whether a particular thing-entity exists (e.g., my laptop, Sancho Panza, etc.) or whether there are entities of a possible subkind of thing-entities (e.g., unicorns, tables, etc.): they do not report whether they are real entities in a metaphysical sense (as entities in a real world). It seems clear that, according to this, to assert that (1) '*this particular* (for anyone) *thing-entity exists*' is not to assert that (2) '*such a particular thing-entity exists in a real thing world*'. For Carnap, a claim like (1) is merely internal and has factual cognitive (or empirical) content, but a claim like (2) is external and lacks cognitive content. Hence, (2) does not have a truth value, while (1) is true or false. We can say that it is possible to answer internal questions about *how the world appears to us* thanks to empirical experience and given a language to express that empirical experience. However, we cannot answer external questions about whether the thing world *really* exists (in a metaphysical sense) or the thing-entities *really* exist (in that metaphysical sense), since any answer to an external question is not really factual, although it seems to be, because there is no way to test

[11] See Carnap (1950a [1956, pp. 210 and 217]).

it (i.e., to confirm or disconfirm it). What we confirm or refute are internal claims such as (1). They are confirmable (or refutable) in accordance with *how the world appears to us,* but not depending on how the world really is (in a metaphysical sense). Claims such as (3) '*there is a system of things*' or (4) '*there really are thing-entities*' are assertions in the external sense and, hence, lack cognitive content. We are not able to confirm or disconfirm (3) or (4), because they are assertions about entities in the real world (metaphysical claims): they are not about how to describe our empirical experiences within a language introduced for that purpose. Anyway, we must keep in mind that, according to Carnap's view, a claim such as (5) '*there are things-entities*' is considered analytic in the internal sense once the thing-language has been accepted.

It should be clear that the crucial reason why Carnap claims that answers to external questions lack cognitive content is because, although they appear to be factual, they are not in fact (directly or indirectly) testable: they cannot be confirmed or disconfirmed. Once we have accepted that, in order to have cognitive content, a statement has to be either analytic or synthetic, it becomes clear that the answers to external questions have no cognitive content (given that they are neither analytic nor synthetic). An ontological claim in the external sense is not formulated as analytically true or false (i.e., by virtue of the mere meaning relations of its component expressions). On the contrary, it seems to demand something of the world, something that must be the case in order for the claim to be true (or false, if the demand is not met). However, according to Carnap, this is what cannot happen, and for this reason such a claim has no truth value.[12]

Returning to the beginning of this section, it is important to remember that we started by saying that Carnap, in his "Empiricism, Semantics and Ontology", adopted a form of agnosticism with regard to the real existence of abstract entities. Now, it is clear that this position does not prevent Carnap from speaking of "the existence of abstract entities" in the internal sense of a linguistic framework or from "designating abstract entities of a certain kind" given a linguistic framework that incorporates a general term for that kind. Once a framework with, at least, a general term for a kind of abstract entities has been accepted, and so have variables for these entities, and perhaps names for some of them, then the existential claims about these entities are trivially true because of their analyticity (though some of them could be difficult to understand). In such a case, a negative answer in the internal sense cannot be seriously given with regard to the entities of that kind; e.g., one the framework of numbers has been accepted, "nobody who meant the question 'Are there numbers?' in the internal sense would either assert or even seriously consider a negative answer." (Carnap, 1950a [1956, p. 209]).

---

[12] Sidelle (2016, p. 77 n. 30) suggests we distinguish between: (i) the *specific inability to understand what is required of the world* for us to be able to attribute *truth* conditions to ontological questions in the external sense, and (ii) the fact that ontological assertions in the external sense cannot be tested. But it is not clear to us (and we think that it is not for Carnap either) how to understand what it is required of the world for us to be able to attribute truth conditions to ontological assertions in the external sense without confirming or disconfirming them.

However, we also have to keep in mind that Carnap, in that same paper and in subsequent works, also adopted a form of agnosticism about the *real* existence of thing-objects (and about the thing world). For him, the existential answers concerning thing-objects, considered as answers to external questions, were also devoid of cognitive content. A different issue, of course, is the consideration of existential answers (in the internal sense) about this or that possible thing-object (e.g., 'the Eiffel Tower exists') or about a possible subgroup of the thing-kind (such as, 'there are cows' or 'there are unicorns'). These can be true or false because of their empirical character (taking into account how the world appears to us). The correct answer in these cases obviously requires empirical inquiry.

2.3 Carnap Against Instrumentalism and Realism

In later works, Carnap's position on the controversy between instrumentalism and realism in science is analogous to his position on the controversy between nominalism and Platonism. Instrumentalists and realists disagree about the real existence of supposed entities that are designated by theoretical terms (which are part of the language of a theory). Carnap (1966 [1974]) defends a form of agnosticism on the *real* existence or non-existence of theoretical entities, and so stands against both positions: instrumentalism and realism.[13]

The instrumentalist approach is presented by Carnap as a view that holds that scientific theories are mere tools for systematizing observational phenomena in order to predict new observables. Moreover, the theoretical terms of a theory are considered 'convenient symbols' and the postulates which contain them are adopted because they are useful. In this sense, theoretical terms have no meaning and theoretical statements have no truth value. Hence, for the instrumentalist "theories are not about 'reality'" (Carnap, 1966 [1974, p. 255]).

The opposing, realist approach is characterized by Carnap as a view that takes theoretical terms of (correct) scientific theories as referring to entities that *really* do exist (in the hard ontological sense). Faced with metaphysical realism, Carnap defends two adequate ways of using the word 'real', to which we will return later: the common-sense use and the scientific use. However, Carnap also mentions an argument adopted by advocates of the realist approach which starts by assuming that there is no sharp line between an observable entity and an unobservable one.[14] The argument goes on to add that some entities which were non-observable at some moment in the past, and designated by (what then were considered) theoretical terms, are now observable thanks to technical advances (e.g., an amoeba that is currently observable by means of a light microscope). So, the argument continues, we

[13] The idea that Carnap is against both positions, instrumentalism and realism, is also defended by Uebel (2010, p. 305).

[14] That there is no such sharp line is assumed by Carnap (1966 [1974, p. 225 ff.]).

have to accept that entities which were unobservable until a given moment can become observable—though not always observable by (direct) perception—thanks to the development of new observation instruments, and hence there is no reason to reject the idea that entities that are non-observable at some moment (and which are considered by a confirmed theory) are as real as entities that are perceptually observed.[15,16]

After presenting the controversy between the instrumentalist and realist approaches, Carnap (1966 [1974]) briefly moves on to consider a position similar to that adopted when faced with questions about the existence of abstract entities, namely:

> I believe that the question should not be discussed in the form: 'Are theoretical entities real?', but rather in the form: 'Shall we prefer a language of physics (and of science in general) that contains theoretical terms, or a language without such terms?' From this point of view the question becomes one of preference and practical decision. (Carnap, 1966 [1974, p. 256])

The paragraph ends with a footnote in which he says:

> In my view greater clarity often results if discussions of whether certain entities are real are replaced by discussions of preference of language forms. This view is defended in detail in my 'Empiricism Semantics, and Ontology' [...]. (*Ibidem*, n. 4)

It is easy to conclude that Carnap, in one of his later works, continues to take the same position regarding external ontological questions (in a metaphysical sense), even when they have to do with theoretical entities: they lack cognitive content and are only legitimate when replaced by practical questions about the fruitfulness of a theoretical linguistic framework (or of the theoretical language of a theory).[17] This conclusion, in fact, entails the idea that Carnap is rejecting both instrumentalist and realist approaches.

Anyway, let us return to the realist argument based on the inclusion of previously unobservable entities within the realm of observable ones, thanks to technical developments. It is surprising that Carnap did not supply a direct answer to this issue. We can certainly imagine two types of answer that Carnap could have provided. The first would consist of establishing, for practical purposes, a distinction between observable and theoretical, and accepting within the realm of observable only what is directly observable by the senses and what is measurable by relatively simple techniques. This is what Carnap himself did, despite accepting that there is a continuum between what is observable and what is not observable.[18] However, this answer does not seem very satisfactory, as it provides no reason why what can only be observed thanks to technical instruments would not be real.

For the second answer, we should start by remembering that, as we saw in Sect. 2.2 above, for Carnap sentences about the real existence of observable things (in a

---

[15] See Carnap (1966 [1974, pp. 255–256]).

[16] An argument along these lines was provided by Maxwell (1962).

[17] This point of view regarding theoretical entities is already found in Carnap (1956, pp. 44–45).

[18] See Carnap (1966, p. 226).

metaphysical sense) do not have cognitive content. The new answer would now hold that ultimately, regardless of which entities we consider to be observable (in a more narrow or broader sense), there is no room for questions about their real existence, as these are *external* questions. If, in contrast, the questions about the existence of these or those observable entities were posed in the *internal* sense, then the affirmative answers would no longer be those of the metaphysical realist, but rather deflationary answers that require empirical investigation. In any case, Carnap should not be worried about the above realist argument if he is assuming, as we saw above (Sect. 2.2), that the existential claims for observable entities have cognitive content only when they are considered in the internal sense. In this case, we think that it should not be problematic for Carnap to accept the existence of these or those observable entities (in the internal sense), according to an expanded notion of "observable" that is not restricted to what is observable by perception. Nevertheless, we will proceed in the rest of this paper to use 'observational' in the narrow sense, because it seems to be the sense which Carnap engaged with (even in his 1966 [1974][19]). So, in agreement with this, only existence sentences for observable entities (things and events) in the internal sense would be admissible, but these are not the existential claims which the metaphysic realist is interested in.

Be that as it may, scientific theories have descriptive terms that do not account for observable entities (e.g., 'electron', 'electromagnetic field', 'molecule', etc.). Such terms are what Carnap calls theoretical terms. As we have seen, Carnap also rejects the negative answers of instrumentalists concerning the real existence of such theoretical entities, primarily because they are answers to external questions. But there are other reasons why Carnap rejects the instrumentalist point of view.

Here we are faced with the instrumentalist consideration that theoretical terms are just convenient symbols without meaning and that the postulates which contain them are adopted because they are instrumentally useful. In response to this, Carnap (1966 [1974]) assumes that the theoretical terms have a partial empirical interpretation acquired through correspondence rules and, because of this, the postulates (and other theoretical sentences) have an empirical interpretation and a truth value. These assumptions have to be put in relation to the synchronic characterization of an empirical theory that Carnap defended after publishing "The Methodological Character of Theoretical Concepts" (1956). There, an empirical theory is represented by a finite list of (theoretical) postulates, T, and a list of rules of correspondence, C. From the mere (theoretical) postulates (or fundamental laws) of a theory, only other theoretical sentences (other theoretical laws) can be inferred. It is the conjunction of (theoretical) postulates and correspondence rules that allows us to infer (either in a deductive or in a probabilistic way) observational statements—generalized and singular ones. As is well known, after his 1956, Carnap

[19] In fact, Carnap was previously speaking of a broad use of "observable", and yet he opted for a narrow use by saying: "What physicist often call 'observable magnitudes,' e.g., mass, position, velocity, energy, frequency of waves, and the like, are not 'observable' in the sense customary in philosophical discussions of methodology, and therefore belong to the theoretical concepts in our terminology." (Carnap 1956, p. 49)

differentiates two levels of language in empirical theories. On the one hand, he has the observational language, L_O, whose descriptive statements only contain observational terms as their descriptive terms; sentences of L_O can also include individual variables and logical expressions typical of classical first-order logic. On the other hand, he has the theoretical language, L_T, whose descriptive statements only contain theoretical terms as their descriptive terms; in addition, sentences of L_T can also include variables of different types and logical expressions typical for modal and higher-order logics and mathematical expressions needed for T. Observational terms have the usual meanings they are given in an everyday language, designating observable properties (of events or things) or observable relations (between those events or things). In addition, L_O has to satisfy requirements such as that of nominalism (the values of the variables must be concrete observable entities) or of finitism (in the weakest form of this requirement: L_O has at least one finite model). All this entails that there is no problem with existential claims made in an internal sense within L_O: as we saw before, these sentences require empirical investigation. Existence sentences, asserted for some observable entities in the internal sense, allow us to establish that those observable entities are 'real' in one of the uses acceptable for Carnap; namely, the common-sense use. This common usage is also appropriated for the question of the 'reality' of possible observable events, as Carnap indicated in a passage clearly reminiscent of the Tarskian analysis of truth:

> The statement that a specified possible observable event, e.g., that of this valley having been a lake in earlier times, is real means the same as the statement that the sentence of L_O which describes this event is true, and therefore means just the same as this sentence itself: 'This valley was a lake'. (Carnap, 1956, p. 45)

Regarding theoretical terms, they obtain empirical interpretation thanks to correspondence rules. These are sentences which directly connect some theoretical terms of an empirical theory with observational terms. The rest of the theoretical terms, which are not components of the correspondence rules of a theory, acquire empirical interpretation indirectly, since they are connected through the postulates of the theory to theoretical terms directly interpreted via correspondence rules. The empirical interpretation of each theoretical term in a theory is partial and open because it is always possible to incorporate new correspondence rules into the theory that will affect the meaning of the term.[20] In this way, the meaning of a theoretical term is relative to a theory insofar as it is represented by the postulates and correspondence rules of that theory. Postulates alone are an uninterpreted calculus; but postulates alone do not represent an empirical theory. To perform this function, the postulates need to go hand in hand with a list of correspondence rules. One may think that there would be a parallelism between the use of 'real' in connection with L_O (or common use) and its use related to L_T. In fact, Carnap stated that this is the case for

[20] The way in which new correspondence rules in a theory affect the meaning of a theoretical term should not be understood in the sense of a radical variation of meaning; that surely would require a new theory with new postulates and correspondence rules for the theoretical term (syntactically considered).

events described in theoretical terms (e.g., the reality (in the internal sense) of a cloud of electrons moving in a specified way here and now): "to accept a statement of reality of this kind is the same as to accept the sentence of L_T describing the event" (Carnap, 1956, p. 45). However, the parallelism breaks down in some sense (as Carnap recognized) when the question of the reality is not about a specified theoretical event (or some specified theoretical events), but about the theoretical entities in general designated by a given theoretical term such as 'electron' or 'electromagnetic field'. The questions concerning the reality of electrons (in general) or about electromagnetic fields, or about theoretical entities of any other kind, were considered ambiguous by Carnap. It is usual (at least in philosophical contexts) to think of the affirmative answer to these questions as a claim made by a metaphysical realist who is accepting entities of a *kind*, where that kind is the result of 'carving nature at its joints'. Nevertheless, Carnap envisioned an acceptable way to answer these questions affirmatively that does not take this interpretation but does correspond to the scientific use of 'real'. In this regard, he noted:

> … if we agree to understand the acceptance of the reality, say, of the electromagnetic field in the classical sense as the acceptance of a language L_T and in it a term, say 'E', and a set of postulates T which includes the classical laws of the electromagnetic field (say, the Maxwell equations) as postulates for 'E'. For an observer X to "accept" the postulates, means here not simply to take T as an uninterpreted calculus, but to use T together with specified rules of correspondence C for guiding his expectations by deriving predictions about future observable events from observed events with the help of T and C. (Carnap, 1956, p. 45)

Through such a comment, Carnap (unsurprisingly) presents the acceptable answers about the reality—the existence—of theoretical entities of a certain kind as sentences in the internal sense; that is, once the L_T of an empirical theory and the theory itself (represented by T and C, or briefly TC) have been accepted. It is obvious from the previous comment that, once an empirical theory has been adopted and with it its theoretical language, the continued acceptance of the theory depends on empirical inquiries into the observational consequences that are inferred from that theory (given certain initial observational reports). In any case, bearing in mind this internal sense, it seems that, in the same way that variables of L_O have observable entities as values, it can be said that variables of L_T for an accepted theory have, in addition to mathematical entities,[21] theoretical entities as values. Nevertheless, while it is relatively easy to clarify the point of view that the observable entities are not metaphysical entities, but rather those that allow us to describe the perceptual experience (given L_O), it is more difficult to clarify the idea that the supposed theoretical entities are not metaphysical entities.

Let us see how Carnap proceeded in his paper of 1956. In the same way that he contemplated certain requirements for the values of the variables of L_O (as we saw

[21] Mathematical entities are needed because empirical theories may incorporate mathematics and according to Carnap (1956), given the conditions assumed for L_O, the mathematical language of an empirical theory is only introduced via L_T.

above), he also proposed certain conditions—he called them conventions—for the domain, *D*, of entities admitted as values of the variables in L_T:

> *C1*. *D* includes a denumerable subdomain *I* of entities.
> *C2*. Any ordered n-tuple of entities in *D* (for a finite n) belongs also to *D*.
> *C3*. Any class of entities in *D* belongs also to *D*. (Carnap, 1956, p. 43)

Thus, *D* is a domain of different types based upon the entities of the subdomain *I*. In order to ensure the denumerability of the subdomain *I*, Carnap proposed we consider its elements as natural numbers. This is certainly correct in order to have of "all kinds of entities needed in the purely mathematical part of L_T" (Carnap, 1956, p. 43); but what about the theoretical entities?

Carnap (1956, pp. 47–48) proposed basing the theoretical entities on a particular space-time coordinate system, where the space-time points *are ordered quadruples of real numbers*, and a space-time region is a class of these space-time points. Any scientific system (e.g., a physical or a psychological system) can be conceived of as ascribing properties, relations or quantitative magnitudes either to points of space-time, or to a space-time region as a whole, or to certain space-time regions or a class of them. The extensions of those properties, relations and quantitative magnitudes are mathematical constructs based upon the previous space-time coordinate system. Surprisingly, thus presented, we would have that all theoretical entities of different kinds, in an accepted empirical theory, would be mathematical entities.

Of course, this way of presenting theoretical entities avoids considering theoretical entities as metaphysical entities (once it is assumed that claims about the existence of mathematical entities only have a place in the internal sense, as Carnap had already defended in his (1950a [1956]) work and was still defending in his 1956). In this regard, Carnap (speaking about the appropriate way—i.e., in the internal sense—to make claims about the reality of theoretical entities in general) proposes the use of the expressions "*O*", "*O*´", "*O*´´", etc., as designating natural numbers—i.e., the entities of the subdomain I. And he adds:

> To say that "*O*" designates the number zero, "*O*´" the number one, etc., gives merely the psychological help of connecting these expressions for the reader with useful associations and images, but should not be regarded as specifying part of the interpretation of L_T. All the interpretation (in the strict sense of this term, i.e., observational interpretation) that can be given for L_T is given in the *C*-rules, and their function is essentially the interpretation of certain sentences containing descriptive terms, and thereby indirectly the interpretation of the descriptive terms of V_T. On the other hand, the essential service that the expressions "*O*" etc. give, consists in the fact that they represent a particular kind of structure (viz., a sequence with an initial member but no terminal member). Thus, the structure can be uniquely specified but the elements of the structure cannot. Not because we are ignorant of their nature; rather, there is no question of their nature. But then, since the sequence of natural numbers is the most elementary and familiar example of the sequential structure here in question, no harm is done in saying that those expressions designate entities and that these entities are the natural numbers, as long as we are not misled by these formulations into asking metaphysical pseudo questions. (Carnap, 1956, p. 46)

We interpret Carnap as adopting a deflationist account of the entities designated by the expressions "*O*", "*O*´", "*O*´´", etc. In this regard, it is relevant to keep in mind what he says about the designation of these expressions. His proposal is that they designate natural numbers as a way of merely giving the "psychological help of connecting these expressions for the reader with useful associations and images" (*ibid.*). He also adds that "no harm is done in saying that those expressions designate entities and that these entities are the natural numbers, as long as we are not misled by these formulations into asking metaphysical pseudo questions". This comment can be extended to any theoretical entity, as designated by theoretical terms, given that the theoretical entities become, as a result of this proposal, mathematical constructs obtained by *C2* and *C3*. In any case, the previous paragraph contains another comment in which Carnap stresses that the relevant service of the expressions is that they represent a structure: "a sequence with an initial member but no terminal member". Psillos (1999, pp. 47–48) interprets this as a manifestation of the early structuralism allegedly expounded by Carnap, adding:

> [...] the appeal to structuralism seems to free Carnap from any explicitly realist account of existential implications of scientific theories. More importantly, it seems to free him from any internal commitments to physical unobservable entities as the referents of theoretical terms. For it implies that what really matters is the logical-mathematical structure of the domain of the theory and not what the nodes in this structure may be. (Psillos, 1999, p. 48)[22]

We have, however, a different interpretation. We agree that Carnap rejected "internal commitments to physical unobservable entities as referents of theoretical terms", although we think that the reason is another: for Carnap it makes no cognitive sense to speak of internal commitments *to physical unobservable entities in general*. For him, to think of the reference of theoretical terms (for accepted theories) as physical unobservable entities means slipping into the realm of metaphysics or hard ontology. Carnap, after the mid-1950s, considered that theoretical terms (of accepted theories) have a reference; that is part of the semantics he accepts. In the internal sense, it is right to speak of theoretical entities as the referents of theoretical terms for a confirmed theory. But they must not be identified as physical entities, since in that case there would be no difference with their consideration in an external sense: they would be entities in the real world (i.e., in a metaphysical sense). Of course, Psillos is aware that the scientific sense of 'real' that Carnap assumes as appropriate is not the external sense, but he seems to demand of Carnap's scientific sense of 'real' more than it can provide. According to our interpretation, the scientific sense of 'real' applied to theoretical entities, as considered by Carnap, is fully deflationary relatively to the accepted language. Given this, theoretical entities cannot have a substantive (i.e., metaphysical) nature as they would have if they were identified with physical entities.[23]

What about the idea that Carnap (1956) propounds an incipient structuralism? Maybe that is true. However, it may be the case that the only point that Carnap wanted to highlight was that this structure is what is needed in order to have all the

[22] Friedman (2011, p. 254), like Psillos, also interprets Carnap's (1956) approach as "already envisioning a kind of structuralist view".

[23] Surprisingly, Psillos, in the quote reproduced above, is announcing Carnap's deflationary position. But Psillos presents that deflationary position as if Carnap surrendered to instrumentalism. See also comments in Psillos (2006, pp. 77–78) in relation to Ramsey sentences.

theoretical entities which are possible referents of the theoretical terms of an empirical theory in a deflationary sense; i.e., without dealing with questions about their nature or their being independently of a theoretical language. In any case, it should be noted that this is the structure of the subdomain I (where I is the basis for the entities that are referenced by the theoretical terms of a theory), and not the structure of the total domain, D, for a theory as Psillos says.[24] In this regard, it is important to keep in mind the idea that the variables of L_T do not range over the elements of I, but rather over the elements of D.

In any case, it seems to us that the important issue for Carnap is to secure referents for the theoretical terms of accepted empirical theories, without attending to the nature of the elements of D. Taking this into account, some of Carnap's remarks acquire full relevance, such as his comment that the appeal to mathematical entities as possible referents of theoretical terms "should not be taken literally but merely as a didactic help by attaching familiar labels to certain kinds of entities" (Carnap, 1956, p. 46).

2.4 Carnap and the Ramsey Sentence

Carnap (1958 [1975] and 1966 [1974]) introduces a new aspect of the entities designated by theoretical terms by incorporating the Ramsey sentence (henceforth: *RS*) into his analysis. The *RS* of an empirical theory is due to Ramsey (1929 [1978]).[25] The problem of the reference of theoretical terms is addressed by the last of Carnap's works in relation to his renewed attempt to explicate the analytic/synthetic distinction for any axiomatized scientific theory using its *RS*.[26] The application of the Ramsey method, also called Ramseyfication, to a theory requires the theory to be finitely axiomatized. In a *RS*, the theoretical terms of an empirical theory, *TC* (the conjunction of its theoretical postulates and its correspondence rules), are replaced by higher-order variables of the appropriate arity and type, bound by existential quantifiers at the beginning of the resultant new expression, while the observational terms are preserved. According to this, given a theory, *TC*, whose theoretical terms are $\tau_1 \ldots \tau_n$ and whose observational terms are $o_1 \ldots o_m$, its formulation is represented by $TC[\tau_1, \ldots, \tau_n, o_1, \ldots, o_m]$. The *RS* of *TC*, or ^{R}TC, is: $\exists X_1 \ldots \exists X_n TC[X_1, \ldots, X_n, o_1, \ldots, o_m]$.[27] The peculiar feature of ^{R}TC is that it has the same observational consequences as *TC*.[28] Apparently, Ramsey's substitute for the original theory circumvents the problem of the meaning of theoretical terms by dispensing with them.

---

24 See Psillos (1999, p. 47).

25 Psillos (1999, pp. 48–51) provides an interesting historical account of how Carnap came to include the Ramsey sentence in his approaches.

26 For Carnap's sense of "explication" see Carnap (1950b, pp. 1–18).

27 See Carnap (1966 [1974, pp. 249–251]).

28 For a proof of this see Stegmüller (1970).

And, in fact, in some sense it does, because if the theoretical terms are not in the *RS*, then questions about their meanings do not arise. In any case, it was not Carnap's purpose to elude the problem of theoretical terms.[29] First, because for him, as we have already indicated, the theoretical terms of an empirical theory, *TC*, have meaning in the context of *TC*, and acquire (partial and open) empirical meaning via the correspondence rules.[30] Second, because, strictly speaking, ^{R}TC does not say the same as *TC*. It is true that both ^{R}TC and *TC* have the same observational sentences as consequences, and hence that they have "the same *explanatory and predictive power*" (Carnap, 1966 [1974, p. 252]), but as Carnap recognized:

> The [...] language [of physics] presupposes that theoretical terms, such as 'electron' and 'mass', point to something that is somehow *more* than what is supplied by the context of the theory itself. Some writers have called this the 'surplus meaning' of a term. When this surplus meaning is taken into account, the two languages are certainly not equivalent. (Carnap, 1966 [1974, p. 254])

In fact, Carnap used the *RS* for another purpose: to offer a new solution to the problem of the analytic/synthetic distinction in a theory. The earlier application of this distinction to theories by the neopositivists (for non-logical–mathematical statements) entailed accepting that the empirical content was guaranteed by the different observational statements and hence the synthetic part was derived from them. Meanwhile, the analytical part had to consist of meaning postulates for the non-observational (or theoretical) terms, which would make their content clear, to the extent that they were reducible to the observational terms. But these earlier attempts to make those meaning postulates explicit for the different theoretical terms were shown to be flawed. It is not possible to make isolated meaning postulates explicit for certain theoretical terms in such a way that some meaning postulates provide the full semantic content for non-observational terms by reductive procedures.[31] Contrary to that account, it should be borne in mind that, if the correspondence rules for such terms only provide part of their empirical content, then they cannot be considered as analytic sentences.

---

[29] Nor was this the aim of Ramsey himself, but rather to demonstrate the indispensability of theoretical terms (at least if we consider the interpretation of Braithwaite (1953)).

[30] Uebel (2011, p. 74) claims that "Carnap had no intention of amending his view of theoretical concepts as only partially interpreted by observational ones".

[31] In Carnap (2000), a talk discovered and published by Psillos (2000b), there is a proposal for the definition of each theoretical term of a theory, *TC*, using Hilbert's ε-operator. It is pertinent to clarify that the definition (or meaning postulate) that is provided for each theoretical term t_i ($1 \leq i \leq n$) of *TC* by means of that operator is not fixed without the set of conditions of the Ramsey sentence ^{R}TC, so that a semantic holism for the theoretical terms is not avoided. Moreover, this procedure provides only *partial* definitions, since the postulates *T* and *C* can always change; and furthermore, for each theoretical term t_i of *TC* the definition takes *an indeterminate representative value* of one of the sequences of values that satisfy the Ramsey-variables of ^{R}TC (for details, see Carnap (2000, pp. 170–172) and Psillos (2000b, 155–157)). Therefore, this type of definition does not provide the *full semantic content* of theoretical terms by reductive procedures. In addition, this final observation preserves the deflationary character of the reference of a theoretical term.

Now, the *RS* contains all the observational consequences of an original theory and so, according to Carnap, it provides the syntactic content of that theory with two peculiarities: (i) it fulfills this function with the help of an expanded observational language, including expressions for "embracing virtually the whole of mathematics"—among them variables of different types—(Carnap, 1966 [1974, p. 253]); and (ii) the syntactic content of the original theory is established holistically by the *RS*. The synthetic part of a theory (at a given moment) is a single complex sentence. For Carnap, ^{R}TC provides the synthetic content of *TC*, and *TC* (jointly) encapsulates its synthetic and analytic content. Given this, the expression which in conjunction with ^{R}TC has *TC* as a consequence must provide the analytic content of *TC*. Hence, such analytic content is the material conditional $^{R}TC \rightarrow TC$ (i.e., $\exists X_1 \ldots \exists X_n T[X_1, \ldots, X_n, o_1, \ldots, o_m] \rightarrow TC[\tau_1, \ldots, \tau_n, o_1, \ldots, o_m]$). This is called the Carnap sentence. For Carnap, the conditional $^{R}TC \rightarrow TC$ is factually empty. It does not say anything about the world. The Carnap sentence simply establishes that: if ^{R}TC has some realization (i.e., if there is at least one realization that satisfies ^{R}TC), the n-tuple of entities named by the theoretical terms $\tau_1, \ldots, \tau_n$ realizes *TC*; or in the words of Carnap: "if the Ramsey sentence is true, we must then understand the theoretical terms in such a way that the entire theory is true" (Carnap, 1966 [1974, p. 270]). This conditional, so considered, is always true: when the antecedent is true, the consequent is true, and so the conditional is true; when the antecedent is false, the conditional is also true. Furthermore, Carnap's notion of $^{R}TC \rightarrow TC$ expressing the analytical content of an empirical theory supposes accepting the theoretical terms as expressions with meaning: "It is a purely analytic sentence, because its semantic truth is based on the meanings intended for the theoretical terms." (Carnap, 1966 [1974, p. 270]) In this sense, Carnap's use of the *RS* to establish the holistic analytic content of a scientific theory with the conditional $^{R}TC \rightarrow TC$ shows, once more, that he was not trying to dispense with the theoretical terms.[32]

Irrespective of this, returning to the issue of the denotations of theoretical terms, it is important to keep in mind that, if the antecedent of $^{R}TC \rightarrow TC$ is true, then the theoretical terms (contained in the consequent) have to denote (since the consequent is true). In addition, it is assumed that if ^{R}TC is true, then the variables that substitute these theoretical terms in ^{R}TC also have to denote, insofar as they are variables bound by existential quantifiers. However, none of these variables (Ramsey-variables, henceforth) refers to any particular class;[33] what satisfies each Ramsey-

[32] Friedman (2008, p. 396) indicates that, with regard to choosing between taking the theoretical terms themselves (as constants) of a physical theory or the existentially quantified variables that replace the former in *RS*, Carnap's suggested proposal "is to retain theoretical terms as genuine primitive constants, while relegating the use of Ramsey-sentences to a subsidiary role–that of allowing us clearly to distinguish between the analytic and the properly empirical components of our axiomatically formulated physical theories." This choice is justified for Carnap in the practical difficulty that the use of a *RS* would involve when it comes to formulate a simple statement attributing a definite mass to a certain object; in this sense, for Carnap it is more convenient the ordinary discourse of physics (see Carnap, 1966 [1974, p. 254]). On the holism of the Carnap sentence, see Uebel (2011, p. 74).

[33] As Psillos (1999, p. 64) says, Ramsey-variables have *multiple realizations*.

variable remains undetermined (see Carnap (1966 [1974, p. 251])). That said, once again it might be surprising to see that Carnap (1966 [1974, p. 253]) chose to consider natural numbers and natural number constructs (e.g., classes of natural numbers, classes of classes of numbers, etc.) as the referents of such variables, as he did in his 1956. His reasons for proceeding in this way should be the same as in his (1956): the referents of theoretical terms or of Ramsey-variables should not be identified with *physical* entities that exist outside in the real world since, if they were, there would be no difference with their consideration in an external or metaphysical sense. Carnap's concern in this regard is clear when, introducing the *RS*, in his (1966 [1974]) he asks:

> How can the right of a scientist to speak of theoretical concepts be justified, without at the same time justifying the right of a philosopher to use metaphysical terms? (Carnap, 1966 [1974, p. 248])

And, under this interpretation, it is not surprising that in his reply to Hempel[34] about the ontological commitments of higher-order variables bound by existential quantifiers in a *RS*, Carnap says:

> I agree with Hempel that the Ramsey-sentence does indeed refer to theoretical entities by the use of abstract variables. However, it should be noted that these entities are not unobservable physical objects like atoms, electrons, etc., but rather (at least in the form of the theoretical language which I have chosen [...]) purely logical-mathematical entities, e.g., natural numbers, classes of such, classes of classes, etc. Nevertheless, ^{R}TC is obviously a factual sentence. It says that the observable events in the world are such that there are numbers, classes of such, etc., which are correlated with the events in a prescribed way and which have among themselves certain relations; and this assertion is clearly a factual statement about the world. (Carnap, 1963b, p. 963)

Be that as it may, another interpretation along instrumentalist lines has been offered by Psillos (1999, p. 51 ff.) and by Lavers (2016, p. 215 ff.). The latter came to assert:

> Just as Carnap sought to allow empiricists to overcome their nominalistic scruples, he himself should have more fully overcome his instrumentalist scruples. (Lavers, 2016, p. 217)

And Psillos (1999, ch. 3) presented Carnap's ontological considerations in relation to the *RS* as an attempt to remain neutral between instrumentalism and realism (in a metaphysical sense), but in fact shifting at times towards the former and at times towards the latter. According to Psillos (1999, pp. 55, 57), when Carnap eludes commitments to theoretical entities and adopts mathematical entities as what the higher-order variables of the *RS* range over, in some sense he is succumbing to instrumentalism.[35] This instrumentalist interpretation would seem more plausible if

[34] Hempel (1958, p. 51) says that Ramsey sentences do not provide a satisfactory way of avoiding theoretical concepts, given that the references of Ramsey-variables are not avoided.

[35] We say "a way of succumbing *in some sense* to instrumentalism" because: (i) Psillos (1999, pp. 59–60) clarifies that Carnap does not assume an instrumentalism *tout court* which includes *negative assertions* about the existence of theoretical entities (and about a theory being true or false); and (ii) Psillos (1999, ch. 3), as we will see later, tries to present Carnap (in the 50s and 60s) as adopting (without great success) neutralism between instrumentalism and realism. Nevertheless, the final picture that Psillos provides of Carnap's approach is that, adopting the *RS* method, his

we considered only the final part of the quote above from Carnap's reply to Hempel. However, we must bear in mind that previously, as we have seen, Carnap (1956) had indicated that resorting to mathematical entities as theoretical term referents was nothing more than *a psychological aid*. Such help is required because, once certain theoretical terms have been introduced within a confirmed empirical theory, these terms (and the Ramsey-variables) designate entities; but, on the other hand, Carnap refused to present these entities as if their nature were determined in the external way of metaphysics. Perhaps this is what Carnap warns us about when, in the quotation from his reply to Hempel, he says that the theoretical entities referred to by the Ramsey-variables are mathematical entities *in the theoretical language chosen by him.* In this sense, it should be noted that, when speaking in the quotation of the language chosen by him, Carnap refers to a footnote (note 41 to his (1958 [1975])) work in *Dialectica*. In that paper, the theoretical language L_T is incorporated in a way that is similar to how Carnap already did so in his (1956): expanding the language of mathematics with theoretical terms, with the peculiarity that the most basic entities of this language are provided by a subdomain, now called D^0, about which Carnap points out:

> Let the structure of L_T be such that it contains a type-theoretic logic with an infinite sequence of domains D^0, D^1, D^2, *etc.* D^n is called the domain of the *n*th level. Each variable and each constant belongs to a definite level. A variable of the *n*th level has D^n as its range of values and a constant of the *n*th level denotes an element of D^n. The elements of D^0 are the members of the infinite sequence *0, 0′, 0″*, etc. *Officially, no meaning is given to these logical expressions.* Their use follows from the rules of the language. *We shall, however, in order to relate these expressions to familiar concepts, unofficially regard* D^0 *as the domain of natural numbers with '0' denoting the number* 0, *'0′' the number* 1, etc. (Carnap, 1958 [1975, p. 76]; emphasis added)

However, Carnap's unofficial use of the entities *0, 0′, 0″*, etc. and of constructs derived from them—mathematical entities—can lead us to think that sentences about such entities are mere mathematical statements, or at least non-synthetic statements.[36] Carnap (1958 [1975, pp. 80–81]) avoids such confusion through an analogy between statements (1) 'the cardinal number of the planets is 9' and (2) 'the mass of the body *k* is 5'. The first statement is expressed by '$n_P = 9$', where n_P expresses 'the cardinal number of the planets'. And the second is expressed by '$F(\alpha) = 5$', where '$F(\alpha)$' expresses 'the mass of α', and 'α' stands for an ordered quadruple of real numbers which is assigned to the body *k* (as representative of a space-time point in a space-time coordinate system). So, 'n_P' is a descriptive expression, although it speaks of a number, and the statement '$n_P = 9$' is synthetic (in fact, it is a false statement, taking into account the new definition of 'planet', according to which Pluto is now considered to be a dwarf planet). In an analogous way, '$F(\alpha)$' is a descriptive expression (a physical magnitude or function) and the statement '$F(\alpha) = 5$' is synthetic. The last statement is synthetic, even though the admissible

---

empiricism collapses to structural realism, without in fact being a realist (in the metaphysical sense) about the existence of theoretical entities (see Psillos, 1999, pp. 67–69).

[36] Obviously, the sequence *0, 0′, 0″*, etc. is isomorphic to the natural numbers.

arguments of F are mathematical entities and the values of F are also mathematical entities, because the arguments of F stand for spatial positions (or regions) and by means of them, finally stand for bodies (thanks to a correspondence rule between coordinates and observable entities). According to Carnap (1958 [1975, p. 81]), it is important to establish that "F is identical with a mathematical function of the same logical type", and that there is a similar identity with any other theoretical term and a mathematical expression. In order to assess Carnap's consideration of these identities, we ought to unofficially adopt mathematical entities as the referents of theoretical terms (and of corresponding existential quantified variables in ^{R}TC). Thus, it is assured that, for the undetermined theoretical entity designated by a theoretical term, there is a mathematical entity designated by the mathematical expression which is identical to that theoretical term. And so, it is ensured that we can always, as *psychological help or unofficially*, connect the theoretical terms to mathematical entities for the semantics of the language of a theory, *TC*.

In the same paper, in answer to questions such as "How should we conceive these things which we can never directly observe?", Carnap (1958 [1975, p. 80]) says that: "These difficulties do not enter for the form of theoretical language described above." And the reason for this is that he is only assuming two types of entities: "observable things and their observable properties and relations", and "mathematical objects of the types of systems described earlier [in the same paper], namely the natural numbers, classes of these and so forth." But we have to remember that, for Carnap, associating theoretical entities to mathematical entities is just a way of ensuring denotations for the theoretical terms of a theory, *TC*, (and for variables substituting them in the corresponding *RS*) without adopting commitments to unobservable entities (i.e., in an external sense), and so avoiding pronouncements about metaphysical entities.

In any event, it should be noted that avoiding such pronouncements should not be understood as if Carnap held that *there are no such unobservable entities*. By adopting such a position, Carnap would have propounded a negative claim of existence in the external sense about unobservable entities, which according to him (as we have seen previously) would lack cognitive content (just as much as any positive claim of that kind would).[37] His point of view is against realism (in a metaphysical sense), but also against an instrumentalist view when it includes *negative assertions* about the existence of theoretical entities.

As we have argued, Carnap maintained an agnostic position regarding the real existence of theoretical entities, as he dismissed any kind of (positive or negative) answer to external questions. And his position in relation to the realism vs. instrumentalism debate was agnostic as well: he rejected both. We think that to describe Carnap's position as neutralist, as Psillos does,[38] is highly problematic. Being

[37] Psillos (1999, p. 58) says: "It seems as though Carnap's neutralism is difficult to maintain: every attempt to restore an empiricist equidistance between scientific realism and instrumentalism makes him fall towards one of these positions".

[38] See Psillos (1999, pp. 40 and 58 ff.) Lavers (2016, p. 217) and Friedman (2011, p. 252) adopt a similar view of Carnap's neutralism. Carnap's position should not be confused with Feigl's (1950)

neutral means not aligning oneself with any position in a certain controversy, but also not expressing oneself against any part of the controversy. Again, it seems to us that Carnap is against both instrumentalism and realism (when we interpret them in a metaphysical sense), since both parts try to give answers to external existence questions concerning theoretical entities. In the case of instrumentalism, these are always negative answers; in the case of realism, they are sometimes positive and other times negative answers (depending on the existential questions). Carnap rejects both kinds of answers to existential questions because both lack cognitive content. In this sense, Carnap is not maintaining an equidistant position between instrumentalism and realism. So, we prefer to say that Carnap is agnostic. In fact, as Uebel (2010, p. 305) has highlighted, Carnap's "own stance was not one of indifference between the ontological commitments of the instrumentalist and the realist, but a position that even rejects the ontological commitments both of them share, namely, to observables".[39]

So far, we have tried to show that Carnap's distinction between external and internal questions analogously allows him to avoid nominalism and Platonism, on the one hand, and instrumentalism and realism, on the other. However, various aspects of Carnap's previous approach have been pointed out as being problematic. Thus, the analytic–synthetic distinction has been widely questioned in the literature since Quine's critique. Despite this, as indicated by Thomasson (2015, p. 52) "Quine's arguments against the analytic/synthetic distinction are far from decisive". Without entering into a debate on this issue, let us grant Carnap the legitimacy of the distinction, since the controversy cannot be considered to have been resolved against his stance.[40]

Another source of attacks is the assumption that the verificationist conception of meaning is what is behind Carnap's view that only existential sentences in the internal sense have meaningful content. We think such attacks are ultimately invalid. Carnap had already rejected verificationism and embraced a weaker position called confirmationism in his 1936 work. A confirmationist conception of meaning only requires expressions (terms and sentences) that are used with a purportedly descriptive function; and cannot be applied to expressions that are not used with this purported descriptive function. Non-descriptive expressions can have meaning in a broad sense and can provide, or contribute to, intelligible sentences. This confirmationist position was assumed by Carnap after the mid-1950s. It is not a general approach to the meaning of any expression; it is only a restricted conception to show whether a purportedly descriptive expression actually contributes to a sentence obtaining either direct confirmation/disconfirmation (when it is observational) or confirmation/disconfirmation through the observational consequences which follow

semantic realism either, as Psillos (1999, pp. 44–45) does. We think that Uebel (2010, p. 301) is right in his considerations regarding this issue.

[39] See our comments about thing-language in Sect. 2.1 of this paper.

[40] Demopoulos (2008, p. 365) considers that "the essence of Carnap's conception of the analytic is that it is *non-factual*—not on *meaning* or *synonymy*", and that "[f]actual content is explicated in terms of the observational vocabulary of a theory".

from an empirical theory thanks to that expression. The confirmationist condition for the meaning of descriptive expressions is easier to accept than a general verificationist conception of meaning, and even than a totally reductive verificationism for non-observational purportedly descriptive expressions. Furthermore, if the confirmationist conception of meaning is sometimes presented as a form of anti-realism, this is completely incorrect as far as anti-realism is attributed to Carnap (as we have been defending).[41] Once again, we must reiterate that Carnap considers that both affirmative and negative responses to external questions lack cognitive content, and this is so because they cannot be confirmed or refuted, despite appearances to the contrary. Carnap adopted an agnostic position in the face of realism and anti-realism (in a metaphysical sense).

But there is another source of problems, directly connected to Carnap's use of the *RS*, which is related to what is known as 'the Newman objection'. This is precisely what we consider in the next section.

2.5 The Ramsey Sentence and the Newman Objection

The mathematician M. H. A. Newman (1928) established a critique of Russell's (1927) suggestion that all knowledge about the unobservable world has to do with its structure, which is in addition to perceptual (or observational) knowledge of the world. That structure provides only logical–mathematical properties, so that, according to Russell's suggestion, no knowledge of unobservable entities (properties and relations), at least no knowledge of their nature, is possible. In this respect, non-observable entities are undetermined.[42] Newman's critique consists of showing that it is *trivial* to assert that an empirical theory, whose observational consequences are true (i.e., a theory which is empirically adequate), only provides *knowledge* of the purely logical–mathematical structure of the unobservable world without constraints regarding its components (except for the requirement that the cardinality of its domain allows it to contain all the entities that are needed for that structure).[43] Hence, if knowledge about the unobservable world is only about its structure, then once the cardinality of the domain has been established, the structure is obtained by mere set-theoretic procedures. As Demopoulos and Friedman (1985, p. 188) say:

> … on this view, only cardinality questions are open to discovery! Every other claim about the world that can be known at all can be known a priori as a logical consequence of a set of α-many objects.

---

[41] See comments by Hirsch (2011, p. 220) and Thomasson (2015, p. 35).

[42] For a comprehensive analysis of Russell's suggestion in his (1927) and the Newman objection, see Demopoulos & Friedman (1985). In that paper, Demopoulos and Friedman recovered the Newman objection from oblivion.

[43] See Psillos (1999, pp. 61–65) and Demopoulos (2003, pp. 387–388, 2007).

Insofar as ^{R}TC for an empirical theory *TC* replaces the theoretical terms of the latter by Ramsey-variables and has the same observational consequences as *TC*, it can be said that ^{R}TC provides, in addition to those observational consequences, a logical–mathematical structure. Along these lines, Maxwell (1970) connected Russell's considerations about the structure of the unobservable world with the *RS* and, on such a basis, proposed that we understand the latter in the light of a version of structural realism; that is, the *RS* for an empirically adequate theory would provide what Russell (1927) suggested was knowable from the world.[44] Psillos (1999, ch. 3) takes a further step by presenting Carnap's use of the *RS* as orienting his empiricism towards structural realism.[45] Let us clarify that initially Psillos does not seem strictly to be claiming that Carnap, by adopting the Ramsey method, is assuming that the structure provided by an empirically adequate theory is a real structure (in a metaphysical sense) of the material world (as in Russell's structural realism). However, Psillos orients his analysis in that direction, since, in considering Carnap's use of the *RS*, he appeals to the connection established by Maxwell between the *RS* and Russell's structural realism. And in this way, Psillos proceeds to attack, on the one hand, Carnap's supposed neutralism given that he makes use of the *RS* and, on the other hand, those who adopt structural realism based on the *RS*. Psillos finally assumes (1999, p. 69) that Carnap, in adopting the Ramsey method, is paying "the price of collapsing his empiricism to structural realism".

In any case, Psillos clearly states that Carnap's use of an empirically adequate *RS* encounters a problem in the Newman objection, namely, what Psillos (1999, p. 64) calls the problem of *trivial realization*. In our opinion, regardless of Psillos's purpose, this problem must be distinguished from the question of whether the assumption of structural realism can be attributed to Carnap.

The Newman objection seems to be intensified by Carnap's use of the *RS*. The reason for this is that Carnap assumes, as we have seen, the infinite sequence $0, 0', 0''$, etc., to be members of a basic subdomain (D^0) of the extended observational language (L'_O) of the *RS*. This subdomain ensures set theoretic constructs are obtained that can realize the structure provided by any consistent *RS*. So, its basic subdomain allows for all the entities that are needed for classical mathematics to be contained. There is no constraint to be considered with regard to cardinality, since the cardinality needed is trivially ensured given the potentiality of the logical–mathematical language of the *RS* as used by Carnap. In this case, a cardinality constraint would be vacuous. Hence, if the *RS is* empirically adequate and the original theory is consistent, its realization is trivial, given the infinite sequence of the basic subdomain. The relational structure provided by the *RS* is thus guaranteed *a priori*, given that we assume the consistency and empirical adequacy of *TC*. It follows that existential assertions about theoretical entities *become trivial and a priori*.[46]

[44] For an interesting comparison of Russell's and Ramsey's proposals, see Demopoulos (2003, pp. 372–384).

[45] See also Psillos (2000a).

[46] See Demopoulos (2003, pp. 387–388, 2007, pp. 263–265). See also Uebel (2010, p. 303; 2011, pp. 75).

Since these considerations by Psillos, some other interesting comments have been made regarding the question of how Carnap's use of the *RS* is affected by the Newman objection. Among those who have participated in this debate, we want to highlight the contributions by Demopoulos, Friedman and Uebel.

Demopoulos (2003, pp. 371–372, 382) indicates that Ramsey himself, with his *RS* proposal, "refrains from offering a positive account of the character of our knowledge of theoretical relations along the lines of Russell's structuralism". However, in the case of Carnap, Demopoulos (2003, p. 393) assumes that by adopting the *RS* reconstruction he is (implicitly) sharing Russell's theory of knowledge developed in his (1927). Despite adopting such an interpretation (which, as we will see, we do not share), Demopoulos (2003, §§4 and 5) provides an interesting analysis of the connections between what he calls the Carnap–Ramsey reconstruction and, on the one hand, Putnam's model-theoretic arguments (Putnam 1977) and, on the other, van Fraassen's (1980) semantic view of theories and constructive empiricism. Regarding the latter, Demopoulos extends application of the Newman objection to it. Finally, Demopoulos (2003, §7), motivated again by the Newman objection, advocates a realist point of view with regard to the referents of theoretical terms of 'true' scientific theories and not for mere structural realism (which he attributes to Carnap's approach). Nevertheless, there is some change in Demopoulos's paper of (2007), since he accepts that some sort of restriction would suffice in order to stop trivialization. This restriction must not be of a realist one, as it could in turn appeal to artificial kinds. Demopoulos's point (2007, pp. 264–265) is to show that there is a difference between the *RS* and *TC*: while *TC* contemplates *specific entities as referents of theoretical terms* establishing specific restrictions of a synthetic nature on the observational consequences allowed, the truth of the *RS* is ensured almost analytically (assuming it is empirically adequate). By adopting this new position, Demopoulos clearly surpasses his initial realistic orientation, regarding the referents of theoretical terms as an alternative to the Newman objection to Ramseyfication. At the same time, he assumes a certain deflationary possibility for the reference of such terms, as the original theory with its theoretical terms designating specific entities (even if they are not real in a metaphysical sense) lacks the triviality of the *RS* and the obscurity of appealing to natural kinds (properties or relations).[47]

Friedman (2011, p. 252) assumes that Carnap's view has much in common with structural realism. But Friedman (2011, pp. 251–252) also advocates a close connection between Carnap's conception of the *RS*, as providing the empirical content of a scientific theory, and what according to van Fraassen matters in a scientific theory: the fact "that observational phenomena are embeddable into an abstract model for the theory".[48] From these connections between Carnap's view and both van Fraassen and structural realism (perhaps with some nuances, as we will see),

---

[47] See Demopoulos (2007, pp. 265, n. 17) for his own contrast to Psillos's position appealing to natural kinds. See also Demopoulos (2008, p. 376).

[48] In a similar way Psillos (1999, p. 55) speaks of Carnap's use of the *RS* as providing "nothing but mathematical models in which observable phenomena are embedded". But Psillos uses this com-

Friedman asserts that "Carnap's attempt at neutrality may have succeeded after all" (*Ibidem*, 252). Nevertheless, Friedman adds:

> [Carnap] may in fact have articulated a version of structuralism that recognizes the strengths of both instrumentalism and realism while simultaneously avoiding the philosophical 'pseudo-questions' on which they appear substantively to differ. (*Ibidem*, 252)[49]

Later, in the same paper, Friedman (2011, p. 256) qualifies these considerations saying that Carnap "is in agreement with van Fraassen but not with contemporary scientific realists like Psillos". And he then immediately points out that: "The Newman problem is no objection to Carnap" (*Ibidem*). Friedman reaches these conclusions after establishing that the *RS* provides an inductive systematization, in addition to the deductive one, of observational statements (in line with what Hempel (1958) proposed).[50]

The connection between Carnap's use of the *RS* and van Fraassen's constructive empiricism (as suggested by Demopoulos and Friedman) is certainly interesting. This is especially so if we focus on van Fraassen's thesis according to which to accept a theory only involves to believe that it is empirically adequate, and not that it is (theoretically) true. But we must not forget that, for van Fraassen, the theoretical part of a theory is literally true or false (in the sense of a metaphysical realism), although this cannot be determined. This is not Carnap's position at all; for Carnap it is not only that the truth value of the theoretical part of a theory (in a realistic sense) cannot be determined, but the mere idea that it could be determined (in a metaphysical realistic way) lacks cognitive content. That is, if van Fraassen adopts a standard semantics that leaves open the possibility that the theoretical part of a theory is true, Carnap adopts a deflationary semantics in the sense that, for him, the truth or falsity of a theoretical sentence is only internal. In fact, Demopoulos's analysis is oriented towards attacking what he considers to be common between Carnap's use of the *RS* and van Fraassen's view, by extending the Newman objection to both. The reason for this is that Demopoulos understands Carnap's position along the

---

ment to present Carnap's position as part of his oscillating between instrumentalism and realism (in a metaphysical sense); in this case as a way of approaching instrumentalism.

[49] Part of Friedman's argument is to show that a proper understanding of Carnap's view on theoretical terms must be linked to Carnap's *Wissenschaftslogik*. For an analysis of the latter, see Friedman (2008).

[50] Friedman (2011, p. 256) intends to reinforce his comments with a passage in which he says that for Carnap theoretical terms "are semantically uninterpreted: we assign no designata to them in our semantical meta-language, and so Tarskian semantics (as Carnap understands it) literally assigns no truth values at all to purely theoretically sentences." He adds that theoretical terms have methodological meaning, because of the inductive connections between observational sentences. The idea that, according to Carnap, no designata are assigned to theoretical terms is not consistent with what we have already seen. In fact, Carnap's approach, as we have pointed out, must be understood along the lines of a deflationary semantics (of Tarskian orientation). Another thing is that the designations of theoretical terms are not presented as non-observational entities (as we have considered here in Sects. 2.2 and 2.3), which justifies Friedman's rejection of any agreement between Carnap's view and those of realists like Psillos.

lines of structural realism; whereas, as an alternative, Demopoulos defends (metaphysical) realism about theoretical entities.

Friedman's point of view is not exactly the same. He tries to show in his 2011 that, beyond what the approaches of Carnap and van Fraassen have in common, for Carnap the truth of the theory is not different from its empirical adequacy; while for van Fraassen, the former is indeed quite different from the latter.[51] Moreover, Friedman seems to sympathize with Carnap's stance against both instrumentalism and realism,[52] without demanding—as Demopoulos (2003) initially did—that Ramsey's variables be constrained in accordance with a realistic orientation. And, as we have seen, Friedman asserts that the Newman problem is not an objection to Carnap's view. The reason for this, according to Friedman, seems to have to do with Demopoulos's initial use (and Psillos's use) of the Newman objection to defend a realistic proposal (in the metaphysical sense) of the theoretical terms, in opposition to Carnap's view.[53] But, is this reason enough? We will try to answer this question shortly. In another respect, it is not clear whether Friedman (2011) is assuming that Carnap's use of the *RS* leads to structural realism (as advocated by Psillos and Demopoulos). His comments on Carnap's position against realism might suggest the contrary, but it seems that Friedman confines such comments to the question of theoretical entities (not to the question of the structure provided by the *RS*).

Finally, Uebel (2010, p. 306 and 309; 2011, p. 76 ff.) brings a brilliant distinction to this debate. He talks of two modes of the Newman objection that are present in the literature on Carnap's use of the *RS*: *the ontological* and *the epistemic* versions. The ontological reading of the Newman objection emphasizes the insufficiency of the *RS* according to Carnap's approach, since the assertions of the existence of the Ramsey-variables are trivially satisfied. As we have said before, the existential assertion about theoretical entities, that is supposedly not trivial at all, becomes trivial and *a priori*: "ramseyfications trivialise existence claims associated with theoretical statements" (Uebel, 2011, p. 76). A clearly representative passage of such an ontological version is found in the following comment by Psillos:

> … provided that the Ramsey-sentence is empirically adequate, we can rely just on a priori reasoning in order to discover what entities realise the theory, i.e. what unobservable entities populate the world. No empirical investigation is necessary. In the end, if no constraints are imposed on the range of the variables of the Ramsey-sentence, it is a *trivial and a priori true assertion* that there are electrons, etc. And this is clearly absurd. (Psillos, 1999, p. 62)

[51] See Friedman (2011, p. 257).

[52] Though Friedman speaks of Carnap's neutralism several times, he recognizes in a footnote that Carnap aims to defend his view "against the metaphysical excesses of both 'realism' and 'instrumentalism'" (Friedman, 2011, p. 256, n. 7).

[53] Friedman (2011, p. 256) makes this claim after stating that "Carnap is not supposing […] that an abstract theory have any synthetic or factual content *beyond* its empirical adequacy (both deductive and inductive), and, in this respect, he (as I have suggested) is in agreement with van Fraassen but not with contemporary scientific realists like Psillos". Once Friedman has rejected the Newman problem as an objection to Carnap, he indicates in a footnote that: "I cannot follow Demopoulos (2003, 2007) in considering (Demopoulos's generalization) of the Newman problem to be a fatal objection to Carnap's view of theoretical terms" (Friedman 2011, p. 256 n. 7).

On the other hand, the epistemic reading raises the problem with Carnap's use of the *RS* that, beyond its empirical adequacy and consistency, the truth of the *RS* is obtained "by what is virtually a purely logical argument" (Demopoulos, 2003, p. 388). Since obtaining an abstract model for the *RS* is *logically* ensured, then the *RS* (and so the original theory) would become almost an analytic truth with no empirical surplus over and above its observational consequences already tested. However, it is unacceptable for the theoretical part of the *RS* to be almost analytic. Hence, it could be concluded that Ramseyfications "misrepresent the empirical nature of the theoretical statements at issue" (Uebel, 2011, p. 76). This seems to entail that, contrary what is supposed to be the case, the theoretical structure in Ramseyfications is not constraining observational sentences, if these are tested and shown to be correct, and there would be no other synthetic content in the *RS* than the observational consequences.

Uebel (2011, p. 76) considers Psillos to be representative of the ontological reading, and Demopoulos and Friedman to be representatives of the epistemic reading. Attending to our previous comments, the latter attribution is not clear enough until Demopoulos (2007, 2008) and, perhaps, Friedman (2011). In fact, in these papers the authors reject, as we have seen, the claim that the ontological reading can be applied to Carnap.[54] In any case, the attributions of the readings are less important here although the distinction itself may be significant. Let us clarify that even if the distinction between the two readings is relevant, there is a sense in which the epistemic is more basic than the ontological version: once it has been established that the Newman objection allows us to obtain an abstract model that makes the *RS* of a theory true, assuming its consistency and empirical adequacy, then in order to identify an ontological problem it is necessary to add that for scientific theories to be true we should claim that there are intended interpretations of theoretical terms that capture entities of the concrete world as their referents.[55] But this supposes a realist approach to the reference of theoretical terms.

After these remarks, we now want to distinguish several issues in these considerations expressed by Psillos, Demopoulos, Friedman and Uebel: (i) whether Carnap's use of the *RS* leads to structural realism; (ii) whether the Newman objection affects Carnap's use of the *RS* in such a way that we should adopt realism (in a metaphysical sense) about the theoretical entities in our best theories; and (iii) whether the

[54] Uebel (2011, p. 78) adopts this interpretation of Friedman's claim that: "The Newman problem is no objection to Carnap" (Friedman 2011, p. 256).

[55] According to the ontological reading, logically ensuring an abstract model *M* for *RS* (with *TC* being the original theory), if we choose *M* as having the same cardinality as a model *W* whose domain is the 'intended domain' *of TC*, entails that it is possible to obtain an interpretation of the theoretical terms of *TC* in *W* which is a mere result of an isomorphism from *M* to *W* (see Demopoulos 2003, p. 387). Demopoulos (2003, p. 388) concludes: "The Carnap–Ramsey reconstruction is unacceptable because it implies that the existence of an *abstract* model for [*TC*] suffices for its truth over [*TC*'s] *intended* domain. What the reconstruction misses is a preanalytic intuition that governs our conception of truth, and this is missed even when we set aside any difficulty in fixing [*TC*'s] *intended* domain." (Let us just add that in this conclusion the preanalytic intuition about the truth is understood in terms of a realist approach.)

Newman objection has an impact on Carnap's use of the *RS* to the point of invalidating Carnap's project.

Regarding (i), our point of view is that Carnap's use of the *RS* should not be understood as implying that for him the structure given by a *RS*, if its observational consequences are correct, would provide an unobservable *real* structure of the world (as would be advocated by a structural realist). There is no reason to limit Carnap's anti-metaphysical attitude to the issue of theoretical entities, and not also to apply it to the theoretical structure provided by the *RS*. This does not mean that that abstract structure is not important for Carnap; it just means that, Carnap cannot conceive of making judgments about whether or not such a structure is real.[56] In addition, we have to keep in mind that Carnap's use of the *RS* for a theory *TC* is not intended as a means of replacing *TC*,[57] but—as we argued above—of providing explications of the synthetic and analytic parts of the theory. Hence, it is not there to provide knowledge of the *real* structure of any part of the world (in addition to the observational knowledge). Carnap agrees with Hempel's comment about the references of Ramsey-variables and states:

> I do not propose to abandon the theoretical terms and postulates, as Ramsey suggests, but rather to preserve them in L_T and simultaneously to give an important function to the Ramsey sentences in L'_O [the extended observational language]. Their function is to serve in the explication of experiential import and, more importantly, in the explication of analyticity. (Carnap, 1963b, 963)

With respect to (ii), if it is not possible to present Carnap's use of the *RS* as accounting for real structure, there is no reason to present his proposal as a weak form of realism, in the sense of the ontological reading of the Newman objection. That interpretation simply undervalues Carnap's deflationary approach, without providing a sound argument against Carnap's thesis according to which realist claims, in general, lack cognitive content. We must add that Carnap's deflationary approach does not depend on his use of the *RS*, but is prior to this (as we have seen previously in Sects. 2.2 and 2.3), even in its application to the descriptive vocabulary of science (not only to the theoretical but also to the observational vocabulary), as verified in his works of 1950 [1956] and 1956. Thus, the motivation and justification for such a deflationary approach would be maintained independently of the *RS* and of the problems that the Newman objection may entail for it.

Be that as it may, such issues could be irrelevant if the Newman objection was applicable to Carnap's use of the *RS* (with true observational consequences) in the sense of rendering the claim that there is a structure realizing the *RS* vacuous, since in such a case elucidation of the analytic/synthetic distinction for an empirical theory would have failed. Although Carnap was surely unaware of this difficulty, it is important to clarify whether it threatens his project, which brings us to (iii).

---

[56] See Uebel (2011, pp. 74–75).

[57] Friedman (2011, p. 257) clearly raised this question.

In this regard, we must acknowledge Uebel's (2011) insight into the search for a solution to what he has called the epistemic reading of the Newman objection.[58] He starts by considering that the epistemic reading of the Newman objection posits that the *RS* is almost an analytic truth *with no empirical surplus* over and above the observational consequences of *TC* being true. However, it is in fact the case that those observational consequences are obtained thanks to a structure that inherits its inductive systematizations from *TC*, in addition to the deductive ones. Hence, this structure accounts for the observational events expressed by all those possible consequences (explaining and predicting them).[59] At any rate, this does not overcome the epistemic version of the Newman objection if this reading of the objection is seen as establishing that there is no empirical surplus over and above empirical adequacy. Uebel's insightful proposal to overcome the problem starts by differentiating two perspectives supposedly involved in Carnap's analysis of theories. The first considers a theory as *a-historical* (in an idealized way) and that perspective, according to Uebel, is provided by the *RS* representation of a scientific theory *TC* with its postulates and its correspondence rules fixed as complete (i.e., with no changes in its conditions; specifically, without considering additions of new correspondence rules), as too are all the observational consequences that follow from those conditions. The second perspective considers a theory as *historically situated*, and so with possible further developments due to (at least) adding new correspondence rules and possibly obtaining new observational consequences; Uebel assumes that this is provided for by *TC* (Carnap's initial representation of a scientific theory). A qualification about the sense of a "complete" theory in the former case is opportune: this designation should not be confused with that of a "final" theory in a diachronic process about the subject matter of that theory. However, that is not all. According to the former perspective: on one hand, the *RS* "not only claims that there exists an abstract structure that embeds the observations made so far, but that it embeds all the possible observations that follow from the theory" (Uebel, 2011, p. 79); and, on the other hand, we have to keep in mind that from the a-historical perspective a "completist" approach means that, in addition to its conditions being fixed as complete, speaking of the *empirical adequacy* of that *RS* has to be understood in the sense that all the corresponding "observational consequences are taken to have been confirmed" (*Ibidem*, 80). As regards the latter perspective, it is not assumed that all the observational consequences of *TC* "have been fathomed, let alone tested" (*Ibidem*). Given this, empirical adequacy has to be understood in a more realistic sense of those observational consequences that have already been taken into consideration (and not all the possible observational consequences that

[58] Uebel (2010, pp. 306 ff.) tries two ways of accommodating the Newman objection in order to maintain an explication of the analytic/synthetic distinction in the formulation of a scientific theory without looking to Ramseyfications. These ways are analyzed attending to Carnap's remarks previous to his use of the *RS*, but, as Uebel himself recognizes, they are not appropriated.

[59] Keep in mind that, as Hempel (1958, §9) indicated, Craig's theorem applied to empirical theories provides a substitute for the original theory, dispensing with its theoretical terms, although without preserving its inductive systematizations.

follow from that situated theory, and less so from posterior developments of that theory).[60] So, we can speak of two different notions of empirical adequacy: one, from the a-historical perspective, which we will call "empirical adequacy-ah"; and the other, from the historically situated perspective, which we will identify by "empirical adequacy-hs". It is important to keep in mind that for Uebel, given Carnap's use of the *RS*, this *RS* reconstruction of a theory embodies the a-historical perspective; while *TC*—at least as it is usually managed by Carnap and specifically in the Carnap sentence—corresponds to a historically situated perspective. (Remember, in any case, that this is only Uebel's opinion.)

Attending to these distinctions, the point for Uebel (2011) is that the "almost analytic" objection would be appropriate if it could be applied to a historically situated theory; i.e., to the representation *TC* of an empirical theory. Nevertheless, the fact is that it is not applicable in this case, since all the observational consequences of a historically situated theory are not tested and hence in this case it is not possible to adopt the requirement of empirical adequacy-ah; i.e., as if all the observational consequences had been tested. Remember that the requirement of empirical adequacy-hs is more modest and less idealized: it has to do with the confirmation of observational consequences already taken into consideration, not all the possible ones. So, it can be assumed that the historically situated theory *TC* has an empirical surplus which can be used to fathom new observational consequences (over and above the observational content already tested and confirmed), as is hoped. The claim that the *RS* is "almost analytic" is not a problem: to say that the *RS* (insofar as it embodies the a-historical perspective) has no empirical surplus is not a relevant objection given the idealized notion of empirical adequacy-ah which is at stake in this case. Hence, we think that Uebel's analysis undermines the epistemic reading of the Newman objection.[61]

Nevertheless, Uebel's idea that the *RS* embodies the ahistorical perspective, while *TC* embodies the historically situated perspective—at least in Uebel's interpretation of Carnap's approach—is odd. It entails that, for Carnap, there is an asymmetry between the *RS* and *TC*. Uebel tries to give some elaborate arguments to convince us of this.[62] But there is not, in our opinion, any indication that for Carnap the *RS* has any other difference from *TC* than the substitution of theoretical terms for existentially quantified Ramsey-variables: otherwise, they are presented as symmetrical. Furthermore, despite Uebel's arguments concerning the Carnap sentence ($^{R}TC \rightarrow TC$), it does not seem reasonable that, if according to Carnap that sentence provides the analytic content of the theory, then its antecedent (^{R}TC) should be

[60] Uebel (2011, pp. 79–80) established a connection between the a-historical perspective and the fact that Ramseyfications do not assume underdetermination by the same data, as English (1973) concluded and Demopoulos (2008) correctly demonstrated. Faced with such a connection, from a historically situated perspective a theory is underdetermined by the evidence, as Duhem (1906 [1914]) pointed out.

[61] See Uebel (2011, pp. 79–80).

[62] See Uebel (2011, p. 81 ff.).

understood according to the ahistorical perspective while we understand its consequent (*TC*) according to the historically situated perspective.

The conditional $^{R}TC \rightarrow TC$ is presented by Carnap as establishing that, if the antecedent is true, the consequent is true. But this leaves open the possibility that the antecedent is false. And such a possibility is closed for the a-historical perspective, given that, from this perspective, the empirical adequacy is considered for all the observational sentences as tested and confirmed. (And the situation is yet worse, given the trivial truth of the theoretical structure according Newman's objection.)

Hence, we have to reject the idea that different perspectives are appropriate for the *RS* and for *TC* as these are considered in Carnap's approach. We think that the same perspective, namely the historically situated one, is appropriate both for *TC* and the *RS* in Carnap's approach. Moreover, if Carnap's use of the *RS* is historically situated (like *TC*), then the Newman objection does not apply to it, or to Carnap's approach. But what about all the arguments relating the Newman objection to the *RS*: are they inappropriate? No, they are appropriate, if we work with a different use of the *RS* from Carnap's; i.e., they apply to a use of the *RS* in which it is understood as an ahistorical, idealized perspective. The mistake was to think that this is Carnap's use; and it seems to us that this is a mistake to which Psillos, Demopoulos and Friedman have contributed (as has Uebel too, though his work provides the key to correcting it). The mistake has its pre-origin in the connection established by Maxwell between the *RS* and the Newman objection; not because such a connection cannot be considered appropriate (as long as the *RS* is understood from the ahistorical perspective), but because later it was connected with Carnap's use of the *RS* (by Psillos, Demopoulos and Friedman) without realizing that he was adopting the historically situated perspective. If this point of view is correct, the conclusion is that Carnap's use of the *RS* is not affected by the Newman objection, though the objection does indeed affect an ahistorical *RS* (connected with Russellian structural realism, but not with Carnap's approach). So, we can conclude that Carnap's approach is not affected by the Newman objection, either in its ontological reading (given Carnap's ontological deflationism) or in its epistemological reading (given that his project is not connected with Russellian structural realism). Of course, if the Newman objection in its epistemic version remains, then *we should opt to maintain Carnap's agnosticism without the Ramseyfication project*. But it seems to us that the epistemic version has also been confuted.

Nevertheless, the idea that a deflated ontology for theoretical terms (or for Ramsey-variables) has to do with mathematical entities designated by those expressions, as proposed by Carnap, is somehow strange. And it is particularly so if we take into account an intuitive way of conceiving the alleged entities designated by such terms. From this point on, given the connection between Carnap's deflationist ontological and semantic proposals (and others in a similar line) we will talk about his deflationist ontosemantics and other related approaches.

2.6 A Possible Way out: A Neo-Carnapian Approach to Ontology

In this section we end by proposing a neo-Carnapian alternative to Carnap's own proposal of adopting mathematical entities for the deflationist ontosemantics of the theoretical terms. But first let us qualify different aspects of Carnap's proposal.

The first of the qualifications was previously announced (see Sects. 2.2 and 2.3 above) by pointing out that Carnap's use of mathematical entities in this sense should be understood as a psychological and didactic aid, and as a way to avoid positions that are typical of metaphysical realism. We will insist no further on this.

The second qualification that we must take into consideration was already highlighted by Psillos (1999, pp. 55–57): Carnap's own indication that the mathematical entities are just extensionally equivalent to the entities designated by the theoretical terms of a theory. But the *intensional content* of a mathematical expression is not the same as the *intensional content* of a theoretical term. From this perspective, the value-intensions of the higher-order variables would be *putative* (not metaphysically real) theoretical entities, though the value-extensions would still be mathematical entities. Recall also what has been said about Carnap's unofficial use of mathematical entities (see Sect. 2.3 above; and also Carnap, 1958 [1975, p. 76 and pp. 80–81]).

A third qualification (which complements the previous one) is that, for Carnap, the *RS* expresses a factual sentence not only because of its observational content, but also because the *RS* "says that the observable events in the world are such that there are numbers, classes of such, etc., which are correlated with the events in a prescribed way, and which have among themselves certain relations; and this assertion is clearly a factual statement about the world" (Carnap, 1963b, p. 963). In asserting this, as Uebel (2010, p. 303) has highlighted, "Carnap [...] had mathematical physics in mind where space-time points are represented by quadruples of real numbers and physical properties like electrical charge density or mass density are represented as functions of such quadruples of real numbers". Hence, the role of the numerical values in an empirical theory can be viewed just as a way of representing the putative referents of the theoretical terms as *intensionally* determined.

Given this, the intensional content of a theoretical term would also help to constraint, in a synthetic way, the corresponding postulated referent of the term (and this without any metaphysical realist assumption, as no real ontological commitment would be made regarding that referent). Anyway, we think we can abandon the idea of identifying theoretical entities with purely logical–mathematical entities, though still accepting, as Carnap does, that theoretical entities cannot be alleged spatiotemporal objects. But how?

Yablo (1998) proposed a way of considering Carnap's "Empiricism, Semantics and Ontology" project as detached from the analytic/synthetic distinction, given the problematic nature of the latter. Nonetheless, Yablo's considerations subsist even if, as we have argued above (see Sect. 2.4) there are no decisive reasons to reject the analytic/synthetic distinction.[63] Yablo's proposal involves relating the distinction between internal and external statements to the distinction between statements in make-believe games and statements raised outside of such games, or, more generically, between metaphorical and literal statements.

Disregarding some peculiar considerations by Yablo in his interpretation of Carnap (1950a [1956]),[64] such a connection between the internal/external distinction and the metaphorical/literal distinction seems to be enlightening. Talking about make-believe games[65] can be understood as excluding any commitment to the supposed entities of such games. However, talking about metaphorical statements places us in an arena—that of metaphors—that is broader and more slippery than the one we intend to consider. So, we opt to dispense with the metaphoric/literal distinction and we prefer to make a connection between internal and external statements, on the one hand; and statements about situations such as make-believe games or works of fiction, and literal statements, on the other. But we must add the following clarifying remarks:

1. Statements made in situations and works of fiction are to be understood within the terms of a fictional realism (of characters, of kinds, of places, etc. of fiction);
2. We interpret such a fictional realism as committing us to abstract artifacts (in a similar sense to that of Thomasson, 1999), not to Platonic entities;[66]
3. In such a scenario, the putative theoretical entities of scientific theories (considered correct or abandoned) would be abstract artifacts, such as: numbers, fictional situations (sometimes called fictional worlds), fictional systems, possible worlds, etc.;
4. Works of fiction, scientific theories, scientific models, scientific laws, propositions, concepts and other similar entities are also abstract artifacts, although of a different character from those considered in '(3)';[67]
5. Abstract artifacts such as those listed in '(3)' are possible deflationary referents for an object language in which expressions about such fictional entities, situations and systems or about possible worlds can be used;
6. The abstract artifacts listed in '(4)' are the *intensional contents* for modes of representation that make use of material supports (such as models, diagrams, sentences and terms of an everyday or technical language, and even of a formalized language) to express such content, and they are just *deflationary referents* for metalanguages within which to speak about such contents.

---

[63] Recall Thomasson's (2015, p. 52) claim, which we introduced in Sect. 2.3 above: "Quine's arguments against the analytic/synthetic distinction are far from decisive".

[64] Yablo (1998) developed an interpretation of Carnap (1950a [1956]) in which he tried to show that it is possible to consider Carnap's project on ontology as complementary to Quine's. However, there is an important difference insofar as Quine contemplates literal external readings of theoretical sentences which are relevant to determining what really exist (in a metaphysical sense) while Carnap rejects this possibility, since it would lack any cognitive content.

[65] Walton (1990) introduces the Make-Believe Theory.

[66] For "abstract objects" see Falguera, Martínez-Vidal and Rosen (2022), especially §3.7 (3.7.1).

[67] See De Donato-Rodríguez and Falguera (2016, 2020).

We would be remaining faithful to Carnap's agnosticism if we took the putative referents of theoretical terms (that belong to empirical theories considered to be correct or even to theories that have been rejected) to be fictional and hence abstract artifacts (in a sense similar to Thomasson, 1999) rather than mathematical entities as (apparently) proposed by Carnap (though mathematical entities are also abstract artifacts in our proposal, but in any case different from putative theoretical entities). These abstract artifacts would be:

(a) Abstract in the sense of not being spatiotemporal, though they would be temporal insofar as they depend on theoretical conditions established by scientists (the entities in question are characterized by theoretical conditions laid out in the corresponding theory), and
(b) They could correspond, or not, to spatiotemporal entities (we leave open the possibility that these abstract artifacts, which are the direct referents of theoretical terms, may in fact (approximately) correspond to spatiotemporal entities (individuals, properties, relations, etc.)); but, as we are dealing with a deflationary view, this is something that could be neither asserted nor denied.

Adopting this position does not involve abandoning the agnostic approach, since the real existence of theoretical entities (in the metaphysical sense) is being neither affirmed nor denied. Our point is that we may introduce quantification over abstract artifacts without taking a particular stance on any external ontological question (of course, this was Carnap's position too). So, in properly Carnapian style, every legitimate ontological commitment is just internal, not external. We can say that, trivially, quantifiers involve ontological commitment to the entities over which they quantify in the internal sense. Quantification over abstract entities, or over any kind of entity (including observational entities), is always justified when the linguistic framework in which we are using the relevant expressions (allegedly referring to these fictions for putative theoretical entities as abstract artifacts) turns out to be useful for our purported intentions; that is, for pragmatic reasons. Each scientific theory would then involve two kinds of commitments:

- *Existential (internally considered), and*
- *Empirical: the empirical adequacy of the theory* (this is related to a *deflationary conception of truth* for empirical as well as for theoretical sentences).

The existential commitment (internally considered) would include both a (deflationary) commitment to the (abstract) mathematical entities that are referred to by the mathematical expressions used in the formulation of the theory and a (deflationary) commitment to the (abstract) theoretical entities that are the direct referents of the theoretical expressions. The question of the real existence of theoretical entities (of correct theories) has no answer.

Anyway, we have to distinguish between theories which we consider correct, since they are highly confirmed (up to the present), and those that have been rejected, since they have been refuted. The former have the peculiarity that their theoretical entities are 'real' in the scientific sense accepted by Carnap—i.e., in the sense of internal existential answers—but we cannot say that they are, or they are not, real *in the metaphysical sense*. Meanwhile, the latter theories are such that the putative theoretical entities designated by their theoretical terms are abstract artifacts (like the characters of a work of fiction), and at least some of them are not real either in the metaphysical or even in the scientific sense that Carnap contemplates in the case of theories that are highly confirmed. This is because the formulations of the theories in which they appear have in fact been disconfirmed.

2.7 Conclusion

Our final proposal is not the same as Carnap's solution for entities designated by theoretical terms (and for entities for Ramsey-variables), but it maintains a deflationary character with regard to the ontosemantics of the formulations of scientific theories (and specifically, of their theoretical terms and Ramsey-variables) which is similar to that advocated by Carnap. Our proposal has the advantage of not introducing mathematical entities (even in an unofficial sense) as the referents of theoretical terms, which, as we have said, sounds decidedly odd.

Acknowledgements The research for this article was supported by the Ministerio de Ciencia, Tecnología e Innovación Productiva (Madrid, Spain, ES) under project PID2020-115482GB-I00 "Deflationary conceptions in ontology and metaontology/Concepciones deflacionarias en ontología y metaontología" and by Fundación BBVA (Madrid, ES) under project "Ingeniería conceptual y desacuerdo epistémico: una perspectiva ontosemántica deflacionista/Conceptual engineering and epistemic disagreement: a deflationary ontosemantic perspective – INCODE".

References

Braithwaite, R. B. (1953). *Scientific explanation*. Cambridge University Press.

Carnap, R. (1934 [1937]). *Logische syntax der Sprache*. Springer. [English version: *The logical syntax of language*. Routledge & Kegan Paul.]

Carnap, R. (1942). *Introduction to semantics*. Harvard University Press.

Carnap, R. (1947 [1956]). *Meaning and necessity: A study in semantics and modal logic* (2nd ed.). University of Chicago Press.

Carnap, R. (1950a [1956]). Empiricism, semantics and ontology. *Revue Internationale de Philosophie, 4*, 20–40. [Used version in: Carnap, R. (1956). *Meaning and necessity* (pp. 205–221). The University of Chicago Press.)]

Carnap, R. (1950b). *Logical foundations of probability*. The University of Chicago Press.

Carnap, R. (1956). The methodological character of theoretical concepts. In: H. Feigl & N. Scriven (Eds.), (1956) *Minnesota studies in the philosophy of science* (Vol. I, pp. 38–76). University of Minnesota Press, .

Carnap, R. (1958 [1975]). Beobachtungssprache und theoretische Sprache. *Dialectica, 12*, 236–248. (Vers. Util.: Observational language and theoretical language. In Hintikka, J. (ed.) (1975), *Rudolf Carnap, logical empirist. Materials and perspectives* (pp. 75–85). Reidel.

Carnap, R. (1963a). Intellectual autobiography. In P. A. Schilpp (Ed.), *The philosophy of Rudolf Carnap* (Library of living philosophers) (Vol. XI, pp. 1–84). Open Court.

Carnap, R. (1963b). Replies and systematic expositions. In P. A. Schilpp (Ed.), *The philosophy of Rudolf Carnap* (pp. 859–1013). Open Court.

Carnap, R. (1966 [1974]). *Philosophical foundations of physics*. Basic Books.

Carnap, R. (2000). Theoretical concepts in science. *Studies in History and Philosophy of Science, 31*(1), 158-172.

De Donato-Rodríguez, X., & Falguera, J. L. (2016). On fictions, theoretical entities, and ideal objects: Applying Zalta's abstract objects theory to scientific theories. In G. Borbone & K. Brzechzyn (Eds.), *Idealization XIV: Models in science* (Poznan studies in the philosophy of the sciences and the humanities, 108. Idealization) (pp. 13–42). Brill-Rodopi.

De Donato-Rodríguez, X., & Falguera, J. L. (2020). The nature of scientific models: Abstract artifacts that determine fictional systems. In J. L. Falguera & C. Martínez-Vidal (Eds.), *Abstract Objects. For and Against* (Synthese library, Vol. 422, pp. 151–171). Springer.

Demopoulos, W. (2003). On the rational reconstruction of our theoretical knowledge. *British Journal for the Philosophy of Science, 54*(3), 371–403.

Demopoulos, W. (2007). Carnap on the rational reconstruction of scientific theories. In M. Friedman & R. Creath (Eds.), *Cambridge companion to Carnap* (pp. 248–272). Cambridge University Press.

Demopoulos, W. (2008). Some remarks on the bearing of model theory on the theory of theories. *Synthese, 164*, 359–383.

Demopoulos, W., & Friedman, M. (1985). Russell's analysis of matter: Its historical context and contemporary interest. *Philosophy of Science, 52*, 621–639.

Duhem, P. (1906 [1914]). *La théorie physique son objet et sa structure* (2nd ed.: 1914). Marcel Rivière.

Eklund, M. (2013). Carnap's metaontology. *Noûs, 47*(2), 229–249.

Eklund, M. (2016). Carnap's legacy for the contemporary Metaontological debate. In S. Blatti & S. Lapointe (Eds.), *Ontology after Carnap* (pp. 165–189). Oxford University Press.

English, J. (1973). Underdetermination: Craig and Ramsey. *Journal of Philosophy, 70*, 453–462.

Falguera, J. L., Martínez-Vidal, C., & Rosen, G. (2022). Abstract objects. In E. N. Zalta (Ed.), *The Stanford Encyclopedia of Philosophy* (Summer 2022 ed.). https://plato.stanford.edu/archives/sum2022/entries/abstract-objects/

Feigl, H. (1950). Existential hypothesis: Realistic versus phenomenalistic interpretations. *Philosophy of Science, 17*, 35–62.

Friedman, M. (2008). *Wissenchaftslogik*: The role of logic in the philosophy of science. *Synthese, 164*, 385–400.

Friedman. (2011). Carnap on theoretical terms: Structuralism without metaphysics. *Synthese, 180*, 249–263.

Hempel, C. G. (1958). The theoretician's dilemma: A study in the logic of theory construction. In H. Feigl, M. Scriven, & G. Maxwell (Eds.), *Concepts, theories and the mind–body problem* (Minnesota studies in the philosophy of science) (Vol. 2, pp. 173–226). University of Minneapolis Press.

Hirsch, E. (2002). Quantifier variance and realism. *Philosophical Issues, 12*, 51–73.

Hirsch, E. (2009). Carnap and the ontological pluralism. In D. M. Chalmers & R. Wasserman (Eds.), *Metametaphysics: New essays on the foundations of ontology* (pp. 130–156). Oxford University Press.

Hirsch, E. (2011). *Quantifier variance and realism: Essays in Metaontology*. Oxford University Press.

Lavers, G. (2015). Carnap, Quine, quantification and ontology. In A. Torza (Ed.), *Quantifiers, quantifiers, and quantifiers: Themes in logic, metaphysics, and language* (Synthese library) (Vol. 373, pp. 271–299). Springer.

Lavers, G. (2016). Carnap on abstract and theoretical entities. In S. Blatti & S. Lapointe (Eds.), *Ontology after Carnap* (pp. 200–219). Oxford University Press.

Maxwell, G. (1962). The ontological status of theoretical entities. In H. Feigl & G. Maxwell (Eds.), *Minnesota studies in the philosophy of science* (Vol. III, pp. 3–27). University of Minnesota Press.
Maxwell, G. (1970). Structural realism and the meaning of theoretical terms. In M. Radner & S. Winokur (Eds.), *Minnesota studies in the philosophy of science* (Vol. 4, pp. 3–34). University of Minnesota Press.
Newman, M. H. A. (1928). Mr. Russell's causal theory of perception. *Mind, 37*, 137–148.
Psillos, S. (1999). *Scientific realism. How science tracks truth*. Routledge.
Psillos, S. (2000a). Carnap, the Ramsey-sentence and Realistic empiricism. *Erkenntnis, 52*, 253–279.
Psillos, S. (2000b). Rudolf Carnap's 'theoretical concepts in science'. *Studies in History and Philosophy of Science, 31*(1), 151–172.
Psillos, S. (2006). Ramsey's *Ramsey sentences*. In M. C. Galavotti (Ed.), *Cambridge and Vienna: Frank Ramsey and the Vienna Circle* (pp. 67–90). Springer Verlag.
Putnam, H. (1977). Realism and reason. *Proceedings of the American Philosophical Association, 50*, 483–498.
Ramsey, F. P. (1929 [1978]). Theories. In F. P. Ramsey (Ed.), *The foundations of mathematics* (pp. 212–236). Humanities Press. (Used version: In: *Foundations: Essays in philosophy, logic, mathematics, and economics* (pp. 101–125). Routledge & Kegan Paul. 1978).
Russell, B. (1927). *The analysis of matter*. Kegan Paul.
Sidelle, A. (2016). Frameworks and deflation in 'empiricism, semantics, and ontology' and recent Metametaphysics. In S. Blatti & S. Lapointe (Eds.), *Ontology after Carnap* (pp. 59–80). Oxford University Press.
Stegmüller, W. (1970). *Theorie und Erfahrung*. Springer.
Thomasson, A. L. (1999). *Fiction and metaphysics*. Cambridge University Press.
Thomasson, A. (2015). *Ontology made easy*. Oxford University Press.
Uebel, T. (2010). Carnap and the perils of Ramseyfication. In M. Dorato, M. Redei, & M. Suárez (Eds.), *EPSA epistemology and methodology of science* (pp. 299–310). Springer.
Uebel, T. (2011). Carnap's Ramseyfications defended. *European Journal for Philosophy of Science, 1*, 71–87.
Van Fraassen, B. C. (1980). *The scientific image*. Clarendon Press.
Walton, K. L. (1990). *Mimesis as make-believe. On the foundations of the representational arts*. Harvard University Press.
Yablo, S. (1998). Does ontology rest on a mistake?. *Aristotelian Society Supplementary, 72*(1), 229–283.

Chapter 3
Against Easy Realism

Tobias Alexius

Abstract According to some contemporary philosophers, call them *easy realists*, traditional ontological question, such as those concerning the existence of *objects, properties, events, numbers*, (etc.) can be fully resolved, in favor of realism, by inquiring into the semantics of ordinary language.

Easy realism builds on the assumption that ordinary talk of the aforementioned kinds of entities is guaranteed to be true because it directly reflects the meaning of the involved expressions. In this paper, I provide an argument against this claim. By pointing to changes in speech-patterns across counterfactual situations where speakers' beliefs vary, I argue that that ordinary talk of the relevant entities do not reflect facts about meaning, but rather widespread metaphysical beliefs. As a result, contra easy realism, inquiry into the semantics of ordinary language does not suffice to show that such talk is true. My focus is mostly on Amie Thomasson's version of the easy realism (see e.g., her *Ontology made easy*. Oxford University Press, Oxford, 2015) but I also briefly discuss how the argument might generalize to the wider easy-realist school.

Keywords Easy realism · Ordinary talk · Counterfactual situations · Ontological Deflationism · Easy Ontology

3.1 Introduction

According to some contemporary philosophers, call them *easy realists*, traditional ontological questions, such as those concerning the existence of *composite objects, properties, events, numbers*, (etc.) can be fully resolved, in favor of realism, by

T. Alexius (✉)
Uppsala University, Uppsala, Sweden
e-mail: tobias.alexius@filosofi.uu.se

X. de Donato-Rodríguez et al. (eds.), *Deflationist Conceptions of Abstract Objects*, Synthese Library 516, https://doi.org/10.1007/978-3-031-98717-5_3

inquiring into the semantics of ordinary language. Crucially, this excludes the need for "epistemically metaphysical theorizing", i.e. inquiry which probes reality, beyond the empirically verifiable, using non-empirical theoretical virtues.[1] In this sense, easy realism encapsulates a form of *methodological deflationism*, related in spirit to Wittgensteinian/pragmatist attempts to resolve metaphysical disputes by paying attention to ordinary language. Chief among contemporary easy realists is, arguably, Amie Thomasson (2007, 2009, 2015), but views in the vicinity have also been defended by e.g., Schiffer (1996, 2003), Hirsch (2011), Rayo (2013), Hale and Wright (2001, 2009), and Cameron (2009, 2020).[2]

In this paper, I develop an argument against easy realism, focusing primarily on Thomasson's version of the view. In Sect. 3.2, I show that Thomasson's easy realism is built on a metasemantic account which emphasizes the importance of ordinary, widespread linguistic practices for determining/revealing the meanings of natural language expressions. More specifically, I show that Thomasson depends on this account when she argues that ordinary uses of philosophically contentious kind-terms such as "table", "property", "event" and "number" are analytic, allowing us to know that they are true/correct through linguistic inquiry alone. In Sect. 3.3, I argue that this metasemantic account fails to distinguish between analytic and synthetic uses of expressions, and so must be rejected. In Sect. 3.4, I propose an improved metasemantic account according to which the analytic uses are those speakers are disposed to accept across counterfactual situations where their beliefs vary from their actual beliefs. In Sect. 3.5, I argue that this improved account classifies the ordinary uses which Thomasson claim are analytic, as synthetic, meaning that their truth cannot be known through linguistic inquiry alone. I discuss some salient ways in which a Thomassonian might respond, but conclude that none looks promising. In Sect. 3.6, I briefly discuss how my argument threatens to generalize to other forms of easy realism.

Note that, in what follows, I use "easy realism" to denote Thomasson's versions of the view, unless otherwise noted.

3.2 Easy Arguments and Linking Premises

Easy realism centers around certain *key inferences*. For current purposes, these can be represented as conditionals in which the antecedent is an uncontroversially true statement and the consequent is a statement which either straight up asserts the existence of philosophically contentious entities, or otherwise refers to such entities

[1] For a useful overview of theoretical virtues, see Keas (2018).

[2] For a historical overview of easy realist views see Thomasson (2020).

(such that the truth of the consequent demands their existence). Here are some examples:[3]

Composite object:

1. If there are particles arranged table-wise, then there is a table.

Event:

2. If P was born on D, then P's birth occurred on D.

Property:

3. If the house is blue, then the house has the property of blueness.

Number:

4. If the ns and the ms are equinumerous, then the number of ns = the number of ms.

I will follow Thomasson in calling these conditionals *linking premises.* The name befits them because they serve as key premises in easy realist arguments for the existence of the contentious entities featuring in their consequents. These arguments can, somewhat simplified, be modeled as instances of modus-ponens:

Composite object:

(i) If there are particles arranged table-wise, then there is a table
(ii) There are particles arranged table-wise
(iii) From (i) and (ii): There is a table

Event:

(iv) If P was born on date D, then P's birth occurred on D
(v) P was born on D
(vi) From (iv) and (vi): P's birth occurred on D

Property:

(vii) If the house is blue, then the house has the property of blueness.
(viii) The house is blue.
(ix) From (vii) and (viii): The house has the property of blueness.

Number:

(x) If the cups and the saucers are equinumerous, then the number of cups = the number of saucers.
(xi) The cups and the saucers are equinumerous
(xii) From (x) and (xi): The number of cups = the number of saucers.

[3] Close variants of these are embraced by Thomasson (2015, especially chapter 3). Other easy realists differ on how they prefer to formulate these key inferences (some preferring, for instance, biconditional formulations), and which they embrace.

Following Thomasson, I call such arguments *easy arguments*.[4] Since easy arguments are deductively valid, and the second premises (or instances of those premises) are philosophically and empirically uncontroversial, the spotlight of controversy naturally falls on the linking premises.

Thomasson argues that linking premises are true by claiming that their antecedents reflect sufficient *application conditions* for the contentious kind-terms found in their consequents. By "application conditions" she means:

> [...]certain basic rules of use that are among those that are meaning-constitutive for the term. In the case of sortal terms, these establish certain basic conditions under which the term will succeed or fail in referring[...] (2015, pp. 89–90)

On this view, a linking premise such as "if there are particles arranged table-wise, then there is a table" reflects a semantic rule for the kind-term "table" which determines its meaning by specifying that it is correctly applied (i.e., successfully refers) whenever there are particles arranged table-wise. Given this view, it is part of the *meaning* of "table" that it refers (and so that there are tables) whenever there are particles arranged tablewise.[5]

This view has an important epistemological upshot: In order to know that a linking premise is true, it suffices to know that its antecedent reflects application conditions for the kind-term in its consequent. Thomasson claims that this sort of knowledge—i.e. knowledge about the application conditions of the relevant kind-terms—in turn, can be made available through linguistic inquiry alone. This claim is core to her deflationary approach to traditional ontology.

Now, many have objected to Thomasson-style views on meaning, reference and existence. For instance, semantic holists claim that uses of expressions cannot be divided into those that do, and those that don't, reflect semantic rules. (See e.g., Quine, 1951, Block, 1994, 1998, Brandom, 1994, Pollock, 2020). Others oppose the idea that uses that reflect semantic rules are guaranteed to be true (e.g., Horwich, 2001, Eklund, 2002, Båve, 2018).[6] Other, still, cast doubt on the view that the contentious kind terms in linking premises play the syntactical role of referring to objects (Button, 2020, Schwartzkopff, 2019, Hofweber, 2005, Felka, 2014).

For current purposes, however, I am willing to look past these objections and grant Thomasson's basic view of meaning, reference, and existence. What I want to argue, instead, is that given certain plausible restrictions on views of this kind,

---

[4]The first two arguments concern *composite objects* and *events*, and so, strictly speaking, require additional premises stating that tables are composite objects and that that births are events. I take such premises to be implicit and philosophically innocent.

[5]Critics might deny that it follows from the fact that a kind term "K" applies/refers that there are entities of that kind. But Thomasson defends, at length, the view that, for any kind K, Ks exist iff the application conditions actually associated with "K" are fulfilled. She takes this principle itself to be analytic for "exists", and hence knowable through linguistic inquiry alone (Thomasson, 2015, chapter 2). Other easy realists make analogous commitments.

[6]Many of these objections focus on *concepts*, not linguistic expressions. But on the natural assumption that the meanings of linguistic expressions are mediated by concepts, they carry over naturally to linguistic expressions.

linking premises do not reflect application conditions for their contentious kind-terms (or any other sort of semantic rule), meaning that their truth cannot be known through linguistic inquiry alone.

Some quick terminological matters: Because Thomasson takes the truth of linking premises to be guaranteed by the fact that they reflect semantic rules, she is fond of calling them *analytic*. I will go along with this. More precisely, I will call any use *S* (e.g., a sentence or inference) involving some expression *E* in a language *L an analytic use of E* so long as *S* reflects a semantic rule for *E* in *L*. (The uses and expressions we are interested in will predominantly be *sentences/inferences* in which *kind-terms* play a referential role.) Furthermore, I assume that all analytic uses are true/valid because they reflect semantic rules.[7] I will call uses of expressions that do not reflect semantical rules *synthetic* uses.

3.2.1 The Epistemology of Analytic Uses

How can we know whether a particular use is analytic? Even though Thomasson's view centers on the claim that linking premises are analytic, she never produces a fully systematic answer to this question. As a result, nailing down Thomasson's epistemology of analyticity requires a bit of rational reconstruction.

Let's start with what is absolutely clear: Thomasson takes observations of ordinary use—i.e., observations of something like collectively shared use-patterns among competent speakers of a language—to be a reliable guide to analyticities. Call this view *Wittgensteinianism*. Note that I do not intend for the notion of "competence" here, or elsewhere, to do any heavy lifting. I take linguistic competence to be nothing other than the ability to engage in fluent communication with other members of the linguistic community.

That Thomasson is committed to (some version of) Wittgensteinianism is evident throughout her work. For instance, in her (2007) she argues that shared use-patterns function as tacit/implicit definitions that fix…

> … [semantic] rules governing proper use of an expression—when it may be successfully applied, what sorts of inferences we may make between and among claims of different types, and so on. (2007, p. 32)

The same view shines through when she talks about language acquisition:

> […] application conditions are among the semantic rules of use for the terms we master as we acquire language. Infants may learn which rules govern which terms as they observe the use of various terms, attempt to use the terms themselves, have their uses approved or corrected or misunderstood by others, and so on […] (2015, p. 90)

---

[7] Neither I, nor Thomasson, take this to imply that analytic uses are *made true* by their meaning alone. Instead, analytic uses are better understood along the lines of semantic stipulations, in which an expression is given a meaning by the role it plays in ensuring the truth of a core set of sentences or inferences. For details, see Boghossian (1996, 2003).

This mode of acquisition clearly presupposes that the linguistic practices of the community constitute a reliable guide to the semantic rules of the language.

But nowhere is Thomasson's commitment to (some version of) Wittgensteinianism more explicit than in her (2015) discussion of Williamson's (2007) argument against epistemic analyticity. Roughly, a use is epistemically analytic, in Williamson's sense, iff all competent speakers must assent to it. Williamson's argument goes as follows (with some purely presentational details altered for current purposes): Imagine two people, *Peter* and *Stephen*, both of whom are competent English speakers, but both of whom refuse to assent to sentences of the following type:

5. Every dragon is a dragon.

They refuse to so assent because they hold deviant theories of the semantics of the English universal quantifier. Peter thinks that universal quantification in English is existentially committing, meaning that (5) entails the existence of a dragon, and so is false. Stephen thinks that what counts as a dragon is *vague*, and that therefore, according to his preferred view of vagueness, (5) is neither true nor false.

Williamson points to Peter and Stephen (as well as other examples) to inductively conclude that, for any use *S* of any expression *E* in any natural language *L*, it is always possible for there to be competent speakers of *L* who consider *S* to be an incorrect use of *E*, and so refuse to assent to *S*. The full details Williamson's argument are irrelevant for current purposes. What is relevant is Thomasson's reply:

> [...] it doesn't seem plausible to think that Peter and Stephen are merely making mistakes in their semantic theories for English – taking these as factual proposals for understanding how our actual terms work. They do not seem to be offering a theory of the truth-conditions for English expressions involving 'every' in the same sense that a linguist would. Being the intelligent people they are, if that were what they aimed to do, they would notice that their 'theories' don't fit the data of English usage very well. They would also notice that when they follow these theories and refuse to accept that every [dragon is a dragon], other competent speakers look bewildered or correct them. Yet they persist just the same. If we take them to be offering a theory of how the English expression works, they must know that their 'theory' leads to a poor fit with usage. But – unlike real linguists – they don't seem to care. (Thomasson, 2015, p. 243; bracketed text added.)

In other words, it is so clear that Peter's and Stephen's semantic theories are unjustified—because those theories don't match the broad use found among ordinary speakers—that, to construe Peter and Stephen as being intelligent, we must think of them as proposing semantic revisions, rather than semantic theories. This response signals strong commitment to the view that semantic rules are tied to something like collectively shared use-patterns among competent speakers.

A quick note: Thomasson often talks about the analytic uses of an expression as being those that competent speakers are *justified* to accept by virtue of their linguistic competence (2015, chapter 7). Whatever to say about this view (see Eklund, 2016 for criticism), it is compatible with everything I'm going to argue in this paper. My quarrel with Thomasson concerns *how we can come to know which uses are analytic*, not with whether or why we are justified to assent to them.

3.2.2 Clarifying Thomasson's Epistemology of Analyticity

For the sake of clarity, I will regiment the above "Wittgensteinian" view as follows:

MEANING AS COLLECTIVE ACCEPTANCE
A use *S* of an expression *E* in a language *L* is analytic for *E* iff competent speakers of *L* are collectively disposed to accept *S* as a correct use of *E*.

Before putting this principle to the test, a few further notes are in order. First, one can read it as either a *metaphysical* or an *epistemic* principle. On a metaphysical reading, it states what metaphysically grounds that *S* is analytic for *E* in *L*. On an epistemic reading it states what *counts as evidence* for the fact that *S* is analytic for *E* in *L*. For current purposes, we are interested in the epistemological reading. However, since the metaphysical reading plausibly entails the epistemic reading (if *S*'s being a collectively accepted use grounds *S*'s being analytic, then the fact that *S* is collectively accepted is also evidence of the fact that *S* is analytic) I will not bother to distinguish between the two readings.

Second, there are two ways of collecting data about collectively accepted use-patterns. One can either do so *empirically*, by observing/asking how speakers use expressions in various circumstances, or one can do so by consulting competence-based intuitions about correct use. Thomasson leans heavily on the latter. I have no principled objection to this. However, note that because such intuitions can only be reflections of one's historical observations of the linguistic community, they are effectively round—about empirical observations.

Third, there are two importantly different readings of MEANING AS COLLECTIVE ACCEPTANCE:

Universal reading

A use *S* of an expression *E* in a language *L* is analytic for *E* in *L* iff *all competent speakers of L are disposed to accept S as a correct use of E.*

Non-universal reading

A use *S* of an expression *E* in a language *L* is analytic for *E* in *L* iff *some privileged subset of the competent speakers of L are disposed to accept S as a correct use of E.*

Easy realists must reject the universal reading. To see why, imagine the following dispute between a mereological nihilist and a realist about the existence of tables. The former believes only in the existence of fundamental particles, and so believes that "if there are particles arranged tablewise, then there is a table" is false. The latter believes that there are composite objects, like tables, in addition to fundamental particles, and so that the same conditional is true.

Now, if, as Thomasson claims, the above conditional is analytic for "table" in English, it follows from the universal reading that the nihilist must either be an incompetent English speaker or be speaking some other language (presumably an idiolect of English which assigns an alternative meaning to "table", "there are" or "there are tables").

Both results are bad for Thomasson. The first is bad because, as an argument for realism, it is merely an instance of *ad hominem,* and so carries no argumentative weight.[8] The second is bad because it implies that there's no substantive dispute between the realist and the nihilist. For if they assign different meanings to "table", "there are" or "there are tables", their claims are compatible, and so their dispute merely verbal (and hence not resolvable by pointing out that the conditional is analytic in the realist's idiolect). In either case, easy realism fails to resolve any ontological dispute.[9]

By contrast, on a non-universalist reading, the nihilist can be classified as a linguistically competent English speaker, even on the assumption that the above conditional is analytic. For the nihilist would then simply fail to be a member of the set of speakers whose use determines/reveals meaning. Furthermore, non-universalism renders the nihilist/realist dispute substantive, i.e., it renders it a dispute over the truth of a conditional with a shared meaning. Consequently, the dispute is resolved by pointing out that the relevant conditional is analytic for "table" in the shared language. For these reasons, it is clear that easy realists must be non-universalists.

Of course, as a non-universalist, Thomasson faces the problem of specifying the set of semantically privileged speakers. One salient option here is to identify it with the set of all experts in the relevant area of inquiry. On this view, we should look towards universally accepted use-patterns among astronomers to discover the analytic uses of e.g., "star" and "galaxy", towards physicists for e.g., "atom", and "proton", and so on.

Two problems with this view. First, expert communities are typically in disagreement about what constitutes correct use of even basic expressions within their discipline (and any restriction to a particular set of experts risks being *ad hoc*). Second, focusing on experts, rather than ordinary speakers, is antithetical to the Wittgensteinian spirit of easy realism.

For these reasons, easy realists are better off identifying the privileged set with the set of speakers whose use aligns with the broad majority (where *broad* does not

---

[8] Thomasson agrees (Thomasson, 2015, chapter 4).

[9] Here's an alternative conception of traditional ontological inquiry: Proponents of different views are proposing alternative languages, not quarreling over the truth of sentences in a shared language. Whatever merits this view has, it is not conducive to the goals of easy realism. For if ontological disputes are modeled as disputes over which language to speak, pointing out that some uses are analytic in *one* language does nothing to resolve the dispute. What's needed are arguments for why one language is superior to the other, and easy realism is of no help here. That being said, some easy realists *do* model traditional ontological disputes as being merely verbal in this sense (Hirsch, 2011, chapter 6, and Thomasson, 2015, chapter 4 seems to indicate agreement). But note that these easy realists *also* holds that *if* those debates are construed as taking place in ordinary English, then the realist wins, because realist uses of the contentious kind-terms reflect their meaning in English. It is this last claim that is the main target of this paper.

necessarily equal *universal*). On the resulting view, the meaning of expressions is determined/revealed by the use-patterns made by the overall linguistic community, but without there being a demand that *everyone* partake in the use-patterns. Reformulated accordingly, the principle looks as follows:

> MEANING AS BROAD ACCEPTANCE
> A use *S* of an expression *E* in a language *L* is analytic for *E* in *L* iff *the broad majority of competent speakers of L are disposed to accept S as a correct use of E.*

MEANING AS BROAD ACCEPTANCE involves no ad-hoc restrictions, stays true to the Wittgensteinian spirit, and remains non-universalist. Furthermore, it plausibly classifies linking premises as analytic, since it is safe to assume that the broad majority of English speakers *are* disposed to accept linking premises as correct uses of the relevant contentious kind-terms (i.e., are disposed to apply "table", "birth", "number", etc., in the circumstances described by the antecedents of the relevant linking premises). These features make the principle a charitable reconstruction of Thomasson's epistemology of analytic uses.

A final point. As it stands, MEANING AS BROAD ACCEPTANCE identifies the analytic uses with those competent speakers are broadly *disposed* to accept. But now consider *normativism*—a view according to which the semantic rules of an expression, that is, the rules that determine how speakers *ought* to use the expression—can come apart from the way speakers are *disposed* to use it.[10] Thomasson explicitly claims that hers is a normativist view (2015, chapter 7).

Now, *prima facie*, normativism poses a threat to MEANING AS BROAD ACCEPTANCE because it invalidates direct inferences from "the broad majority of competent speakers are disposed to accept *S*" to "*S* is analytic". However, the threat should not be overstated. Possible normativist views fall along a spectrum. At one end, speakers' actual use-dispositions and the semantic rules diverge greatly, at the other end they converge. MEANING AS BROAD ACCEPTANCE is rendered useless by normativist views on the divergent end of the spectrum. But given convergent normativism, the principle remains a useful heuristic, as collective use-dispositions remain a reliable indicator of analyticity.

Since divergent normativism entirely severs meaning from use, it not only contradicts the Wittgensteinan spirit of Easy Realism, but also looks independently implausible (what *could* determine/reveal meaning other than *something* like widely accepted use?) Therefore, any plausible form of normativism, at worst, threatens to turn MEANING AS BROAD ACCEPTANCE into a useful heuristic, rather than an all-or-nothing biconditional. For current purposes, this makes no substantive difference, so I will gloss over the normativist/non-normativist distinction.

[10] For a recent overview of normativist views, see Glüer and Wikforss (2022).

3.3 Overgeneration

In this section, I argue that MEANING AS BROAD ACCEPTANCE massively misclassifies synthetic uses as analytic, and so fails to be a reliable guide to analytic uses. This undercuts Thomasson's justification for claiming that linking premises are analytic.

Consider the term "dinosaur". Most English speakers believe that creatures such as dimetrodons, plesiosauruses, and pterodactyls are dinosaurs. Consequently, the broad majority are disposed to accept the following conditional as a correct use of "dinosaur":

6. If x is a dimetrodon/plesiosaurus/pterodactyl, then x is a dinosaur.

Furthermore, most English speakers believe that birds are not dinosaurs, and so are broadly disposed to accept the following conditional:

7. If x is a bird, then x is not a dinosaur.

As a result, MEANING AS BROAD ACCEPTANCE classifies both conditionals as analytic for "dinosaur". But they are not analytic. Paleontologists have long known that the three first mentioned species are not dinosaurs and have recently come to agree (more or less) that birds are avian theropods (specifically, maniraptorans).[11] Consequently, MEANING AS BROAD ACCEPTANCE overgenerates.

Of course, defenders of the principle may claim that paleontologists are using "dinosaur" as a technical expression, call it "dinosaur*", with a non-colloquial meaning, allowing that (6) and (7) stay analytic for the colloquial expression.

But this response is problematic. Consider a counterfactual scenario in which the broad majority of English speakers have learned contemporary paleontology. They would then be disposed to use "dinosaur" as contemporary paleontologists already do, and so be disposed to reject (6) and (7).[12] This use-synchronization given belief-synchronization indicates that current divergences are due to differences in beliefs about dinosaurs, rather than differences in the meaning of "dinosaur".

Of course, defenders may attempt to explain the use-synchronization in terms of *both* groups having switched from the colloquial "dinosaur" (for which (6) and (7) are analytic), to "dinosaur*" (for which the same are synthetic). But this explanation quickly runs into trouble. For upon learning contemporary paleontology, speakers may reasonably react by saying things such as: "Oh, and I used to think dinosaurs went extinct!", "I thought I'd never see a real dinosaur!", or "Before the discovery that dinosaurs and dimetrodons fall on different sides of the sauropsid-synapsid

[11] For an overview on the bird-dinosaur connection see O'connor et al. (2011).

[12] My reasoning here is straightforward: Paleontologists are just ordinary speakers who are currently in the same situation as the broad mass of ordinary speakers would be in the specified counterfactual situation. Since the former are disposed to reject (6) and (7), it stands to reason that ordinary speakers would also be so disposed if they learned contemporary paleontology.

split, we thought that dimetrodons were dinosaurs". But if we interpret these retroactive statements as involving "dinosaur*", they come out *false* as speakers didn't have prior beliefs about dinosaurs*. We get the same result if we interpret them as involving the colloquial "dinosaur", since colloquial dinosaurs did go extinct, will never be seen by ordinary speakers, and fall on both sides of the sauropsid-synapsid split.

By contrast, if ordinary speakers and paleontologists are interpreted as using the same term "dinosaur", for which (6) and (7) are synthetic, before as well as after synching their paleontological beliefs, we have no problem rendering such retroactive statements true. They then simply reflect a change in speakers' beliefs about dinosaurs.

We should conclude, therefore, that MEANING AS BROAD ACCEPTANCE has mistakenly classified synthetic uses of "dinosaur" as analytic.[13] Since the dinosaur case is but an instance of a common kind (think of any instance where ordinary speakers fail to update their use of a term to accord with scientific developments), the resulting overgeneration is quite massive. Consequently, MEANING AS BROAD ACCEPTANCE cannot be a reliable guide to analytic uses, and Thomasson's argument for the view that linking premises are analytic is flawed.

At this point, easy realists may object that MEANING AS BROAD ACCEPTANCE gets the dinosaur case wrong only because "dinosaur" has a world-deferential semantics (whilst claiming that the principle still correctly identifies semantic rules for terms with non-world-deferential semantics). Here's the thought: A kind-term has a world-deferential semantics if its semantic rules refer to worldly facts that may be hidden to speakers (in the case of "dinosaur", these would presumably be various phylogenetic facts). Yet speakers still have *descriptions/theories/stereotypes* associated with such terms to guide their use. If these descriptions (etc.) fail to match the entities picked out by the semantic rules—in the case of "dinosaur", the entities that stand in the right phylogenetic relations—we end up with description-extension mismatches, and thereby use-extension mismatches. Such mismatches are the reason MEANING AS BROAD ACCEPTANCE gets the dinosaur case wrong.

By contrast, a kind-term with a non-world deferential semantics has its extension *determined* by the descriptions/theories/stereotypes speakers associate with it, meaning there's no room for description-extension mismatches of the sort just described. So, the argument would go, MEANING AS BROAD ACCEPTANCE gets the semantics of such terms right.

The problem here is that a lack of description-extension mismatches does not entail a lack of use-extension mismatches. For *use* is never just governed by the descriptions speakers associate with an expression, but also by speakers' beliefs

---

[13] The choice here is whether to render the original statements or the retroactive statements true. I go with the latter because semantic theories should prioritize *well informed* uses of expressions over *less informed* uses. Without such a prioritization, we risk ending up with the view that all, or the vast majority of changes in use-patterns entail a change in meaning. But that is clearly a non-starter. See Hirsch (2011, chapter 10) for discussion on this point.

about which situations/entities are picked out by the descriptions. Since such beliefs can be false, a semantic theory which focuses on widespread use will be unable to differentiate between synthetic uses that result from widespread, but false beliefs about description-satisfaction, and analytic uses. Thus, the risk of overgeneration remains for expressions with a non-world deferential semantics.

3.4 Revising Meaning as Broad Acceptance

To avoid this threat of overgeneration, easy realists must find some way of making MEANING AS BROAD ACCEPTANCE sensitive to speakers' use-dispositions across counterfactual situations where their beliefs differ from their actual beliefs (so that only uses which speakers remain disposed to accept under belief change are classified as analytic).[14] Here's a revision along these lines[15]:

MEANING AS COUNTERFACTUALLY STABLE USE (MaCSU)
A use *S*, of an expression *E* in a language *L* is analytic for *E* in *L* iff…

> … *the broad majority of competent speakers of L are disposed to accept S as a correct use of E, including in all counterfactual situations where their beliefs differ from their actual beliefs.*

This version gets the dinosaur case right because the relevant uses of "dinosaur" fail to be counterfactually stable under changes to speakers' paleontological beliefs.

Unfortunately, MaCSU instead *undergenerates* analytic uses. To illustrate, suppose that "if *x* is a bachelor, then *x* is unmarried" is analytic for "bachelor" in English. Now consider a counterfactual situation in which English speakers come to believe a false semantic theory according to which "bachelor" applies to *x* iff *x* falls under a certain behavioral stereotype *B*. Suppose further that it is common knowledge that there are married persons who fall under *B*. Speakers would then be disposed to reject "if *x* is a bachelor, then *x* is unmarried" as an incorrect use of "bachelor", and MaCSU would misclassify it as non-analytic. This problem generalizes: For any use *S* of any expression *E* in any natural language *L* one can conjure up a counterfactual situation in which the broad majority of competent speakers refuse to accept *S* as a correct use of *E* due to having embraced some deviant semantic theory.[16]

---

[14] A precursor to this view is found in Carnap (1955).

[15] If one read "disposed to accept" in MEANING AS BROAD ACCEPTANCE as *disposed to accept regardless of belief*, then the principle was already sensitive to variations in beliefs across counterfactual situations. However, if one reads the same clause as *disposed to accept given current beliefs*, then MEANING AS BROAD ACCEPTANCE is not so sensitive. In previous sections, I presupposed the former reading. But if one prefers the latter reading, consider this section an effort to make explicit, and deal with, some implications of that reading.

[16] This is a version of Williamson's argument from Sect. 3.2.1, except that it targets non-universalist, as well as universalist views.

To fix this, the set of counterfactual situations witnessed by MaCSU must exclude those where speakers have beliefs about the semantics of expressions in their language that differ from their actual beliefs:

MEANING AS COUNTERFACTUALY STABLE USE II (MaCSU II)
A use *S*, of an expression *E* in a language *L* is analytic for *E* in *L* iff…

> … *the broad majority of competent speakers of L are disposed to accept S as a correct use of E, including in all counterfactual situations where their beliefs are different from their actual beliefs, but disregarding those where their semantic beliefs concerning L differ from their actual beliefs.*[17]

This restriction is quite natural. By analogy, imagine how bizarre it would be to attempt to identify the rules of English grammar by probing how English speakers would be disposed to use expressions in counterfactual scenarios where they have non-actual beliefs about the grammatical rules of English.

Unfortunately, MaCSU II still undergenerates. Return to the bachelor case and consider a counterfactual scenario in which speakers come to believe, for some reason other than having deviant semantic beliefs, that there is a married bachelor (insert any creative sci-fi explanation, such as e.g., belief-changing cosmic rays). In such a situation, speakers would be disposed to reject "if *x* is a bachelor, then *x* is unmarried", meaning that MaCSU II classifies it as non-analytic. Again, this generalizes: For any use *S* of any expression *E* in any natural language *L* simply imagine a scenario where speakers of *L*, for whatever creative reason, come to believe the negation of *S*, and so become disposed to reject *S*.[18]

To fix this, I suggest the following restriction: Exclude all counterfactuals where speakers' non-semantic beliefs have been formed on the basis of anything other than *responsible inquiry*:

MEANING AS COUNTERFACTUALY STABLE USE III (MaCSU III)
A use *S*, of an expression *E* in a language *L* is analytic for *E* in *L* iff…

> … *the broad majority of competent speakers of L are disposed to accept S as a correct use of E, including in all counterfactual situations where their non-semantic beliefs are (a) different from their actual beliefs, and (b) based in responsible inquiry, but disregarding those where their semantic beliefs concerning L differ from their actual beliefs.*

We can leave the notion of "responsible inquiry" vague and undefined. There is no shortage of paradigmatic examples—physics, biology, mathematics, astronomy, etc.—and these grant sufficient content to the notion for current purposes.

This version of the principle has a bunch of good-making features. First, it does not (obviously) undergenerate. Take e.g., the bachelor case. There are no salient

---

[17] Note that "semantic beliefs" here is not to be read as *analytic beliefs* (that would be circular), but as *beliefs that explicitly concern the meaning/correct use of expressions.*

[18] It may be objected that, if "if x is a bachelor, then x is unmarried" is analytic for "bachelor", one cannot express/hold the belief that x is a married bachelor without changing the meaning of "bachelor". But easy realists must reject this view. For, on the assumption that linking premises are analytic, this view would turn traditional ontological disputes into merely verbal disputes, as disputants are forced into different languages (see Sect. 3.2.2).

counterfactual scenarios where responsible inquiry into the non-semantic world would yield evidence for the existence of a married bachelor, or anything else that would make English speakers disposed to reject "if x is a bachelor, then x is unmarried". Indeed, it is entirely unclear what kind of responsibly precured data about the non-semantic world could, even in principle, form the doxastic basis for such a disposition. As a result, MaCSU III classifies the conditional as analytic.[19] We plausibly get the same result for e.g., "if x is a mammal, then x is an animal", "if x is colored, then x is extended", and "if x is square, then x is not round". There simply are no salient counterfactuals in which responsible inquiry into the non-semantic world would make English speakers believe in the existence of non-extended colored things, non-animal mammals, square circles, or anything else that would make them disposed to reject those counterfactuals.[20] So they are classified as analytic.[21]

Now, one may object by pointing to counterfactual situations in which a hyper-reliable agent, say a powerful A.I., tells English speakers that there are e.g., married bachelors. If trusting such an A.I. counts as basing one's beliefs in responsible inquiry, this would undermine the analyticity of "if x is a bachelor, then x is unmarried", and so show that MaCSU III undergenerates.

However, the key question isn't whether speakers who trust the A.I. have engaged in epistemically responsible inquiry, but whether *the A.I.* has done so. If it hasn't, then there is a clear sense in which the beliefs of speakers who trust the A.I. are *not* based in responsible inquiry. In other words, a belief is "based in" responsible inquiry, in the relevant sense, only if the *root inquiry* from which support for the belief is drawn is an instance of responsible inquiry. (What we are trying to find out is not whether *some agent can trick us*, but whether inquiry into the non-semantic world could ever give us reason to revise our use-dispositions.)

Given this, the question of whether a counterfactual situation in which speakers are told by a reliable A.I. that there are e.g., married bachelors is witnessed by MaCSU III can only be answered by checking whether the A.I.'s root process of inquiry was an instance of reliable inquiry. To do that, we must spell out and

[19] Nothing in this paper hinges on this actually being the *correct* view on the semantics of "bachelor". The core point I'm making is more general.

[20] When employing MaCSU III, one cannot consider *every* counterfactual witnessed by the principle. In application, the principle only produces inductive hypotheses about analyticity. This isn't a weakness. For barring fully reliable rational insights, or some other methodological mysticism, that's the best we are ever going to achieve.

[21] Some commentators have objected that MaCSU III classifies obviously true, but presumably synthetic sentences, such as "the earth is round" as analytic. But it is easy enough to imagine a far-out counterfactual scenario in which responsible, worldly inquiry reveals that the round-earth-theory was, after all, a NASA plot. In such a scenario, English speakers would be disposed to reject "the earth is round", so it is not classified as analytic. Another objection that's been raised is that MaCSU III classifies uses that are presumably *syntehtically a priory* (e.g. mathematical statements) as analytic. But even assuming that this objection is right, my overall argument need not be threatened. For MaCSU III can be reformulated as a necessary, though not *sufficient* condition on analyticity. The problem for the easy realist, then, is that her linking premises fails to meet a necessary demand on analyticity.

evaluate *that* process of inquiry. But now we are back where we started: We find ourselves unable to formulate counterfactual situations in which we (or a powerful A.I., or anyone else) uncovers reason, by way of responsible inquiry into the non-semantic world, to believe in the existence of a married bachelor, or anything else that would make us disposed to reject "if x is a bachelor, then x is unmarried". As a result, MaCSU III classifies that conditional as analytic, and, hence, does not undergenerate.

Second good-making feature: The principle handles the dinosaur case. For that case involved changes in paleontological beliefs formed via responsible inquiry, so the fact that the relevant uses of "dinosaur" failed to be stable under such belief change means that MaCSU III does not classify them as analytic.

Third good-making feature: MaCSU III lines up with the traditional assumption that analytic uses are in some sense true/valid independently of what the world is like (and hence are necessarily true/valid). For MaCSU III identifies the analytic uses with those we remain disposed to accept regardless of how carefully we investigate the non-semantic parts of the world.

Finally, MaCSU III stays true to the Wittgensteinain spririt of connecting meaning to ordinary use (albeit ordinary use that remains stable across relevant counterfactuals).[22]

For these reasons, MaCSU III looks like a solid improvement on MEANING AS BROAD ACCEPTANCE. The question now is: Does it classify linking premises as analytic?

3.5 MaCSU III and Linking Premises

Unfortunately, MaCSU III does *not* classify linking premises as analytic. To see why, consider a counterfactual situation in which the broad majority of English speakers have come to believe, on the basis of epistemically metaphysical inquiry (i.e., abductive inquiry based on non-empirical theoretical virtues), that there are no entities of the contentious kinds that figure in the consequents of linking premises. Among these would be the counterfactual situations where text-book anti-realist arguments, such as arguments from parsimony, vagueness, metaphysical queerness, the impossibility of colocation etc., have won the popular vote. In such scenarios, English speakers would be disposed to reject linking premises for being incorrect uses of their contentious kind-terms (just as actual anti-realists are so disposed now). As a result, MaCSU III classifies linking premises as non-analytic, indicating that widespread, current acceptance of linking premises reflects commonly held synthetic beliefs, not semantic rules.

[22] For an account of meaning/analyticity along broadly similar lines, though focusing on ideal reasoning rather than responsible inquiry, see Chalmers (2012, chapter 5).

I will now go through four salient responses to this argument. The first is to reject MaCSU III, and propose an alternative epistemology of analyticity. Of course, the viability of this response depends on the merits of the alternative, and I don't have an argument to show that a better epistemology couldn't, in principle, exist. There is, however, good reason for thinking that no plausible replacement principle could deviate enough from the core commitments of MaCSU III to make a difference. For MaCSU III is but a language-focused instance of a more general, and highly plausible epistemology of implicit rules, applicable across multiple, rule-governed social activities. Consider the following scenario.

A person "*A*" is tasked with figuring out the rules constitutive of chess. At her disposal is a group of competent chess players, all of whom have internalized the rules of chess, but none of whom can articulate them explicitly.

A's only option would be to observe how the players move their pieces, hoping to find stable, widespread use-patterns. Once such a pattern has been identified (for instance that rooks never move diagonally), *A* would thereby have evidence for there being a corresponding rule (that rooks are not *allowed* to move diagonally).

Of course, there's always the risk that an observed use-pattern merely reflects wide-spread beliefs about sound strategy. To weed out such possibilities, *A* could ask players how they *would* move e.g., rooks in various counterfactual situations where they face unusual play. If *A* doesn't find any such counterfactual situation in which players would move rooks diagonally, this would be solid evidence that rooks are not *allowed* to move diagonally. By contrast, if *A* does find such a counterfactual situation, this would be evidence that the tendency not to move rooks diagonally reflects widespread beliefs about sound strategy, not a rule.

Of course, in probing such counterfactual situations, *A* must disregard those where players have explicit beliefs about the rules of chess. For those beliefs would influence the way players move their pieces without any guarantee of matching the actual rules of chess. *A* must also take care to disregard counterfactuals where players have other sorts of deviant beliefs, e.g., that losing a game will net them some great reward.

The parallels between this quickly sketched, but highly intuitive epistemology of the rules of chess and MaCSU III are striking, and similar epistemologies look *prima facie* plausible for rules governing other social activities (e.g., grammatical rules, rules of social etiquette, traffic rules etc.). The sheer plausibility of these parallel epistemologies makes it difficult to see how there could be a replacement for MaCSU III that deviates from it to any meaningful degree. (Unless coupled with the view that language is radically different from these other social activities. But then easy realism would be conditional on this presumably contentious view.)

The second response I want to consider is to accept MaCSU III but argue that epistemically metaphysical inquiry is not responsible inquiry.

I think this response has some merit. There are plenty of philosophers who doubt the credentials of epistemically metaphysical inquiry (see e.g., Bennett, 2009; Saatsi, 2017; Huemer, 2009). The problem with this response is that it makes easy realism conditional on the success of such methodological skepticism, and there is nothing in the easy-realist toolbox itself to help the sceptic argue her case.[23] It also makes easy ontology anything but a *deflationary* approach to ontology, as it now includes substantial commitments concerning the ways we can responsibly probe reality, and these commitments clearly cannot be defended through linguistic inquiry alone.

The third response is to add further restrictions to the set of counterfactual situations witnessed by MaCSU III to exclude those where speakers become convinced anti-realists by reading metaphysics literature.

The problem with this response is that it would be *ad hoc* if these restrictions were motivated by the desire to render linking premises analytic. And since MaCSU III already avoids under- and overgeneration issues, and seemingly gets basic cases of analytic truths right, it's difficult to see what other motivation one might cite for such restrictions.

The fourth, and final response I want to consider is to deny that speakers would be disposed to reject linking premises in counterfactual situations where they've been convinced by anti-realist arguments on the grounds that talking about tables, events, properties, numbers (etc.) under the conditions specified by linking premises is simply too useful and convenient to give up.

I think this response is right as far as it goes. Speakers *would* plausibly be disposed to accept ordinary uses of "table", "event", "property", "number", etc., as represented by linking premises, even if they were convinced by traditional anti-realist arguments. That being said, we must distinguish, as is standard, between contexts where speakers accept a use because it has some desired pragmatic effect, and contexts where they accept it because it's taken to be *literally correct*. For instance, a paleontologist visiting an amusement park might point to a statue of a dimetrodon and say, to his group of friends, "we met at that dinosaur statue an hour ago". Even though dimetrodons are not dinosaurs, there is a sense in which the paleontologist has not committed an error. For he is using "dinosaur" merely to facilitate communication about a location, not to state that dimetrodons are literally dinosaurs. By contrast, if the same paleontologist were to say that dimetrodons are dinosaurs in a paleontology seminar, he *would* be in error. For in such contexts, he is (presumably) trying to use "dinosaur" literally.

Now, if we want to identify *the application conditions* of "dinosaur"—i.e., the semantic rules that determine under what conditions the term refers—the non-literal, purely pragmatic uses are clearly *noise*, and only the literal uses are *data*. So, what I am willing to grant is this: English speakers would continue to be disposed to assent to linking premises even if they were convinced by traditional anti-realist arguments, *but only in non-literal contexts*. For instance, they would continue to be disposed to talk of tables in the presence of particles arranged table-wise in contexts

[23] Thomasson sketches an initial defense of such methodological skepticism in her (2019). An evaluation of that view, and other forms of similar skepticism, is outside the scope of this paper. Note, though, that Thomasson does not seem to take the success of her skepticism to be crucial to the success of her easy realism.

where they are concerned to quickly communicate information about where to place a glass of water. However, they would be disposed to reject such talk *in contexts where they are concerned with literal correctness,* and these are the only contexts that are relevant for determining/revealing meaning. Such contexts include philosophy seminars, of course, but also serious dinner conversations, moments of personal reflection, contemplations of serious works of art, etc.

Now, easy realists might attempt to respond to this by embracing a form of semantic contextualism (where semantic contextualism about an expression *E* is the view that *E* shifts its meaning depending on the context of use). This could allow linking premises to stay analytic for their contentious kind-terms *in the context of ordinary use*, but non-analytic in stricter contexts, due to contextual shifts in meaning.[24]

But going contextualist will not help the easy realist. For traditional ontological debates do not take place in ordinary contexts, but in contexts of serious inquiry/thought. Thus, the easy realist who takes the contextualist route would be the equivalent of someone who, upon hearing a paleontologist explain why dimetrodons are not dinosaurs, replies that in *ordinary settings* it is true to say that dimetrodons are dinosaurs. Even supposing that such a person is right, there's no reason why paleontologists, and by analogy, philosophers should care when formulating their paleontological/philosophical views.[25]

Perhaps easy realists will here choose to echo the Wittgensteinian slogan about language going on holiday. That is, claim that the contentious kind-terms have no meaning outside their ordinary contexts of use. The problem with traditional ontology, on this view, is that it is an attempt to use ordinary vocabulary outside of its semantic home turf, meaning it constitutes a failure to use that vocabulary in a meaningful way.

But this does not seem right. Doubts about the existence of e.g., composite objects is a live (and exercised) option in many contexts outside of academic philosophy, and thus fully intelligible and meaningful to ordinary speakers. If proof of this I needed, just ask students taking introductory courses to metaphysics, or

[24] Semantic contextualism is typically defended with regards to indexical expressions such as "here", "I", and "now", and predicates like "is tall" or "is flat" and quantifiers. (See Silk, 2016 for a recent defense of a broad application of contextualist semantics.) But contextualism is not typically defended with regards to kind-terms. The best candidate for a contextualist treatment in the context of easy realism might therefore be the existential quantifier and the existence-predicate. On this view, "there are" and "exists" shift in meaning across contexts, rendering statements like "there are tables" true in some contexts, but false in others. However, the context sensitivity of quantifiers is typically spelled out in terms of domain restriction, and that is not obviously the sort of meaning variance required by a contextualist easy realist. At any rate, this sort of approach is close to the "quantifier variance"-thesis put forth by Hirsch (2011, chapters. 5, 6 and 10 especially), only Hirsch puts his view in terms of different possible languages having different quantifier meanings, not in terms of contextual shifts in quantifier meaning.

[25] The problem here is similar to the one discussed in connection with universalism in Sect. 1.2, only here easy realists and their anti-realist opponents are assigning different meanings to linking premises because they are speaking in different contexts, rather than in different languages/idiolects.

thoughtful non-philosophers in general, if they've ever had issues getting the manifest image to cohere with the scientific image, and, as a result, have doubted the truth of ordinary assertions about tables, numbers, properties and whatnot (spoiler: many of them have). It would be uncharitable in the extreme to interpret all such people as having engaged only in the production of meaningless nonsense. The rules of ordinary language clearly allow such skepticism to be articulated in a meaningful way.

Having considered, and dismissed these responses, I conclude that Thomassonian deflationism about traditional ontology fails to convince.

3.6 Generalizing to Other Forms of Easy Realism

I will end by indicating how my argument threatens to generalize to other versions of easy realism. Due to lack of space, I will focus on the views of Eli Hirsch, Agustín Rayo, and Ross Cameron, and omit many details.

First, these easy realists do not operate with notions like analyticity and application conditions. For instance, Hirsch (2011, chapters 6 and 9) takes the antecedents of linking premises to state truth-conditions for their consequents, Rayo (2013) takes both sides to have *the same content* (i.e., as demanding the same coarse-grained fact), and Cameron think of the antecedents as perspicuous representations of the *truth-makers* of the consequents. (Cameron, 2009, 2020). However, these philosophers all share the view that we can come to know that linking premises are true through linguistic inquiry alone. The core idea is parallel to Thomasson's: We can come to know that the antecedents of linking premises represent truth-conditions/truth-makers for, or demand the same fact as, their consequents, and hence are true, by observing widely shared use patterns among competent speakers. Hirsch, for instance, invokes a well-known principle of charity and argues that those who deny that ordinary uses of "table", "blueness", "birth", etc., are true have…

> […] failed drastically to interpret [English] in a plausibly charitable manner. Their various doctrines seem to be trivially absurd in plain English, as charitably interpreted. (Hirsch, 2011, p. 184, text within brackets added by current author)

In a similar vein, Rayo (2013, chapter 1) argues that because speakers routinely, and without further thought, transition between the two sides of linking premises, the correct semantic theory will assign the same content to both sides. Finally, Cameron writes that:

> On discovering that the thing in my kitchen upon which I place my dinner is mostly empty space, should I conclude that there are no tables? No, because 'table' is used by ordinary speakers to refer simply to things like this (gesturing at the thing on which I place my dinner), and whatever that turns out to be, that is what a table is. (Cameron, 2020, p. 38)

On this view, ordinary uses of "table" are guaranteed to be true because whatever reality happens to be like in the circumstances where that term is ordinarily applied will be assigned as truth-maker for those uses.

The problem with all these views, just as with MEANING AS BROAD ACCEPTANCE, is that they overemphasize the importance of *current/actual* use-patterns. As a result, just like MEANING AS BROAD ACCEPTANCE, they over-generate truth-conditions/truth-makers/content-identities whenever currently/actual widespread patterns of use are the result of widespread synthetic beliefs. Furthermore, just as with MEANING AS BROAD ACCEPTANCE, the only salient remedy is to focus on uses that show stability across counterfactual situations where speakers beliefs vary (holding still their semantic beliefs). But once the dialectic reaches this stage, we inevitably run into those counterfactual situations in which speakers have been convinced by traditional anti-realist arguments to abandon the relevant linking premises, revealing that the wide-spread acceptance of them reflected widespread synthetic beliefs, rather than truth-conditions/truthmakers/content identities. And here it should be fairly trivial to plug in appropriate analogies of Sects. 3.3 and 3.4.

Of course, this is just a quick sketch. To make a truly convincing case, one would have to go into the details of each view individually. But the sketch nevertheless indicates the general threat.

3.7 Conclusion

Thomasson's deflationary approach to traditional ontology relies on the claim that her linking premises—which represent ordinary uses of philosophically contentious kind-terms—are analytic. To support this claim, Thomasson embraces an epistemology of analyticity which identifies analytic uses with those that the broad majority of speakers are disposed to accept. I have argued that this epistemology fails to distinguish between synthetic and analytic uses. I've also suggested that an imporived epistemology should identify the analytic uses with those that competent speakers are disposed to accept across counterfactual situations where their beliefs are (a) different from their actual beliefs, (b) have been formed on the basis of responsible inquiry, and (c) their semantic beliefs are held still. Furthermore, I've argued that this improved epistemology classifies Thomasson's linking premises as synthetic, not analytic, due to the existence of counterfactual situations in which competent speakers would be disposed to reject linking premises because they've been convinced by text-book anti-realist arguments. If this is right, it follows that the truth of linking premises cannot be known through linguistic inquiry alone, and so that Thomasson's deflationary approach to traditional ontology fails. I've also discussed various responses available to Thomasson but argued that none were promising/convincing. Finally, I quickly sketched how the argument threatens to generalize to other versions of easy realism.

I will close by saying that I take the core mistake of easy realism to be rather basic: It defends realism by elevating our current, ordinary and unreflective speech-patterns to the point of making their truth a matter of meaning. By doing so, it fails to accurately capture the degree to which those speech-patterns are a function of widespread metaphysical beliefs, rather than the semantic rules of our languages.

Acknowledgements Thanks to Matti Eklund, Sebastiano Moruzzi, Øystein Linnebo, Einar Bøhn, and Arvid Båve, as well as participants at a research seminar at Uppsala University (2020), at the workshop Abstract Objects: Deflationist Approaches at the University of Santiago de Compostela (2021), and at the PhD-net seminars in Helsinki (2019) for feedback on the material that turned into this paper.

References

Båve, A. (2018). Trivial languages. *Acta Analytica, 33*(1), 1–17. https://doi.org/10.1007/s12136-017-0328-1

Bennett, K. (2009). Composition, colocation, and metaontology. In D. J. Chalmers, D. Manley, & R. Wasserman (Eds.), *Metametaphysics: New essays on the foundations of ontology* (pp. 38–76). Oxford University Press.

Block, N. (1994). An argument for holism. *Proceedings of the Aristotelian Society*, New Series, *95*, 151–169.

Block, N. (1998). Holism, mental and semantic. In E. Craig (Ed.), *The Routledge encyclopedia of philosophy*. Routledge.

Boghossian, P. (1996). Analyticity reconsidered. *Noûs, 30*(3), 360–391. https://doi.org/10.2307/2216275

Boghossian, P. (2003). Epistemic analyticity: A defense. *Grazer Philosophische Studien, 66*(1), 15–35.

Brandom, R. (1994). *Making it explicit*. Harvard University Press.

Button, T. (2020). Deflationary metaphysics and ordinary language. *Synthese, 197*(1), 33–57. https://doi.org/10.1007/s11229-016-1212-z

Cameron, R. P. (2009). How to have a radically minimal ontology. *Philosophical Studies, 151*, 249–262. https://doi.org/10.1007/s11098-009-9442-2

Cameron, R. P. (2020). Easy ontology, two-dimensionalism, and truthmaking. In K. Bennett & D. W. Zimmerman (Eds.), *Oxford studies in metaphysics* (Vol. 12, pp. 35–57). Oxford University Press.

Carnap, R. (1955). Meaning and synonymy in natural languages. *Philosophical Studies, 6*(3), 33–47.

Chalmers, D. J. (2012). *Constructing the world*. Oxford University Press.

Eklund, M. (2002). Inconsistent languages. *Philosophy and Phenomenological Research, 64*(2), 251–275.

Eklund, M. (2016). Review: Ontology made easy. *Notre Dame Philosophical Reviews*. https://ndpr.nd.edu/reviews/ontology-made-easy/. Accessed 03 April 2020.

Felka, K. (2014). Number words and reference to numbers. *Philosophical Studies, 168*(1), 261–282. https://doi.org/10.1007/s11098-013-0129-3

Glüer, K., & Wikforss, Å. (2022). The normativity of meaning and content. In E. N. Zalta (Ed.), *The Stanford encyclopedia of philosophy* (Spring 2022 ed.). https://plato.stanford.edu/archives/spr2022/entries/meaning-normativity/. Accessed 15 July 2022.

Hale, B., & Wright, C. (2001). *The reason's proper study: Essays towards a neo-fregean philosophy of mathematics*. Oxford University Press.

Hale, B., & Wright, C. (2009). The metaontology of abstraction. In D. J. Chalmers, D. Manley, & R. Wasserman (Eds.), *Metametaphysics: New essays on the foundations of ontology* (pp. 178–212). Oxford University Press.

Hirsch, E. (2011). *Quantifier variance and realism: Essays in metaontology*. Oxford University Press.
Hofweber, T. (2005). A puzzle about ontology. *Noûs, 39*(2), 256–283.
Horwich, P. (2001). Stipulation, meaning, and apriority. In P. Boghossian & C. Peacocke (Eds.), *New essays on the a piori* (pp. 150–169). Oxford University Press.
Huemer, M. (2009). When is parsimony a virtue? *The Philosophical Quarterly, 59*(235), 216–236.
Keas, M. N. (2018). Systematizing the theoretical virtues. *Synthese, 195*(6), 2761–2793.
O'connor, J., Chiappe, L. M., & Bell, A. (2011). Pre-modern birds: Avian divergencies in the mesozoic. In G. Dyke & K. Gary (Eds.), *Living dinosaurs: The evolutionary history of modern birds* (pp. 39–154). John Wiley & Sons.
Pollock, J. (2020). Holism, conceptual role, and conceptual similarity. *Philosophical Psychology, 33*(3), 396–420. https://doi.org/10.1080/09515089.2020.1729973
Quine, W. V. (1951). Two dogmas of empiricism. *The Philosophical Review, 60*(1), 20–43.
Rayo, A. (2013). *The construction of logical space*. Oxford University Press.
Saatsi, J. (2017). Explanation and explanationism in science and metaphysics. In M. Slater & Z. Yudell (Eds.), *Metaphysics and the philosophy of science: New essays* (pp. 162–191). Oxford University Press.
Schiffer, S. (1996). Language-created language-independent entities. *Philosophical Topics, 24*(1), 149–167.
Schiffer, S. (2003). *The things we mean*. Oxford University Press.
Schwartzkopff, R. (2019). Singular terms for numbers? *Fudan Journal of the Humanities and Social Sciences, 12*, 371–383. https://doi.org/10.1007/s40647-018-0248-5
Silk, A. (2016). *Discourse contextualism: A framework for contextualist semantics and pragmatics*. Oxford University Press.
Thomasson, A. (2007). *Ordinary objects*. Oxford University Press.
Thomasson, A. (2009). Answerable and unanswerable questions. In D. J. Chalmers, D. Manley, & R. Wasserman (Eds.), *Metametaphysics: New essays on the foundations of ontology* (pp. 444–471). Oxford University Press.
Thomasson, A. (2015). *Ontology made easy*. Oxford University Press.
Thomasson, A. (2019). Replies to comments on *Ontology made easy*. *Philosophy and Phenomenological Research, 99*(1), 251–264. https://doi.org/10.1111/phpr.12619
Thomasson, A. (2020). Easy ontology. In J. Miller & R. Bliss (Eds.), *Routledge handbook of metametaphysics*. Routledge.
Williamson, T. (2007). *The philosophy of philosophy*. Blackwell Publishing.

Chapter 4
From Deflationism About Truth to Deflationism About Abstract Objects

Thomas Schindler

Abstract I present a deflationary account of abstract objects, focusing on numbers and properties. This account can be seen as a natural extension of the deflationary account of truth. Very roughly, the key idea is that (1) our talk about numbers and properties serves a quasi-logical function analogous to our talk about truth, and (2) key questions about the metaphysics and epistemology of numbers and properties can be answered with reference to that function. In contrast to many other deflationary approaches in metaphysics, the present account does not assume that the principles governing our talk about numbers and properties is analytically true.

Keywords Deflationism · Truth · Numbers · Properties · Abstract objects · Analyticity · Easy ontology

4.1 Introduction

Do abstract objects exist? If so, what kinds of abstract objects exist? How do we come to know the truth or falsity of sentences that involve quantification over or reference to abstract objects?

What explains the utility of quantifying over and referring to abstract objects? Why are the laws of arithmetic, say, applicable to the empirical world?

In recent years, a number of deflationary views of the nature of abstract objects have emerged in response to these and other problems, e.g. (Hale & Wright, 2001; Schiffer, 2003; Thomasson, 2015; Båve, 2015; Hofweber, 2016; Linnebo, 2018; Warren, 2020. Very roughly, the core idea behind these approaches (as I understand them) is that questions as the ones raised above can be answered by reference to our

T. Schindler (✉)
University of Amsterdam, Amsterdam, Netherlands
e-mail: t.schindler@uva.nl

X. de Donato-Rodríguez et al. (eds.), *Deflationist Conceptions of Abstract Objects*, Synthese Library 516, https://doi.org/10.1007/978-3-031-98717-5_4

linguistic and conceptual practices. These approaches do not take recourse to the abstract object themselves in order to answer these questions.

While I am generally sympathetic to these views, I take issue with many details. Let me mention the two most important points of disagreement. First, most of these accounts assume that the principles governing our talk about numbers, properties, and so on, are automatically valid or analytically true. However, I do not believe that the axioms of Peano arithmetic, for instance, are analytically true. Second, most of these accounts have a Carnapian bend—they trivialise questions of ontology, i.e. questions about what kinds of entities exist. The Quine-Putnam indispensability argument, for example, plays no role in their considerations. I do not believe that questions of the existence of abstract objects can be answered so easily.

The purpose of this paper is to sketch an alternative deflationary account of abstract objects. There are four main points in which the proposed account will differ from most other deflationary accounts about abstract objects. First, the proposed account of abstract objects will presuppose semantic deflationism, i.e. deflationism about truth, satisfaction, and reference. Second, it is articulated in strong analogy to semantic deflationism. It can be seen as a natural extension of deflationism about truth. Very roughly, the key idea is that our talk about abstract objects such as properties and numbers serves a quasi-logical function analogous to our talk about truth, and all questions regarding abstract objects can be answered with reference to that function. Third, the proposed account does not assume that the axioms characterising truth, numbers, properties, and so on, are automatically valid or analytically true. Fourth, the proposed account has a more Quinean bend—questions of ontology are answered by developing the best overall theory.

Naturally, I believe that my account has various advantages over the other views. It retains the main idea of the other accounts—namely, most philosophical questions about abstract objects can be answered (largely) by looking at our linguistic and conceptual practice—but without relying on the controversial idea that the truths of arithmetic, say, are analytically true. For another, my account provides a better—and more direct—explanation of the utility and applicability of (talk of) abstract objects. Finally, my account better fits the fact that disputes in ontology are taken very seriously by their participants.

Outline of the Paper In Sect. 4.2 I will briefly survey some of the problems that the existence of abstract objects pose. In Sect. 4.3 I will outline some of the ideas behind the recent deflationary approaches to abstract objects in the literature, and how they answer these challenges, focussing on the work of Thomasson (2015). I will also indicate what I take to be the main problem with these approaches. After that, we will move on to the original and constructive part of the paper. In Sect. 4.4 I will outline my version of the deflationary theory of truth. This will be helpful because (i) deflationism about truth provides a neat example of how the function of a term can explain everything of interest about that term, (ii) it provides the semantic background against which we will formulate our deflationary account of abstract objects, and (iii) it provides a blueprint for our formulation of our account of abstract objects. In Sect. 4.5.1 I will outline my deflationary account of properties and in Sect. 4.5.2 I will outline my account of numbers. In Sect. 4.6 I will explain how my account answers the problems surveyed in Sect. 4.2.

4.2 Some Problems Posed by the Existence of Abstract Object

Quantification over abstract objects such as numbers and properties is ubiquitous throughout science and philosophy. By a well-known argument of Quine and Putnam, abstract objects therefore exist, at least in the minimal sense that our variables need to range over such entities. Even if one denies that quantification over abstract objects ontologically commits us to the existence of abstract objects (Azzouni, 2004), the fact that we frequently quantify over abstract objects raises many deep and difficult questions. Since most of the questions that I am interested in still arise if one denies that quantification is ontologically committing, I will, in what follows, neglect the possible gulf between quantification and ontological commitment and often write as if quantification was ontologically committing.

Below I have listed some of the problems that the existence of (or quantification over) abstract objects pose. This list is by no means exhaustive. I have merely listed the problems that I find most urgent or otherwise intriguing.

1. *The epistemological problem*

   We have knowledge of abstract objects. For example, we know that 2 + 2 = 4. We know that there are infinitely many prime numbers. If abstract objects exist, we might expect that such objects play a role in the acquisition of knowledge about them. But this does not seem to be the case. We do not seem to have any causal interactions—direct or indirect –with abstract objects. How do we gain knowledge about them?
2. *The special status*

   Our knowledge of abstract objects (or at least of some abstract objects such as the natural numbers) seems to enjoy a special status—it enjoys a degree of certainty that other areas of inquiry lack. For example, the claim that 7 + 5 = 12 seems less assailable than the claim that there are twelve apples in front of me (I might be dreaming, or I might have miscounted and there are actually thirteen apples in front of me). What explains this special status?
3. *The utility of talking about abstract objects*

   Quantification over abstract objects is ubiquitous in science. What explains this fact? There have been many attempts to eliminate reference to abstract objects in scientific theories. None of them can be called an undivided success (see Burgess and Rosen (1997) for an overview). Why is reference to abstract objects indispensable? Why is it that we can't avoid reference to abstract objects in science?
4. *The application problem*

   Given that abstract objects are causally inert and not located in space and time, why does reference to abstract object help us understanding the empirical world? For example, why can we rely on the laws of arithmetic when we count concrete objects?

4.3 Recent Deflationary Approaches in Metaphysics

In recent years, a number of deflationary accounts of abstract objects have emerged in response to these problems, e.g. (Hale & Wright, 2001; Schiffer, 2003; Thomasson, 2015; Båve, 2015; Hofweber, 2016; Linnebo, 2018; Warren, 2020. Many of these accounts have a conventionalist bend, even if their adherents often do not characterise their position in this way (except for Warren).

Very roughly, the core idea behind these approaches (as I understand them) is that the questions listedin Sect. 4.2 can be answered by reference to our linguistic and conceptual practices. These approaches do not take recourse to the abstract object themselves in order to answer these questions. For purposes of illustration, let us have a brief look at Thomasson's account. I choose this example because her account is more general and easier to present than most of the other accounts.

4.3.1 Easy Ontology

According to Thomasson's easy approach to ontology, 'existence questions [are] easy to answer in the sense that answering them requires nothing more than conceptual and/or empirical work. [...] in many cases ontological debates may be resolved by engaging in trivial inferences from uncontroversial premises' (Thomasson, 2015, p. 127). In slightly more detail, the idea behind Thomasson's approach is as follows.

First, Thomasson claims that general nouns such as 'property' or 'number' have certain kinds of constitutive rules, including so-called application and co-application conditions (p. 86). Application conditions are certain basic rules of use that are among those that are meaning-constitutive for the term (p. 89). They are among the semantic rules that we master as we acquire language (p. 90). They need not take the form of necessary and sufficient conditions, and they need not be statable (p. 91).

Second, Thomasson accepts Quine's criterion of ontological commitment. She is also a deflationist about the term 'exists'. She thinks of the term as a univocal, formal notion governed by certain rules. One of them is this: Fs exist iff the application conditions actually associated with the term 'F' are fulfilled (p. 86).

Finally, Thomasson holds that existence questions can be easily answered by those who have mastered the rules governing the terms employed and who know the truth of certain undisputed claim as follows (p. 232). One starts from an undisputed truth, such as (1) 'The apple is red'. Now if we have mastered the use of property terms, we are entitled to infer (2) 'The apple has the property of being red' and therefore we can infer the existence claim 'There are properties'. What underlies the inference from (1) to (2) is the conceptual truth that if x is P, then x has the property of being P. 'The conceptual truth may be seen as an object-language articulation of a rule of use (in this case, an introduction rule for nominative property talk) that competent speakers have mastered' (p. 234). Hence, the existence of properties follows via trivial inferences from uncontroversial truths. In a similar way, questions about the existence of numbers, sets, mereological sums, and ordinary objects can be easily answered as well.

What we have here is (i) a simple realism about the entities in questions and (ii) a deflationary metaontological approach. The claim that we have a simple realism about properties and numbers means that properties and numbers exist—where the word 'exist' has the same meaning as in 'Dogs exists'. The claim that we have a deflationary metaontological approach means that the easy ontologist considers 'hard' debates about the existence of abstract objects (and certain other kinds of entities) as misguided and pointless. One need not, as the Neo-Quinean does, look to our best total theories to tell us whether numbers or properties exist. Instead, we simply need to look at how we use number and property terms in ordinary language.

4.3.2 How to Answer the Problems Posed by the Existence of Abstract Objects

According to Thomasson's account, we can know that abstract objects exist by deriving this claim from some undisputed truths using trivial inferences only. Let us now turn to the problems, enumerated in Sect. 2, that the existence of abstract objects pose. How are they answered by Thomasson's account? We will confine our attention to the first two problems.

First, let us address the epistemological problem. How do we come to know, for example, the truths of arithmetic? Thomasson is not very explicit on this point, but we can develop an answer based on the fact that Thomasson considers Hume's Principle as a conceptual truth about numbers. Hume's Principle is also the corner stone of the Neo-Fregeans such as Hale and Wright (2001) and Cook (2009), who have developed a detailed account of the epistemology of natural numbers.

Hume's Principle (HP) states that the number of Fs is identical to the number of Gs iff there is a one-one correspondence between the Fs and the Gs. Here, the phrase 'there is a one-one correspondence between the Fs and Gs' is formulated in second-order logic. There is a well known mathematical result concerning HP which is known as *Frege's Theorem*. Frege's Theorem tells us that we can derive the axioms of Peano arithmetic, in second-order logic, from HP and Frege's definition of zero, predecession, and natural number.

Given the assumption that HP is a conceptual truth, and given Frege's Theorem, we can develop an answer to the epistemological question. Very roughly, we can and do have knowledge of numbers, but not by being in causal contact with them, but by deducing truths about them from HP. The latter is of course nothing more than the articulation of the application and co-application conditions for number terms, which every competent speaker of English has mastered.

Second, let us address the question of the special status of our knowledge of (certain) abstract objects. Again, Thomasson is not explicit on this point, but we may imagine a response along the following lines. Very roughly, our knowledge of arithmetic, say, has the special status it does in virtue of the fact that it follows from our knowledge of (second-order) logic and the conceptual truth HP. Again, this is essentially what the Neo-Fregeans propose: '[A]rithmetical knowledge inherits whatever epistemological advantages are enjoyed by standard second-order logic and (HP)' (Cook, 2009, p. 29).

4.3.3 The Problem with Recent Deflationary Approaches

I am very sympathetic to the deflationist's idea that all questions about abstract objects should be answered with reference to our linguistic and conceptual practice. However, I take issue with one of the central assumptions of the deflationary approaches mentioned above. Easy ontologists such as Thomasson claim that principles such as HP are *conceptual truths*. Similarly, Neo-Fregeans claim that abstraction principles such as HP are *analytic*, or at least that they are available without significant epistemological presupposition. And conventionalists such as Warren claim the meaning-determining inference rules for mathematical expressions (say, the rules of Peano arithmetic) are *automatically valid* in that language (Warren, 2020, pp. 58, 190).

I stress at the outset that my worry is not about the intelligibility of the notion of analyticity itself. It is also not about the claim that there are analytic truths. For all I know, there may well be some analytic truths (e.g. 'All bachelors are unmarried'). But what I do not accept is that the (basic) principles characterising our talk about numbers, properties, and so on, are analytic. My reasons for thinking this are not particularly original. I therefore shall be brief.

First, it seems that we can always reject or doubt the existence of some alleged entities, the Fs. But if it is a conceptual truth that Fs exist, then it seems that only semantically incompetent speakers can doubt the existence of Fs. This contradicts the fact that many philosophers, all semantically competent speakers, have doubted the existence of numbers, properties, and so on.[1]

Second, and most importantly, concepts such as *number* and *property* are *law-cluster concepts* in the sense of Putnam. And as Putnam emphasises, no principle involving a law-cluster concept should be considered analytic.

Law-cluster concepts are constituted [...] by a cluster of laws which, as it were, determine the identity of the concept. The concept 'energy' is an excellent example of a law-cluster concept. It enters into a great many laws. It plays a great many roles, and these laws and inference roles constitute its meaning collectively, not individually. I want to suggest that most of the terms in highly developed science are law-cluster concepts, and that one should always be suspicious of the claim that a principle whose subject term is a law-cluster concept is analytic. The reason that it

[1] Thomasson responds to this objection in Thomasson (2015, pp. 172–176). I am not convinced by her response.

is difficult to have an analytic relationship between law-cluster concepts is that such a relationship would be one more law. But, in general, any one law can be abandoned without destroying the identity of the law-cluster concept involved, just as a man can be irrational from birth, or can have a growth of feathers all over his body, without ceasing to be a man. (Putnam, ch. 2, p. 52).

Interestingly, Putnam does not deny that there are *some* analytic truths. For example, he says that we may consider 'All bachelors are unmarried' as an analytic truth. Essentially, this is so because the term 'bachelor' does not play a significant role in science, and therefore there is little reason to believe that we should ever come to the point where we have to reject the claim that all bachelors are unmarried. But there can be no analytic truths about law-cluster concepts.

Third, the laws that one would naturally assume regarding certain abstract objects, such as properties or classes, can easily lead to paradoxes. For example, properties are typically characterised by comprehension axioms, e.g. sentences of the form

For all x, x is an F iff x has the property of being an F.

As is well known, such comprehension principles lead to paradoxes such as Russell's. There have been many different attempts to deal with the paradoxes. For example, we can reject some instances of the comprehension axiom scheme. Or we replace the comprehension axiom scheme by some other principles. Or we can adopt a non-classical logic in which the comprehension axiom scheme does not lead to triviality. Whatever option we choose, I think it is clear that we have to do this by considering our best overall theory.

A related objection by Boolos is that such alleged conceptual truths are not as epistemically innocent as some make them out to be. For example, Boolos is concerned about the 'quite strong content that [Hume's Principle] seems to possess' (Boolos, 1999, p. 304). For HP, when formulated in second-order logic, is equiconsistent with PA2, a theory in which most theorems of classical mathematics can be expressed and proved. Since we cannot be sure that PA2 is consistent, how can we claim that HP is analytic or available without significant epistemological presupposition?

4.3.4 Where to Go from Here

In my view, the claim that the (basic) principles of arithmetic, say, are analytic, or conceptually true, or available without significant epistemic presuppositions, has to be rejected. Let us assume we do that. Where do we go from here?

Note that one cannot easily remove the element of analyticity from the deflationary accounts just discussed without drastically changing them. For instance, easy ontologists claim that existence questions are too easy to answer to warrant serious metaphysical debate. On Thomasson's view, the Neo-Quinean approach, according to which we have to look to our best overall theories to answer existence questions, is confused. Instead, we can answer existence questions by simple conceptual means—by drawing existential consequences from undisputed observations and simple conceptual truths. But this doesn't work if there are no conceptual truths of the desired kind.

In order to make progress, we should ask ourselves:

(i) What is the main insight behind the deflationary approaches mentioned above?
(ii) Why is it that most deflationary views of abstract objects claim that the axioms of arithmetic, say, are analytic or conceptually true?

Let me start by responding to question (ii). I suspect that what underlies this is *the need to find truthmakers for e.g. arithmetical truths*. Since deflationists don't want to take recourse to the abstract objects themselves to account for the truth or falsity of the sentences of arithmetic, they need to explain it in a different way. And what better way than to claim that the axioms of arithmetic are simply true by stipulation (or something along these lines)?

One way to resist the temptation is to claim that the truths of arithmetic are made true by stipulation is to resist the idea that every truth requires a truthmaker in the first place. The deflationary conception of truth is well suited for that purpose. It will form the backbone of our deflationary view of abstract objects.

As to question (i), we have already seen the answer. The main insight behind deflationary approaches is that the problems posed by the existence of abstract objects can and should be resolved by reference to our linguistic and conceptual practice. Again, the deflationary account of truth provides a very nice example of a successful theory of this type. For the deflationary account of truth views the logico-linguistic function of the truth predicate as the key to a proper understanding of truth. We will use this example as a blueprint for formulating our deflationary account of abstract objects.

4.4 Deflationism About Truth

In the next section I will outline an alternative deflationary account of abstract objects. But before I get to this it will be helpful to have a brief look at the deflationary account of truth. As I just mentioned, this will be helpful because:

- it provides a neat example of how the logico-linguistic function of a term can explain everything of interest about that term;
- it provides the semantic background against which we will formulate our deflationary account of abstract objects; and
- it provides a blueprint for the formulation of our deflationary account of abstract objects.

There are many different formulations or versions of deflationism about truth, many of which I reject. In presenting deflationism about truth, I shall seize the opportunity to clarify some points about how I think deflationism about truth is best formulated.

Core Tenets of Deflationism About Truth In my view, the deflationary account of truth is best characterised by the following three claims:

1. The sole reason for having a truth predicate in our (regimented) languages is that it can serve a certain logical or quasi-logical function: namely, to enable us, in combination with our ordinary (first-order) quantifiers, to quantify into the syntactic position of sentences.
2. In order to serve this function, it is necessary and sufficient that every (non-paradoxical) sentence A is (in some sense to be specified) equivalent with its truth ascription $T(\ulcorner A \urcorner)$.
3. All truth-theoretic principles can be explained on the basis of claims (1) and (2) in conjunction with other explanatory principles that do not directly involve the concept of truth.

In what follows I will explain these three claims in slightly more detail. But before I do that, I want to highlight two things.

First, I have started the deflationary account of truth by focusing on the function of the truth predicate. This characterisation differs importantly from that of many other authors who take the equivalences between A and $T(\ulcorner A \urcorner)$ as their starting point. In particular, many deflationists think of the equivalences as 'conceptually and explanatorily basic' (Beall & Armour-Garb, 2005, p. 1), where

> To say that they are *conceptually basic* is to say that there are no 'deeper facts' about the given notion [...] in terms of which it can be defined or in terms of which the given equivalences themselves can be explained.

and

> To say that they are *explanatorily basic* is to say that our 'truth'-talk is explained, where explanation is required, by nothing 'deeper' than such equivalences—or by the inference rules (e.g., true-in, true-out) that yield them (Beall & Armour-Garb, 2005, pp. 1–2).

I do not think that this is the best way of formulating deflationism. In my view, deflationism is more attractive if the function of the truth predicate is taken as conceptually and explanatorily basic.[2] We will see one important applications of this reversal later on.

Second, deflationism (as I understand it) is not an empirical claim about the use of the truth predicate in ordinary discourse, but *a claim about scientific theories* and the language(s) in which these theories are formulated. That is to say, deflationism concerns regimented languages.[3] Of course, when I say that deflationism is concerned with regimented languages, I do not automatically wish to deny that deflationism is correct about the use of the truth predicate in ordinary language as well. For all I know, this may well be the case. But the deflationist need not make such a strong empirical claim; she may remain agnostic about it. Deflationists, as I

---

[2] Azzouni (2005) expresses a similar view.

[3] By a regimented language I do not necessarily mean a formal language. It could also be a fragment of English (perhaps augmented with some formal symbols) that is governed by certain normative rules guiding our scientific discourse.

conceive of them, think that little of philosophical interest hinges on whether claims (1) to (3) are true about ordinary language.

The Function of Truth Talk That the truth predicate is only in our language to serve a certain (quasi-)logical function is a thesis that has been held by many deflationists. However, what that function precisely consists in has been a matter of some debate. There have been at least three different proposals:

- to express infinitary conjunctions and disjunctions;
- to finitely axiomatise infinite sets of sentences;
- to enable us to quantify into the syntactic position of sentences.

In a number of publications, Lavinia Picollo and I have argued against the first two proposals and argued in favour of the third proposal (Picollo & Schindler, 2018a, b, 2022). I shall not rehearse these arguments here, but since it is important to understand what our proposal amounts to,[4] let me illustrate it by way of example.

Consider the claim that knowledge is factive. We can express this claim through the following schema, where K is a sentential operator standing for knowledge:

$$\mathrm{K}A \rightarrow A$$

Now suppose we want to express this as a single claim rather than as a schema that has infinitely many instances. One way to do this is to introduce variables that can stand in the syntactic position of sentences, and quantifiers for binding them. Thus we can express the factivity of knowledge like so:

$$\forall \alpha (\mathrm{K}\alpha \rightarrow \alpha)$$

However, we can express the factivity of knowledge with a truth predicate as well, without introducing these new variables and quantifiers. Given that A is equivalent to $\mathrm{T}(\ulcorner A \urcorner)$, we get from the factivity schema the following equivalent schema:

$$\mathrm{KT}(\ulcorner A \urcorner) \rightarrow \mathrm{T}(\ulcorner A \urcorner)$$

And now we can generalise as follows, using only our ordinary first-order quantifiers:

$$\forall x (\mathrm{KT}(x) \rightarrow \mathrm{T}(x))$$

[4] For example, since many deflationists consider truth to be an 'insubstantial' notion, some philosophers have argued that deflationary theories of truth need to be conservative, i.e. they must not prove new non-semantic facts. This is often seen as an objection to deflationism because most interesting theories of truth are not conservative. In Picollo and Schindler (2018a), we argue that the conservativeness requirement is based on a misunderstanding of deflationism: if it is the job of the truth predicate to emulate higher-order quantification, then we shouldn't expect truth to be conservative because higher-order quantification isn't conservative either. A second important application, concerning the question how deflationists can account for the alleged causal-explanatory role of truth, is given below.

In Picollo and Schindler (2022), we have shown that all inferences that can be carried out in standard higher-order languages can be mimicked within a suitable truth theory.

The Equivalences In order for the truth predicate to serve its function, it is necessary and sufficient that every (non-paradoxical) sentence A is equivalent to its truth ascription $\mathrm{T}(\ulcorner A \urcorner)$. Importantly, the deflationist should *not* claim that the equivalences are automatically valid or analytically true. Now, I do not wish to deny that we may think of the truth predicate, and the principles or laws governing it, as being introduced into our regimented language(s) by some kind of stipulation. On the contrary. But even if the truth predicate, and the laws governing it, are introduced into our language by stipulation, that does not make the laws analytically true. In fact, it doesn't even guarantee their truth.

That the laws of truth are not analytic is best seen by reflecting on the semantic paradoxes. The paradoxes show that the naïve theory of truth (i.e. the set of all Tarski-biconditionals) cannot be correct. They show us that we need to make tough decisions about: what logic we should use; how the equivalences should be formulated (e.g. as Tarski-biconditionals, as introduction and elimination rules, etc.); for what sentences we can accept the equivalences; and so on.

The reader may wonder: What makes the equivalences true, if they are not made true by stipulation? To this I reply the following: In asking that question, one presupposes that truth requires a truthmaker—something that makes it true. I reject that assumption. The deflationist need not answer the question, if formulated in this way. We may ask instead: What are the correct principles governing the truth predicate? And how do we know that they are correct? And the answer is of course: in order to find out what principles govern the truth predicate, we need to look to our best overall theory. After all, theory is our best guide to truth—even to the truth about truth.

Explaining All Truth-Theoretic Principles The deflationist claims that all truth-theoretic principles can be explained in terms of the function of the truth predicate and the equivalences governing it. I will illustrate this by going through three important classes of principles that have caused some concern in the literature.

(a) *Explaining the basic laws of truth.* If one takes the equivalences as conceptually and explanatorily basic then one cannot explain why the equivalences are true. However, by taking the function to be explanatorily and conceptually basic allows us to explain *why* the truth predicate satisfies the equivalences. The truth predicate satisfies the equivalences because it is in virtue of satisfying the equivalences that the truth predicate can play the role it is intended to play – namely, to enable us, in combination with our ordinary quantifiers, to quantify into the syntactic position of sentences. The equivalences are necessary and sufficient conditions on the functioning of the truth predicate in its intended sense. Since the truth predicate is introduced into our scientific theories for its logical function, it is no wonder that it satisfies the principles it needs to satisfy in order to do the work it is intended to do. Of course, how precisely the equivalences have to be formulated depends, among other things, upon which logic is used in our scientific theories. This is precisely why the equivalences are not analytically true. (Again, note that I am not saying that the equivalences are *made true* by the fact we introduce them in this way. It is because our best theories require a device for sentential generalization that we need to introduce a truth predicate satisfying the equivalences. And it I because we have good reason to think that our best theories are true that we have good reason to believe that the equivalences, being part of them, are therefore true as well.)

(b) *Explaining compositional principles of truth.* Both friends and foes of deflationism have wondered how the deflationist can account for so-called compositional principles of truth such as

(C) 'A conjunction is true if and only if both conjuncts are true'

In Picollo and Schindler (2021), we have argued that compositional principles are simply generalisations of schematic consequences of the equivalences. For example, (C) is a generalisation of the schema

$$\mathrm{T}(\ulcorner A \wedge B \urcorner) \leftrightarrow \mathrm{T}(\ulcorner A \urcorner) \wedge \mathrm{T}(\ulcorner B \urcorner)$$

(C) and similar principles can be derived from the equivalences together with certain additional (non-truth-theoretic) principles that the deflationist has every reason to accept.[5] I shall not rehearse this argument here but refer the interested reader to the mentioned paper.

(c) *Explaining other generalisations involving the truth predicate.* The truth predicate is used in the formulation of certain explanatory principles such as:

(S) If your beliefs are true then you are likely to achieve your goals.

This raises two questions: (i) Can a deflationary truth predicate be used in explanatory principles? (ii) How can the deflationist explain the truth of such principles? I will answer (i) in the affirmative. My answer will imply that (S) is not a truth-theoretic principle at all, and therefore its truth need not be explained by the deflationist.

The answer to (i) is that a deflationary truth predicate can be used in explanatory principles as long as it is only used as a logical device. Here is how we can see that the truth predicate occurs only as a logical device in (S): The function of the truth predicate is to enable us, in combination with our ordinary quantifiers, to quantify into the syntactic position of sentences.

[5] These additional principles are known in the literature as *reflection principles*. See Cieslinski (2017) and Horsten and Leigh (2017) for a similar solution.

This allows us to devise a little test for when the truth predicate is used in accordance with its logical function, and when it isn't.[6] Azzouni sees this important point very clearly:

We can (artificially) introduce prosentences, or more precisely, impose an additional prosentential capacity on a pronoun already available in English ("it") so that it can now also appear in sentential positions but still refer back to quantifiers (in English) that—when functioning as they ordinarily do—only accept anaphora from pronouns in nominal positions (e.g., in our artificial English, we can say, "Some sentence although no one can prove it," or perhaps, "Some sentence, it, although no one can prove it"). Whenever "true" is functioning solely as a disquotational device, it should be replaceable by these *anaphorically unrestricted pronouns*. If usage of "true" can't be so replaced in a locution, then that use of "true" isn't functioning as a disquotational device compatibly with [deflationism]. (Azzouni, 2006, p. 23).

So the idea here is roughly as follows. First, let L be our language, which we assume to contain the truth predicate. Second, consider the language L^* that is exactly like L except that it does not contain the truth predicate and instead contains variables that can stand in the syntactic position of sentences, and quantifiers for binding them. Now, consider some sentence of L involving the truth predicate. If we can find a sentence of L^* that, for all intents and purposes, can play the same theoretical role as our L-sentence, then the truth predicate is used in the L-sentence as a logical device only.

If we apply this procedure to the sentence (S), then we can easily see that the truth predicate is used in (S) only as a logical device. For we can replace (S) with

(S^*) For all α, if you believe α, and α, then you are likely to achieve your goals.

This shows that a deflationary truth predicate can be used in explanatory principles. This answers (i).

To answer (ii), just note that (S^*) is not a truth-theoretic principle. Instead, it is a law-like generalisation about how certain believes allow you to achieve your goals. In (S), the truth predicate is merely used to express this law-like generalisation without making use of sentential quantifiers.[7] Like (S^*), (S) is not a truth-theoretic principle. Hence, explaining why it is true does not fall within the purview of the deflationist.

[6] Here we have another example of why it is important to have a proper understanding of the function of the truth predicate.

[7] A similar argument has been worked out in slightly more detail by Wrenn (2023).

4.5 Deflationism in Metaphysics

I am finally in a position to present my deflationary account of abstract objects. In this paper, I will restrict my attention to properties and numbers. However, one can develop analogous accounts for classes and other abstract objects as well. In presenting my accounts of properties and numbers I shall follow the scheme as in the case of truth:

- a specification of the function of the relevant notion (e.g. property, number);
- a specification of what formal rules or axioms are needed for the notion to fulfil its function;

and

- a claim to the effect that all facts about the relevant notion can be explained on the basis of the first two bullet points together with additional explanatory principles that have nothing specifically to do with the relevant notion.

In Sect. 4.5.1 I will formulate the three claims for properties and in Sect. 4.5.2 I will do the same for numbers. Then, in Sect. 4.6 I will explain how my account answers the challenges or problems that we surveyed in Sect. 4.2.

4.5.1 A Deflationary Account of Properties

There are various deflationary accounts of properties in the extant literature. The views that are closest to my own are those proposed by Hofweber (2006, 2016) and Båve (2015).[8] However, there are various points of divergence between us, points that have many ramifications for our outlook on epistemology, metaphysics, and semantics.

Core Tenets of Deflationism About Properties My account of properties consists in the following three claims:

1. The sole reason for having property terms in our regimented languages is that they can serve a certain logical or quasi-logical function: namely, to enable us, in combination with our ordinary (first-order) quantifiers, to quantify into the syntactic position of open formulas.
2. In order to serve this function, it is necessary and sufficient that every (non-paradoxical) formula $\varphi(x)$ is equivalent (in some sense to be specified) with the formula 'x has the property of being a φ'.
3. All facts about properties can be explained on the basis of claims (1) and (2) in conjunction with other explanatory principles that do not directly involve the concept of a property.

[8] See also (Armour-Garb & Woodbridge, 2015, p. 203, fn 15).

The case of property talk is quite similar to the case of truth talk. I will confine myself here to brief comments on claims (1) and (2).

The Function of Property Talk Hofweber seems to think that the function of property terms is to express certain infinite conjunctions and disjunctions (or at least, that a quantification over properties is truth-conditionally equivalent to an infinite conjunction or disjunction). In (Schindler, 2022a, sec. 4.2) I have argued that this cannot be right. In my view, the function of property talk is best understood as a means to quantify into the syntactic position of open formulas. A version of this thesis was, to the best of my knowledge, first proposed by Charles Parsons (1983). Let me illustrate this proposal by way of example. Assume one wishes to express (the following version of) Leibniz's Law as a single claim by universally generalising on φ's position in

$$\forall x \forall y \left(x = y \rightarrow \left(\varphi(x) \rightarrow \varphi(y) \right) \right) \quad (4.1)$$

Using second-order quantifiers, we can write

$$\forall V \forall x \forall y \left(x = y \rightarrow \left(Vx \rightarrow Vy \right) \right) \tag{4.2}$$

However the introduction of property terms satisfying property comprehension affords us the same inferential power. For one can formulate Leibniz's Law as a single claim as follows:

$$\forall v \forall x \forall y \left(x = y \rightarrow \left(x \,\text{has}\, v \rightarrow y \,\text{has}\, v \right) \right)$$

It is quite easy to see that any inference that one can draw in a higher-order language can be mimicked in a suitable (first-order) theory of properties.

The Equivalences The equivalences for properties have traditionally been expressed by comprehension axioms, i.e. x has the property of being an F iff x is an F. As in the case of truth, property talk gives rise to paradoxes (such as Russell's paradox), forcing us to make tough decisions about the choice of logic, about the formulation of the equivalences, and for what open formulas we can accept these equivalences, and so on. Again, this shows that the equivalences cannot be considered as analytic.

4.5.2 A Deflationary Account of Numbers

My deflationary account of numbers will follow the same scheme that we saw in connection with truth and with properties. But in order to specify the function of number talk, and what principles it requires, we need to introduce some terminology.

Throughout this section, we will often refer to *numerically definite quantifiers*. These are ex- pressions of the form '$\exists^n$' and they are contextually defined. If 'Fx' is a formula with one free variable, we define:

$$\exists^0 x Fx =_{Df} \neg \exists x Fx$$

$$\exists^1 x Fx =_{Df} \exists x \left(Fx \wedge \forall y \left(Fy \rightarrow x = y \right) \right)$$

$$\ldots$$

$$\exists^{n+1} x Fx =_{Df} \exists x \left(Fx \wedge \exists^n y \left(Fy \wedge y \neq x \right) \right)$$

$$\ldots$$

We may read '$\exists^n x\ Fx$' as 'There are (exactly) n-many objects satisfying F'. Numerically definite quantifiers allow us to count how many objects satisfy the formula Fx *without referring to numbers* (i.e. without using singular terms referring to numbers). The occurrence of 'n' in '$\exists^n x\ Fx$' is syncategorematic and disappears upon analysis.

In a similar way, we can count how many pairs of objects satisfying some formula $F(x, y)$ with two free variables, the number of triplets satisfy some formula with three free variables, and so on, without referring to numbers. For details I send the reader to (Schindler, 2022b).

Now, let '#' be a variable-binding operator that, applied to an open formula 'Fx', yields a singular term '$\#x.Fx$' in which 'x' is bound. We may read '$\#x.Fx$' as 'the number of Fs'.

Similarly, let '$\#^2$' be an operator that, applied to an open formula '$F(x, y)$', yields a singular term '$\#^2 xy.F(x, y)$' in which the variables 'x' and 'y' are bound. We may read '$\#^2\ xy.F(x, y)$' as 'the number of pairs that fall under F'. Finally, we write '**n**' for the numeral of the natural number n. I assume that numerals are formed via some canonical recursive procedure, e.g. from '**0**' and the successor function symbol.

Core Tenets of Deflationism About Numbers We are now in a position to state my deflationary account of numbers:

1. The sole reason for having number terms in our language is that it can serve a certain logical or quasi-logical function: namely, to enable us, in combination with our ordinary (first-order) quantifiers, to quantify into the index position of a numerically definite quantifier.
2. In order to serve this function, it is necessary and sufficient that for every formula Fx, the formula $\exists^n x\ F(x)$ is equivalent (in some sense to be specified) with the formula $\#x.Fx = \mathbf{n}$ (and analogously for formulas with more than one free variable).
3. All facts about natural numbers can be explained on the basis of claims (1) and (2) in conjunction with other explanatory principles that do not directly involve the concept of a natural number.

The Function of Number Talk As we have seen, numerically definite quantifiers allow us to count how many objects satisfy a formula without referring to numbers. For example, we can say that there are two apples without incurring an ontological commitment to numbers. Moreover, we can prove, using standard first-order logic, that if there are two apples and two bananas, then there are four pieces of fruit.[9] Similarly, we can also prove that if there are two spoons and two forks, then there are four pieces of cutlery. Now, the introduction of number terms enables us, given the equivalences, to generalise these statements. We simple say that $\mathbf{2} + \mathbf{2} = \mathbf{4}$.

The Equivalences In contrast to the Tarski-biconditionals for truth or the comprehension axioms for properties, the numerical equivalences do not lead to a contradiction when formulated in classical logic with a material biconditional. Thus, in what follows, I will assume the following formulation of the equivalences:

$$\# x.Fx = \mathbf{n} \leftrightarrow \exists^n x Fx$$

I will call this scheme *the numerical equivalence scheme* (NES, for short). (This is the version of the NES for formulas with one free variable. We need, in addition, an analogous scheme for formulas with two free variables, and so on. See Schindler (2022b) for details.) Nevertheless, we should not consider the instances of the NES as analytic truths. Given the importance and central role of number terms for our scientific theories, we should always be prepared to revise our principles. For example, we may have to revise our logic in order accommodate the paradoxes of truth and properties. In the wake of this, we may also have to reformulate the NES (e.g. as introduction and elimination rules rather than as a material biconditional).

Explaining All Number-Theoretic Principles Given suitable definitions, one can derive, in classical first-order logic, all true numerical equations and inequalities (involving addition and multiplication) from the NES (Schindler 2022b, sec.4.4). Hence we can easily explain this simple fragment of arithmetic (the diagram of the standard model of Peano arithmetic). What about general claims about the natural numbers? There are various possibilities. Let me mention two. First, it can be shown that the equivalences entail all truths of arithmetic in second-order logic (with the so-called standard semantics, where the second-order quantifiers range over the full power set of the domain of discourse) (Schindler 2022b, sec.4.4). Hence we have an explanation of all number-theoretic principles if we accept second-order logic with standard semantics. However, some deflationists will not accept second-order logic with standard semantics.[10] Alternatively, we may explain general claims about

[9] Strictly speaking, this also requires the background assumptions that no apple is a banana and vice versa, and that apples and bananas are fruits.

[10] Some will reject second-order logic with standard semantics because it is not effective in the computational sense. Others will reject it because the deflationist wants to stay first-order. Indeed, the entire point of talking about truth and properties is to mimic higher-order quantification within a first-order framework.

numbers substitutionally. Thus, we may define that a general statement about the natural numbers is true iff every numerical substitution instance is true. Since we already have an explanation of the non-quantificational fragment of arithmetic, this definition assigns to every arithmetical sentence a determinate truth value.

4.6 How to Answer the Problems Posed by Abstract Objects

We are finally in a position to respond to the problems posed at the beginning of this paper.

The Epistemological Problem *We have knowledge of abstract objects. If abstract objects exist, we might expect that such objects play a role in the acquisition of knowledge about them. But this does not seem to be the case. We do not seem to have any causal interactions—direct or indirect—with abstract objects. How do we gain knowledge about them?*

We have seen how all principles involving a given notion (truth, property, number) can be explained with reference to the basic principles or equivalences governing that notion. The equivalences, in turn, can be explained by reference to the function of the notion. Hence, knowledge of the function gives us knowledge of the basic principles (together with systematic considerations concerning how to best formulate the equivalences in light of the paradoxes). Knowledge of the basic principles, in turn, provides the basis for our knowledge of non-basic principles involving the given notion. For instance, I have said that, given suitable definitions, all true numerical equations and equalities can be straightforwardly deduced, in first-order logic, from the NES. We therefore have a relatively easy epistemic access to this fragment of arithmetic. However, while we can give an easy explanation of our grasp of the elementary truths of arithmetic, this is not the case for general statements about numbers. Earlier I pointed out that we can explain general statements about numbers either through the use of second-order logic or substitutionally. For example, if second-order logic is logic, then every arithmetical truth is a logical consequence of the NES. But this merely provides a *semantic* route from elementary arithmetic to full arithmetic. It doesn't provide an *epistemic* route. This is so because there is no sound deduction system that is complete with respect to the standard semantics of second-order logic. Even though all truths of arithmetic follow from the NES, not all of them can be derived from the NES in a sound deduction system. So the deflationist needs to tell a story how we can get from the knowledge of elementary arithmetic to our knowledge of a larger fragment of arithmetic (say, how we come to know the axioms of first-order Peano arithmetic). In Schindler (2022b) I have outlined several proposals, the most promising of which involves the so-called *regressive method*. The term was coined by Bertrand Russell who assimilated the method of mathematics to that of the ordinary sciences of observation. According to this idea, we start out with some elementary truths as our data, just as the ordinary sciences start with some observational statements. General laws are then obtained by inductive methods, and tested against these data, thereby conferring some degree of probability to the laws (Russell, 1907). On Russell's account, it is not clear how we come to know the elementary truths of arithmetic. However, as we have seen, our knowledge of the elementary truths of arithmetic can be explained on the basis of our knowledge of the NES. And the regressive method then allows us to explain our knowledge of general statements about numbers.

The Special Status *Our knowledge of abstract objects (or at least of some abstract objects such as the natural numbers) seems to enjoy a special status—it enjoys a degree of certainty that other areas of inquiry lack. What explains this special status?*

From what I have said in my answer to the previous question, it should be clear that I do not consider all parts of arithmetic as certain. For example, I have suggested that general claims about numbers are often obtained and justified according to the regressive method, and therefore do not enjoy the special status that some people have thought it does. However, I agree that certain laws of arithmetic—e.g. that $2 + 2 = 4$—have a special epistemic status. Similarly, I do believe that some (but not all) comprehension axioms—e.g. that Fido has the property of being a dog iff Fido is a dog—have a special epistemic status. How can that be explained on my account? Consider first the case of truth. The truth predicate is introduced into our language to enable us to quantify into the syntactic position of sentences of our language. We know that the equivalences (e.g. the Tarski-biconditionals) cannot be analytic because of the paradoxes.

This forces us to make decisions regarding the logic which we should use, how the equivalences are to be formulated, for which sentences we can adopt the equivalences, and so on. But no matter what logic we choose, and how we formulate the equivalences, typically we will accept the equivalences for all sentences that do not contain the truth predicate themselves. Why? Let L be the language without the truth predicate. Now, if we accept the equivalences for the sentences of L, then we will be able to quantify into the syntactic position of the sentences of L. But this is the only reason why we added a truth predicate to L in the first place. Of course, once we have added a truth predicate, we will be able to formulate liar-like sentences. Adopting the equivalences for them may lead to paradox. But these sentences are of little interest when it comes to theorising about the world. We may learn to live without having the equivalences for them. But if we did not accept the equivalences for the sentences of L, the truth predicate would completely fail to serve its purpose. This is why it is very unlikely that we will ever reject the equivalences for the sentences of L. An analogous point can be made for the corresponding instances of the comprehension axiom for properties and the NES for numbers. Since the latter are sufficient to allow us to prove that $2 + 2 = 4$, it is unlikely we will ever reject that, given that we have little reason to reject the NES. Thus, the elementary laws of numbers and properties have a special epistemic status—but they are not analytic.

The Utility of Talking About Abstract Objects *Quantification over abstract objects is ubiquitous in science. What explains this fact? Why is reference to abstract objects indispensable? Why is it that we can't avoid reference to abstract objects in science?*

Talk about abstract objects such as properties and numbers is ubiquitous in science because such talk enables us to formulate certain important high-level generalisations. Reference to abstract objects can't be avoided in this enterprise because there is no other way of formulating such generalisations within a first-order framework. In slightly other words: The function of talking about properties and numbers is to quantify into syntactic positions of our language other than singular terms. Since in first-order languages we can only quantify into the syntactic position of singular terms, there is simply no other way to do this but to introduce singular terms that stand proxy for the relevant categories. For example, if we want to quantify into the position of open formulas (and stay within a first-order framework), then we can do nothing except to turn the occurrences of these expressions (e.g. open formulas) into singular terms—e.g. by correlating the singular term 'the property of being a dog' with the open formula 'is a dog', and then saying 'Fido has the property of being a dog' instead of 'Fido is a dog'.

The Application Problem *Given that abstract objects are causally inert and not located in space and time, why does reference to abstract object help us understanding the empirical world? For example, why can we rely on the laws of arithmetic when we count concrete objects?*

Very roughly, claims about abstract objects such as properties and numbers are applicable to the empirical world because of the equivalences. The equivalences (e.g. the comprehension axioms for properties and the numerical equivalence scheme for numbers) connect our use of terms referring to abstract objects with ordinary statements that concern the empirical world and do not contain reference to abstract objects. For example, let us have a look at a question raised by Dummett (1991, p. 99): How can we explain the fact that we can appeal to the equation '$\mathbf{5} + \mathbf{2} + \mathbf{0} = \mathbf{7}$' in order to justify that there were seven animals in the field from the fact that there were five sheep, two cows and no other animals there? Dummett's question can be answered as follows. Assume there were five sheep, two cows and no other animals in the field (formulated in terms of numerically definite quantifiers). Given the NES, we infer from our premise that the number of sheep = **5**, that the number of cows = **2**, and that the number of other animals = **0**. Next, given suitable definitions—for example, that $\#x.Fx + \#x.Gx = \#x.(Fx \vee Gx)$, provided that the Fs and Gs are mutually disjoint—we conclude that the number of sheep, cows and other animals = $\mathbf{5} + \mathbf{2} + \mathbf{0}$. Now, basic arithmetic tells us that the number of sheep, cows and other animals = **7**.

Finally, another application of the NES yields the desired conclusion, viz. that there were seven animals in the field. Thus, the NES acts as a bridge principle: given some premises involving numerically definite quantifiers, and given some body of (purely) arithmetical truths, the NES enables us to utilise the arithmetical truths to draw consequences involving numerically definite quantifiers. According to the deflationary position outlined here, this bridge principle is at the centre of our number talk and therefore it is not mysterious at all that arithmetic is applicable to the empirical world.

Acknowledgements I thank Marta Campa for many conversations about deflationism and meta-ontology and for her very helpful comments on an earlier version of this paper.

Funding A first draft of this paper was written while the author received support from the Research Council within the project Truth and Semantics (TRUST, grant agreement no 803684). It was finished while the author received support from the European Research Council within the project Generalisation into sentence and predicate positions (GENERALISE, grant agreement no 101076054).

References

Armour-Garb, B., & Woodbridge, J. (2015). *Pretense and pathology. Philosophical fictionalism and its applications*. Cambridge University Press.

Azzouni, J. (2004). *Deflating existential consequence. A case for nominalism*. Oxford University Press.

Azzouni, J. (2005). Anaphorically unrestricted quantifiers and paradoxes. In J. C. Beall & B. Armour-Garb (Eds.), *Deflationism and paradox* (pp. 250–273). Oxford University Press.

Azzouni, J. (2006). *Tracking reason. Proof, consequence, and truth*. Oxford University Press.

Båve, A. (2015). A deflationist error theory of properties. *Dialectica, 69*, 23–59.

Beall, J. C., & Armour-Garb, B. (2005). Introduction. In *Deflationism and paradox* (pp. 1–6). Oxford University Press.

Boolos, G. (1999). Is Hume's principle analytic? In *Logic, logic, and logic* (pp. 301–314). Harvard University Press.

Burgess, J., & Rosen, G. (1997). *A subject with no object. Strategies for nominalistic interpretations of mathematics*. Clarendon Press.

Cieslinski, C. (2017). *The epistemic lightness of truth. Deflationism and its logic*. Cambridge University Press.

Cook, R. T. (2009). New waves on an old beach: Fregean philosophy of mathematics today. In O. Bueno & O. Linnebo (Eds.), *New waves in philosophy of mathematics* (pp. 13–34). Palgrave Macmillian.

Dummett, M. (1991). *Frege: Philosophy of Mathematics*. London: Duckworth.

Hale, B., & Wright, C. (2001). *The reason's proper study*. Oxford University Press.

Hofweber, T. (2006). Inexpressible properties and propositions. In D. Zimmerman (Ed.), *Oxford studies in metaphysics* (Vol. 2, pp. 155–206). Oxford University Press.

Hofweber, T. (2016). *Ontology and the ambitions of metaphysics*. Oxford University Press.

Horsten, L., & Leigh, G. E. (2017). Truth is simple. *Mind, 126*, 195–232.

Linnebo, O. (2018). *Thin objects. An abstractionist account*. Oxford University Press.

Parsons, C. (1983). Sets and classes. In *Mathematics in philosophy* (pp. 209–220). Cornell University Press.

Picollo, L., & Schindler, T. (2018a). Deflationism and the function of truth. *Philosophical Perspectives, 32*, 326–351.

Picollo, L., & Schindler, T. (2018b). Disquotation and infinite conjunctions. *Erkenntnis, 83*, 899–928.

Picollo, L., & Schindler, T. (2021). Is deflationism compatible with compositional and Tarskian, truth theories? In C. Nicolai & J. Stern (Eds.), *Modes of truth. The unified approach to truth, modality, and paradox* (pp. 41–68). Routledge.

Picollo, L., & Schindler, T. (2022). Higher-order logic and disquotational truth. *Journal of Philosophical Logic, 51*, 879–918.

Russell, B. (1907). The regressive method of discovering the premises of mathematics. In B. Russell (Ed.), *Essays in analysis* (pp. 272–228). Allen and Unwin, 1973.

Schiffer, S. (2003). *The things we mean*. Oxford University Press.
Schindler, T. (2022a). Deflationary theories of properties and their ontology. *Australasian Journal of Philosophy, 100*, 443–458.
Schindler, T. (2022b). Steps towards a minimalist account of numbers. *Mind, 131*, 865–893.
Thomasson, A. L. (2015). *Ontology made easy*. Oxford University Press.
Warren, J. (2020). *Shadows of syntax*. Oxford University Press.
Wrenn, C. (2023). Deflating the success-truth connection. *Australasian Journal of Philosophy, 101*, 96–110.

Part II
Themes on Thomassonian Easy Realism

Chapter 5
Easy Ontology Made Easier

Arvid Båve

Abstract Easy Ontology (EO), defended in several recent works by Amie Thomasson, and based on Carnap's famous deflationism about metaphysics, is the view that many ontological questions, like 'Are there numbers?', are at bottom easy, at least when taken in their "internal" sense. Both Carnap and Thomasson take for granted that serious metaphysicians therefore cannot plausibly be interpreted as asking internal questions. Thus, they think they are committed to finding some alternative, special interpretation of metaphysicians' utterances. I argue that none of the alternative interpretations proposed in the literature are viable. More importantly, I argue that Carnap and Thomasson are wrong about this commitment. EO should instead be coupled with the view that metaphysicians ask internal questions, yet have the false belief that they can be answered only through metaphysical speculation. This is a significant modification of EO, in which the notion of an "external question" plays no role, and which results in a simpler and more plausible theory.

Carnap took for granted that metaphysicians cannot be asking internal questions, and claims instead that they use and understand ontological questions in a special, "external" sense, i.e., with a certain lexical meaning (1950/1956, pp. 209, 210, 213). As we will see, however, all extant proposals about this alleged, special sense are untenable. Fortunately for EO, there is no need to identify such a special sense: we can just take metaphysicians to (typically) use and understand ontological questions in their ordinary sense, which is their internal sense. This interpretation of metaphysicians is more charitable and yields a simpler and more innocuous version of EO—indeed, one which makes no use of the notion of an external question. This is a radical departure from the views of Carnap himself and his main contemporary follower, Amie Thomasson. It makes EO easier to state, understand, and defend.

Keywords Carnap · Thomasson · Ontological question · Ontological deflationism · Easy ontology · Internal · External-factual · Quantifier variance · Normative-linguistic

---

A. Båve (✉)
CFUL-Lisbon, Lisbon, Portugal

X. de Donato-Rodríguez et al. (eds.), *Deflationist Conceptions of Abstract Objects*, Synthese Library 516, https://doi.org/10.1007/978-3-031-98717-5_5

5.1 What to Keep

This section lays out what I take to be the central aspects of EO, which I think should be retained, while the next one describes Carnap's and Thomasson's views about how metaphysicians ought to be interpreted, which should be abandoned. The positive view I describe below will also include some additions not found in Carnap's or Thomasson's work, but which are relevant for my overall case.

Now, consider the following question:

Q Are there numbers?

I will throughout use Q as an example of a paradigmatically ontological question. When referring to Q or to an "ontological question", however, I will sometimes refer to a mere string, i.e., an uninterpreted sentence in interrogative form, which is individuated purely orthographically-phonetically. The reason is that I sometimes discuss whether Q might have several senses, which sense(s) it has, etc., which would make no sense if Q is semantically individuated. Similarly, one cannot sensibly discuss the different senses of non-linguistic entities that are expressed by linguistic entities, e.g., question(s) expressed by utterances of Q, or propositions expressed by declarative sentences. Therefore, when I discuss questions below, I will usually mean strings that are interrogative sentences, rather than something expressed thereby. Context will make clear which type of entity is referred to by 'Q' or 'ontological question'.

Ontological Deflationism in general is typically defined as the view that ontological questions are not deep, substantive questions about the extra-linguistic world. Carnap's theory is a subspecies of this view, since ontological questions do not come out as deep, substantive questions about the extra-linguistic world on any one of the senses he distinguishes (as we will see below). But there are also non-Carnapian instances of Ontological Deflationism, e.g., Eli Hirsch's view, discussed in Sect. 5.3.

It is not entirely clear how ontological questions should be distinguished from existence questions that are not ontological, if such there be (see, e.g., Hofweber (2016: Ch. 12) and Korman (2019) for discussion). If one takes all existence questions to be ontological, then it is clear that we cannot say that all ontological questions are easily answered (consider, 'Bosons exist'). Even so, Carnapian deflationists will insist that the difficulties involved in answering some existence questions are not of the kind that worry metaphysicians. They are due to complexity, unavailability of empirical evidence and the like; they are not, in Ted Sider's phrase, "epistemically metaphysical" (2011, p. 187).

To take these complications into account, Thomasson proposes that EO be defined as the conjunctive view that

> all well-formed existence questions may be answered by conceptual and/or empirical work (requiring nothing 'epistemically metaphysical'), and that at least some disputed existence questions may be answered by means of trivial inferences from uncontroversial premises. (2015, p. 128)

Here, "empirical work" does not include pragmatic considerations, e.g., of simplicity. I will follow Thomasson's definition of EO.

Among existence questions, Q would count as one of the easily answered ones (when taken internally). To wit, the answer to Q is Yes, and can be conclusively justified simply by noting, e.g., that 3 and 4 are greater than 2, hence, there are numbers greater than 2, and, hence, there are numbers. I will call this piece of reasoning Simple Reasoning and I will take the claim that Q is "easily answerable" to follow (more or less) from the claim that Q can be given a correct answer purely on the basis of Simple Reasoning.

When it is said that Q is easily answered, this should be understood as the claim that it is easy to *know* the answer, i.e., to know that there are numbers. Naturally, the claim is not that it is easy to produce a token of the word 'Yes' in response to the question. The issue also cannot be taken to concern merely believing the truth, since it can plausibly be easy to make a lucky guess. Even those who claim that it is not easy to make a lucky guess as to whether there are numbers should agree that this is not what the debate is about. What is at issue is knowledge. However obvious, Thomasson's *Ontology Made Easy* actually never makes this explicit. Doing so, as we will see in Sect. 5.7, turns out to reveal some interesting possibilities for EO hitherto unnoticed.

Since many philosophers think the proposed version of EO is a non-starter because it is immediately undermined by the "Bad Company Problem", I should say, very briefly, why such an assessment would be premature. Adherents of EO think that the existence of properties, for instance, can be established by a simple application the unrestricted, "naïve" inference rules for 'property', which are held to be analytic. However, if applied to the property of being a property that does not instantiate itself, these rules also allow us to infer a contradiction. This may seem to show that the existence of properties cannot be known, let alone easily known, by applying these rules.

The reason this problem does not immediately undermine EO is that it is possible for Carnapians to adopt a rather standard form of reliabilism, on which one knows P just in case P is true and one believes P on the basis of a reliable method, and add that believing a proposition on the basis of one's conceptual competence is a reliable (but fallible) method. Thus, when a proposition is believed on such a basis and it is true, then it is known. (One can also argue that if it is easy to come to believe a given true proposition on the basis of one's conceptual competence, this proposition is also easily known. The reliabilist response is therefore not merely "possible", but also has some benefits.)

Some propositions that we can infer by purely conceptual means are false, like the contradictions involving pathological, self-referential properties or sets. But this is consistent with the view proposed, since we already agreed that the relevant method is fallible. Since this line of defence is at least possible, it shows that the examples of pathological properties does not immediately undermine EO. I take this also to be the most promising way to supplement EO and that it forms part of a plausible overall treatment of the Bad Company Problem (which is more complex than suggested by the quick remark above), but the full defence must await another occasion.

5.2 What to Give Up

Let us now turn to the claims by Carnap and Thomasson that I think should be rejected, namely, the idea that metaphysicians understand ontological questions in a special, "external" sense. Carnap is usually interpreted as distinguishing three ways of taking an ontological questions: as internal questions, in which they are answerable by purely conceptual and/or empirical means (and in which some of them, like Q, are easily answered), and as external questions, which are in turn divided into two kinds: factual, language-independent questions, which are "meaningless", or "pseudo-questions", and, secondly, as pragmatic questions about language, in which they are sensible, but normative questions of which "linguistic framework" to adopt.

We can now see more clearly why Carnap's view is a species of ontological deflationism, as defined above: Internal questions are trivial, and, hence, not deep or substantive; factual-external questions are not even meaningful, and so cannot be deep or substantive; and external-pragmatic questions are worthwhile and perhaps deep, but they are not about the extra-linguistic world, but rather about language.

Why did Carnap identify other ways of taking ontological questions than the internal? One reason (probably the main reason, and perhaps the only reason) was to make sense of ordinary, "serious" metaphysicians, i.e., philosophers who take these ontological questions to be important, deep, and difficult. He writes,

> …nobody who meant the question 'Are there numbers' in the internal sense would either assert or even seriously consider a negative answer. This makes it plausible to assume that those philosophers who treat the question of the existence of numbers as a serious philosophical problem and offer lengthy arguments on either side, do not have in mind the internal question (1950/1956, p. 209).

Carnap thought that metaphysicians rather have in mind "external questions". Thomasson (2015, p. 33 f.) cites this passage approvingly. In Sect. 5.6, I criticize Carnap's reasoning in the quoted passage in detail, but before that, in Sects. 5.3, 5.4, and 5.5, I will argue that all of the interpretations of metaphysicians that have been offered on behalf of EO (namely, candidates (a)–(c) below) are untenable. Fortunately, since Carnap's reasoning in the quoted passage can be resisted, EO is not committed to choosing between them, but can without strain be coupled instead with what I call the "internal interpretation" (candidate (d) below).

Here, then, are the four candidate interpretations of metaphysicians that I will be discussing:

(a) Quantifier variance: metaphysicians' disagreements are merely verbal: they (can be taken to) interpret quantifiers and 'exist' differently, and in such a way that each makes claims that are true-in-their-language (cf. Hirsch, 2009, 2011).

(b) External-factual: metaphysicians (attempt to) use ontological expressions, not metalinguistically, but somehow externally to the linguistic framework they belong to. When used in this sense, they are nonsensical and raise mere "pseudo-problems" (cf. Eklund 2013).
(c) Normative-metalinguistic: metaphysicians, while apparently discussing non-normative questions about non-linguistic matters, are really engaged in discussions about which terms or terminologies to use, i.e., normative discussions about expressions or concepts.
(d) Internal: metaphysicians use and understand ontological questions in their ordinary, internal sense, in which, e.g., Q is easily answered.

My critique of (a)–(c) will repeat some familiar complaints, since they form part of my case for (d), but several new ones will be appear, too. As we will see, comparisons between these interpretations will have much to do with interpretive charity, as will my case for (d), presented in Sect. 5.7. (I will not discuss the idea that different metaphysicians may require different interpretations, since the upshot of my discussion will be that, except perhaps in some special cases, (d) is the only plausible interpretation.)

Let me close this section by clarifying and qualifying my positive view. Firstly, (d) must be qualified, since metaphysicians sometimes explicitly claim to use quantifiers or the existence predicate in some alternative, technical sense, or in some alleged ordinary sense, on which they are not answerable by purely conceptual and/or empirical means. This is so, for instance, when philosophers characterize quantifiers or the existence predicate in terms of "joint-carving", "fundamentality", or "domain condition"—see, e.g., Hofweber (2016: Ch. 3). (d) should therefore be taken as concerning only metaphysicians doing ontology without such terminological preliminaries. Even with this qualification (d) is substantive and contentious claim, which is rejected by Carnap and Thomasson, as well as by opponents of EO.

Note that this paper is not exactly about the question of how serious metaphysicians ought to be interpreted, but rather about what adherents of EO should say about this matter. Although I will mostly leave this qualification implicit, it is important to keep in mind when assessing the philosophical significance of the paper. For while it is commonplace to say that metaphysicians usually understand ontological questions in their ordinary, literal senses, it is contentious to say that adherents of EO can and should say this. Indeed, Carnap and Thomasson both reject this claim, and think they are committed to doing so.

Is the question of this paper about what Carnap in particular should say about serious metaphysicians, or what Carnapians more generally should say? There is no clear, significant difference here. In his seminal 1950 paper, Carnap made many different claims that were often obscure and often in tension with each other, resulting in an overall picture that is not clearly different from "Carnapianism in general", on any reasonable understanding. Also, no specific issue arising in my discussion depends on this choice. Nevertheless, since my purposes are philosophical rather than exegetical, I would prefer to say that this paper concerns EO in general, keeping in mind the obvious weight of Carnap himself as the founder of this position.

5.3 Quantifier Variance

Option (a) is loosely connected with Eli Hirsch, whose views has been intensely discussed in the literature. It is something of an outlier in the present context, however, which rather concerns Carnapian metaontology (Eklund (2013) argues that the relationship between quantifier variance and Carnap's views is tenuous at best). While they are both instances of Ontological Deflationism, as defined above, they differ on what is wrong with metaphysical disputes. Hirsch thinks they are merely verbal, in that different metaphysicians mean different things, and thus speak past each other. Carnap held no such view.

Hirsch sometimes formulates his view in terms of what metaphysicians might have meant, rather than what they actually mean. His point is that for each metaphysical sentence, there is a language (in the abstract, Lewisian sense) in which it comes out true, and that the different languages are all metaphysically on a par. It is mainly this last claim that his opponents have criticized (e.g., Sider, 2011). Since the present paper concerns what metaphysicians actually mean, however, we will pay less attention to this more general and abstract question.

One influential objection against (a) is that it is uncharitable. This may seem surprising, since (a) makes metaphysicians' claims come out true (as stressed by Hirsch himself in his (2007)). But, as argued by Brendan Balcerak Jackson (2013, 2014), (a) is still uncharitable, because it makes the contestants' assertions come out as trivial (given what they mean by the relevant expressions). Similarly, John Horden (2014) argues that while the Hirschian interpretation of what metaphysicians are saying is charitable, the accompanying interpretation of what they intend and desire is not. (McGrath (2008) finds several further, more specific, reasons why (a) involves uncharitable interpretations.)

Interpretation (a) is also uncharitable in that it entails that the contestants mistakenly think they have a genuine disagreement, when they don't (cf. Thomasson, 2017a, p. 9). Actual metaphysicians will often reject the notion that they are just talking past each other, each speaking a language in which their respective claims come out true. Thus, while (a) makes their basic first-order claims come out true, certain central higher-order beliefs of theirs come out false, namely, beliefs about what they mean, what they are doing, and what their debates are about. It is arguably less charitable to attribute false beliefs about what one means than about philosophical matters, the latter being notoriously difficult and the former something that should be more transparent to speakers.[1]

---

[1]Analogous objections threaten Yablo's (1998) identification of the internal/external distinction with the fictional/literal distinction. He claims that metaphysicians understand ontological questions literally, whereas laymen understand them fictionally. But since metaphysicians typically take themselves to understand these questions the same way as ordinary speakers do, including themselves, as they speak outside of the seminar room, Yablo's theory commits him to saying that they are wrong about whether they are speaking literally or fictionally, which is uncharitable. This objection presumably targets other forms of "hermeneutical fictionalism" as well, e.g., concerning mathematics (see Eklund, 2019; Båve, 2015, p. 28).

As against this, one might note that it is rather common, even for intelligent, rational, careful, and patient people, given sufficient time to think things over, to mistake a verbal disagreement for a real one. On the other hand, the case of metaphysicians is extreme. It is an understatement to say they have had "sufficient time to think things over", and they cannot plausibly be taken to be lack sufficient intelligence, rationality, care, or patience. In particular, they are surely especially acute with respect to the relevant kind of semantic/conceptual matters. Also, as Balcerak Jackson notes, it is difficult to see, in the case of ontological debates, how the differences in meaning are supposed to have arisen, and how they could have gone unnoticed for so long (2014, p. 43).

While some of the above considerations tell against (a), others may seem to mitigate. The responses on behalf of (a) seem rather weak, but there is a more conclusive point to be made against (a): (d) has several benefits that (a) lacks, and, moreover, it does not have any of the disadvantages of (a) listed above. The overall cost-benefit analysis therefore favours (d) over (a).

5.4 Factual-External Questions

Option (b) derives directly from Carnap's seminal paper and is notoriously obscure. Matti Eklund (2013) identifies several different interpretations of the notion of a factual-external question, and takes the most promising one to be a "language pluralist" one. On this interpretation, external questions are confused because they conflict with a certain trivial way in which ontological matters depend on language. Eklund explains,

> Suppose one of us speaks a language where "there are numbers" is true and one of us speaks a language where "there are numbers" comes out false. Suppose we come to find out that it is so. But then I go on to say "OK, 'there are numbers' comes out true in my language and false in yours. But, language-independently, are there numbers?". This would be confused. What could this supposed further question amount to? (2013, p. 232)

But is it really confused to think there is a further question whether there are numbers, once it has been established that 'There are numbers' is true? If the sentence is not semantically individuated, it seems perfectly fine to think that its being true does not settle whether there are numbers (whether language-independently or not). It seems this is potentially confused only if it is presupposed that the sentence means that there are numbers. Then, thinking the question is not settled amounts to hesitating over an inference of the form 's means that p and is true; therefore, p'.

Such inferences have occasionally been rejected, particularly in the literature on the Liar paradox, so it would be too strong to say that doubting them is just confused. On the other hand, in those discussions, there has always been a reason to doubt them, namely, the fact that rejecting them is a way of avoiding contradiction. So perhaps Eklund's idea is that what is confused is to doubt such an inference without having any special reason to.

Against the idea of interpreting metaphysicians as asking external questions in this sense, one might note, firstly, that doing so is uncharitable. It is also hard to think of any textual evidence for interpreting actual metaphysicians as taking metaphysical questions not to be settled by claims about sentence-truth in the relevant way. Nor can I think of any actual metaphysician whose views can be seen to make more overall sense if interpreted in this way. (I should add that Eklund does not himself take this to be a plausible interpretation of metaphysicians: it is rather a proposal that arises in his Carnap exegesis.)

Since the primary aim of this paper is to discuss how to interpret actual metaphysicians from the perspective of EO, these considerations seem to me sufficient for rejecting (b) on Eklund's interpretation.

I will now consider two further interpretations of 'external-factual question'. It will turn out that metaphysicians may well be taken to ask external questions on each of these two interpretations, but that their doing so is consistent with (d).

A first interpretation takes an external question to concern which framework is the right/best one, where this concerns some kind of fit with reality that goes beyond purely empirical or pragmatic considerations. It is clear that some metaphysicians—notably, Ted Sider (2011)—think this is an important question. But their doing so is consistent with (d). For asking external questions in this sense does not entail using ontological expressions in any particular sense, but rather to have a philosophical view about objective structure, which one can clearly have while using the relevant terms in their internal senses.

Secondly, we might think of an external question as by definition a question that can only be asked if the asker believes that it cannot be settled merely by conceptual or empirical means. To ask an external question in this sense is to ask a certain question and to take it not to be easily answerable (but, perhaps, answerable only in what Thomasson calls "epistemically metaphysical" ways). The claim that metaphysicians ask external questions in this sense is very plausible, but it, too, is consistent with (d) (since someone's believing a question not to be answerable a certain way does not make it so).

Carnap would reject my above claims about that asking external questions in either of these two sense is consistent with (d). As is made clear in the passage quoted above, he thought that if one asks a question in either of these ways, one must *ipso facto* understand it in some other sense than the internal one. But, as I will argue at length below, this was a mistake.

Both of these last two interpretations fail to capture Carnap's claim that external-factual questions are questions used in a certain sense. They also fail to capture his claim that external-factual questions are "meaningless". These two claims of Carnap's are of course hard to reconcile. According to the present paper, both claims are false. The latter claim, moreover, is extremely implausible, and is best seen as an instance of Carnap's and many of his contemporaries' sloppy and exaggerated rejections of various claims as "meaningless". The simpler and more moderate diagnosis advanced in this paper is that metaphysicians understand ontological questions in their ordinary, internal sense, in which they are easily answered, but falsely believe they are difficult.

5.5 Pragmatic-External Questions and Metalinguistic Negotiation

The second understanding of the notion of an external question takes such questions to concern which linguistic framework we ought to use (for certain purposes), how we ought to reform our language, etc. Of course, metaphysicians asking questions like Q cannot well be taken thereby to explicitly ask such normative, metalinguistic questions. But maybe they can be taken to do so somehow implicitly? Amie Thomasson (2017a) proposes that metaphysicians are engaged in metalinguistic negotiation, a kind of discourse which appears superficially to be concerned only with non-linguistic, non-normative matters, but which is in fact about which terminology to use, or how to modify our terminology, or some such normative, metalinguistic matter (see also Price (2011: §13.2) and Belleri (2017)). My discussion of (c) will focus entirely of this idea, since, taken so widely as Thomasson takes it, it seems to cover any possible rendering of the idea that metaphysicians are "implicitly discussing normative, metalinguistic questions".

Metalinguistic negotiation is arguably a real phenomenon.[2] A good example comes from Peter Ludlow (2008, p. 118), and involves a debate in which one party assertorically utters, while the other dissents from, 'The racing horse Secretariat is an athlete'. Thomasson writes,

> Even if we interpret the disputants as using the term 'athlete' in different ways (given their dispositions to apply/not apply it to non-human animals) and each speaking a truth in their own idiolect, we can still see them as involved in a genuine dispute: not at the level of literally asserting conflicting propositions, but rather at the level of pragmatics. For each speaker can be seen as advocating for ways the term 'athlete' should be used: whether it should be applied to non-human animals or reserved for humans. And such disputes may be very much worth having. For how we use words matters, given their relations to other aspects of our conceptual scheme, and to our non-verbal behavior (2017a, p. 10 f.).

Paralleling Hirsch's distinction between metaphysicians actually and possibly speaking past each other, Thomasson distinguishes between the idea that metaphysicians actually engage in metalinguistic negotiation and the idea that it can be fruitful to view their debates in such a way (see her (2017b, 2020)). Given my aims, however, I will focus only on the former.

Thomasson highlights several important advantages of interpreting metaphysicians as engaging in metalinguistic negotiation. Firstly, it is charitable because, contrary to option (a), it takes them to genuinely disagree. This is an advantage, firstly, because they themselves believe that they disagree. Hence, this (higher-order) belief

[2] See Burgess and Plunkett (2013a, b), Plunkett and Sundell (2013), and Plunkett (2015).

comes out true, whereas on (a), it doesn't. Secondly, the fact that they disagree makes broader, practical sense of their conversation: genuine disputes are precisely genuine, whereas verbal ones are not.

Furthermore, this interpretation is charitable because it takes metaphysicians to be engaged in a kind of dispute that is not only genuine, but important. It may be important whether 'athlete' should be used for non-human animals, given the social significance of this label. Similarly, it is obviously of great practical and moral importance whether 'torture' should be applied to water-boarding and whether 'marriage' should be applied to same-sex unions (cf. Chalmers, 2011, p. 516 f.). Closer to the metaphysical case, it is arguably important (as far as philosophy is important, anyway) which terminologies or notations we use, and what virtues they may have (simplicity, manageability, expressive power, etc.).

But (c) also has its problems. The most serious problem is the plain fact that virtually every contemporary metaphysician will deny that they are engaged in metalinguistic negotiation. They will not just say that this is not primarily what they are doing: they will typically say they are exclusively concerned with describing extra-linguistic reality. As with (a), this fact is particularly pressing given that it cannot be explained away by saying that metaphysicians lack the requisite intelligence, conceptual/semantic competence/expertise, care, rationality, patience, or time. It is simply hard to see how metaphysicians, not lacking in any of these respects, could fail to know, when prompted to reflect on the matter, that they are engaged in metalinguistic negotiation, given how this notion was explained. In real-life examples of the phenomenon, like the debate about Secretariat, the contestants would surely realize, if they were sufficiently rational and intelligent, etc., that their disagreement really concerned how a certain word should be used, or at least that this is what they *should* have been debating instead. To accuse metaphysicians of the kind of error that these contestants would be making if they insisted that their disagreement was purely factual is uncharitable.

Still, some uncertainty lingers, due to the lack of a precise definition of metalinguistic negotiation. Here are some things one would like to know in order to properly assess the idea that metaphysicians are engaged in it:

1. Must speakers who engage in metalinguistic negotiation be able to know that they are, given sufficient intelligence, rationality, care, time to ponder, patience, etc.?
2. Do people engaged in metalinguistic negotiation thereby *mean that* so and so is the case (e.g., that some expression ought to be used a certain way)? That is, is metalinguistic negotiation manifested at the level of speaker meaning?
3. Does metalinguistic negotiation consist in participants' conversationally implicating propositions of the relevant kind?

These questions are closely connected, and are reasonably answered the same way. When we consider how they might be answered, however, a problem looms. For if the answer to these questions is Yes, it seems hopeless to argue that serious metaphysicians are engaged in metalinguistic negotiation. But if it is No, it is hard to see what metalinguistic negotiation is supposed to be. The most promising option for interpretation (c), in any case, seems to be to answer questions (1)–(3) in the negative, and then provide a different understanding of metalinguistic negotiation. I will consider two possibilities, but argue that the claim that metaphysicians are engaged in metalinguistic negotiation is implausible on both understandings.

Firstly, we could suppose that although metaphysicians do not actually mean to discuss normative questions about words, they can still, in some sense, be "charitably interpreted as doing so", in the sense that there is some similarity between what they are actually saying, what they actually mean, etc., and something one could be saying, where the latter is reasonable. The interpretation could be seen as charitable because of the reasonability, and it could be called an interpretation of metaphysicians because of the similarity.

I think the distinction between "actually meaning" and "being charitably interpretable as meaning", in the special sense explained, while important in many contexts, is of no help to (c). Note that the claim that metaphysicians are, in this special sense, "charitably interpretable as engaged in metalinguistic negotiation" is uncontroversial, at least to Carnapian deflationists. But it does not even begin to answer the question that is the subject matter of this paper, which is what deflationists should say about what metaphysicians actually mean.

Here, it may be objected that if someone is "charitably interpretable" as meaning p, then that is *ipso facto* reason to take them as actually meaning p. This is true, but the reason here must be seen as merely defeasible. While perhaps decisive in a tie, this kind of charity cannot trump clear, first-person judgments by sufficiently rational, patient, and intelligent speakers, about what they mean. However reasonable we take it to be to engage in metalinguistic negotiation, metaphysicians think it is equally reasonable to engage in purely worldly metaphysics. Even if they are wrong about this (as ontological deflationists maintain), that fact cannot be used, in conjunction with some principle of charity, to show that they didn't actually mean to discuss purely worldly matters.

The second proposal is inspired by Thomasson's characterization of metalinguistic negotiation as a conversational move in which participants push for certain linguistic or conceptual changes. This is an interesting and telling phrasing, for it suggests that metalinguistic negotiation might not best be cashed out in terms of the usual speech acts, but rather in terms of some other type of action. As such, it is consistent with a negative answer to (1)–(3).

Here is a possible clarification. Suppose a person knows, at least inchoately and implicitly, that facts about which sentences are uttered (assertorically and seriously) contribute to determining the extensions of the words therein, and that they also want the extension to change in a certain way. Suppose, more concretely, that they know that utterances of a sentence '*Fa*' contribute to making a part of the extension of '*F*' and that they also want this to happen. Then, uttering 'Secretariat is an athlete' could be seen as an action performed with the intention of making Secretariat a member of the extension of 'athlete' (at some future time). Performers of this action need not take it to lead directly to the intended goal, but may take it to do so indirectly, by leading others to make the same utterance, so that, if enough people do it, the balance is eventually tipped, resulting in the desired change of extension.

While this may be an interesting notion, it does not seem to help (c). Surely, metaphysicians' disputes cannot plausibly be taken to consist of utterances-intended-to-change-extensions of this kind. Metaphysicians instead take their claims to be accurate descriptions of reality, i.e., true, given the actual, current extensions of the terms used. The fact that they would therefore protest against an interpretation on which they instead make their utterances in order to change those extensions seems enough to refute it. Hence, metaphysicians cannot plausibly be taken to engage in metalinguistic negotiations in either of these two senses, and so it becomes even less clear whether there is any sense of 'metalinguistic negotiation' in which metaphysicians could reasonably be said to engage in it. This concludes the first part of this paper, where I argued that the alternative interpretations of metaphysicians that Carnap and Thomasson are forced to choose between are all implausible. I proceed in the next section to Canrap's argument for thinking some such a special interpretation is necessary.

5.6 Carnap's Wrong Turn

The passage quoted in Sect. 5.2 shows that Carnap considered (d) patently false. But this was a mistake, for (d) is both available to deflationists and the most plausible interpretation of metaphysicians they can offer. The positive support for (d) will be deferred to the next section. Here, we will rather consider three possible motivations for Carnap's reasoning in the quoted passage, and argue that each motivation fails. This list of possible motivations and accompanying rebuttals also functions as a list of objections against (d) with replies, and will thus form part of my overall case.

Carnap claims that "nobody who meant the question 'Are there numbers' in the internal sense would either assert or even seriously consider a negative answer". He thus inferred from the fact that metaphysicians take ontological questions to be difficult to the conclusion that they do not understand them in their internal sense. It seems to me that the most obvious explanation of why anyone would reason like this is that they accept

(C) Someone who uses and understands Q in its internal sense *ipso facto* knows that it is easily answerable in the affirmative.

According to the present paper, (C) is false. Rather than trying to refute it, however, I will here consider three possible motivations for it, and argue that they all fail. This should be enough for undermining any initial plausibility of (C).

Firstly, (C) may seem to follow from some plausible principle of the transparency of meaning, i.e., some claim to the effect that one always knows what one means, or in which sense one is using one's expressions. But while there are plausibly true principles of this kind, it is doubtful whether there is any such principle that is both true and supports (C). For even if one knows in which sense one is using an expression, one need not know all meaning-related facts about the expression, taken in that sense. There are many facts about expressions that hold in virtue of their being used in a certain sense by a given person, which that person need not know. For instance, a sentence, when used in a certain sense, might have certain truth-conditions relative to worlds, contexts, and further parameters, that one need not know about, even if one uses it in that sense. This is enough to show that there are meaning-related facts of which competent users may be ignorant.

To take a more relevant example, the sense of a word may be determined by certain meaning-constitutive inference rules even if a speaker using the word in that sense fails to know that it is so determined. This is true independently of whether meaning is determined by normative or non-normative inferential facts about expressions. It may be that competence with the relevant expressions requires a defeasible disposition to infer in accordance with such rules. But such a disposition does not entail knowledge that the meanings of the words are determined by those rules (it does not even entail any ability to conceive of the rules). This is evidenced by the fact that philosophers disagree about which rules are meaning-constitutive—indeed, about whether any rule is—while still using them in the same senses. This example is more relevant because adherents of EO are arguably committed to some form of inferential role semantics (cf. Thomasson, 2015, p. 95, esp. note 13). If one accepts such a semantics, then one is clearly committed—given some obvious empirical facts about what philosophers say and think—to saying that one can use an expression in a sense determined by a given inference rule without knowing that it is so determined.

Now, is easy answerability the kind of meaning-related property one must know *qua* competent speaker? More precisely, if a speaker uses and understands Q in its internal sense, must she then know that it is easily answered? One may be tempted to argue that Carnapians must answer in the affirmative, and for the following reason. Firstly, and to repeat, Carnapians are committed to some version of inferential role semantics. But on this kind of semantics, one can be competent with the expressions of Q only if one will answer Q in the right way, since doing this requires merely inferring in accordance with meaning-determining rules. Thus, Carnapians are committed to taking "easy answerability" to be a meaning-related property one must know of, if one uses the relevant expressions in the internal sense. Hence, the internal interpretation of metaphysicians is false, since they do not recognize any easy answerability.

The problem with this argument is that it presupposes an implausible version of inferential role semantics, namely, one which takes it to be necessary for understanding an expression that one actually infer in accordance with a particular rule. Any rule whatsoever can be rejected by competent speakers, e.g., for theoretical reasons (cf. Williamson, 2007: Ch. 4). The more plausible alternative holds that semantic competence with the relevant expression consists in a defeasible disposition to infer in accordance with the rules (Williamson's argument of course mainly targets this position, but it can be resisted—see Eklund (2007, p. 562ff.) and Båve

(2020)). This is commonplace among non-normativist inferentialists (see, e.g., Paul Horwich (2005, p. 49 f.) and Daniel Korman (2019), who makes this point in relation to Thomasson's theory in particular). That a disposition is defeasible means its manifestation can be blocked by various factors, such as the philosophical belief that certain considerations going beyond the simple inference rules are relevant for answering Q.

On the basis of such a defeasibility view, we can explain how a competent speaker could understand Q in its internal sense without knowing it is easily answered: she may have some theoretical belief leading her to think Q is not easily answered and yet be defeasibly disposed to follow these rules. Her having this defeasible disposition should here be understood as entailing that: were she not to have any such recalcitrant theoretical belief, then she would make the relevant inference. Alternatively, we could simply take this counterfactual, more directly, as a necessary condition for competence with the expression in question, and leave dispositions aside.[3] I conclude that the above attempts to support (C) on the basis of some transparency-of-meaning thesis are unsuccessful. It is unclear whether any such argument for (C) could work.

Having thus criticized the idea of supporting (C) on the basis of some transparency-of-meaning principle, let me now close this section by considering, more briefly, two further possibilities. A second possible motivation for accepting (C) stems from verificationism, in particular, the conditional,

If a sentence has one set of verification conditions when taken in sense S, and a different set of verification conditions when taken in sense S′, then S ≠ S′.

This claim is plausible, perhaps trivial, but note that it does not entail that if two people associate different verification conditions to a given sentence, they must mean different things with it. For one of them might simply be wrong, just like a sceptic might simply be wrong about what justifies belief in external objects, or like two people may disagree about which inference rules are valid. Such disagreements do not immediately entail that the disagreeing parties differ in how they interpret the relevant expressions. Hence, while metaphysicians and Carnapians disagree about what verifies Q, it does not follow that they associate different senses to it. Hence, metaphysicians may well understand Q in its internal sense.

A third argument for (C) assumes that the internal sense is technical and that this fact makes it impossible to attach this sense to Q without knowing that it is easy. The underlying reasoning is that one can only attach this sense to an expression by way of some explicitly made stipulative definition, which would unfailingly make one aware that Q is easily answerable. While this might appear to be a promising argument for (C), it is actually unavailable to Carnapians. For Carnapians cannot say that Q and similar questions are easily answerable only when taken in a certain technical sense. This claim is something metaphysicians and opponents of EO may well accept. That is, metaphysicians can easily grant that Q could be imbued with a

[3] See Båve (2020) for details.

special sense such that Q is easily answerable (on the basis of Simple Reasoning) when taken in that sense. (Metaphysicians merely need to deny that they are themselves using Q in this special sense.) Note that I'm not saying that Q could be imbued with such a sense. The point is rather that it is a claim that opponents of EO can accept, and therefore not contentious in the way it would have to be in order to be identified with an interesting version of EO. Thus, Carnapians must take the internal sense to be a non-technical, "ordinary" sense of Q.

Could we even go further, and say that the internal sense is the ordinary sense of the relevant questions? That would presuppose that the relevant expressions 'there is', etc.) have only one ordinary sense. Carnap and his contemporaries would probably have found this naïve, given their default assumption that natural languages are rife with ambiguity and other kinds of obscurity.[4] However, with Grice's razor and his reduction of lexical meanings by appeal to conversational implicatures, and, subsequently, a host of other putative pragmatic mechanisms, this presumption has increasingly come to be seen as unwarranted. Most philosophers of language now take it as a non-negotiable methodological principle that lexical ambiguity should be appealed to only given special reasons to think that the data cannot be accommodated without it. Ideally, the ambiguity posited should also be independently justified and/or explicable, e.g., justified by intuitions and/or explained causal-historically.

I know of no data that justify positing further senses of 'There are numbers'. Furthermore, the sentence seems to fail the standard ambiguity tests (see, e.g., Zwicky and Sadock's classical (1975)). By the methodological principle described above, we are therefore justified in presuming it to be unambiguous, and, hence, in using the definite, 'the ordinary sense'.[5]

The more important conclusion, however, is that also the third argument for (C) fails, since adherents of EO cannot hold that it is true because the internal sense of ontological questions is a technical one. Hence, all three motivations for (C) fail.

---

[4] Notable philosophers with such views include Frege, Næss, Waismann, Wittgenstein, and the "Oxford philosophers", including Austin, Strawson, and many others.

[5] Hofweber (2016: Ch. 3) appeals to certain linguistic data to justify an ambiguity view of quantifiers. But these data, it seems to me, are better explained by appeal to ordinary quantifier restrictions (pace Hofweber (2016: §3.5.2)). This issue cannot be resolved here, however. A different question is whether quantified sentences like, 'There are numbers' entail corresponding existence claims ('Numbers exist'). There do seem to be reasons for denying this, and, if so, Carnapian ontological deflationists might have been wrong to suppose that existence questions are easily answered in just the same way as quantificational questions like Q. This intuition that quantification and existence come apart is central to neo-Meinongians, but it bears noting that many neo-Meinongians hold views about existence that are in fact congenial to Carnapian ontological deflationism. Richard Routley and Graham Priest, for instance, hold that existence entails concreteness (Priest, 2005, p. 136). But this view is arguably in line with the Carnapian deflationist's claim that Q is easily answered. For it can be argued that if existence entails concreteness, then the question whether numbers exist can be easily answered *in the negative*, and perhaps by partly empirical means, rather than merely conceptually.

5.7 The Positive Case for (d)

The position being advanced here is that metaphysicians who discuss ontological questions without terminological preliminaries typically use and understand such questions in their internal sense, in which they are answerable by purely conceptual and/or empirical means, and, in the case of Q and certain other cases, they are also easily answered. Aside from simplicity, the main recommending feature of this alternative is its interpretive charity. It does take metaphysicians to make a mistake, of course, in thinking Q is not easily answerable. But, firstly, it is clear that a Carnapian deflationist must take them to be wrong in some respect. Secondly, it seems preferable to take them to be wrong about philosophical matters, on which we are often wrong, rather than in the ways implied by the other interpretations.

The respect in which (d) takes metaphysicians to be wrong is in thinking that Q cannot be easily answered. This is a mistake about what kind of considerations may and may not be adduced in answering a certain philosophical question. Attributing this mistake to someone is not particularly uncharitable. (Diagnoses of this philosophical mistake are bound to be plentiful, but a natural suggestion is that it results from over-generalizing from real, substantive questions (cf. Horwich, 1998, p. 89 f.))

By contrast, (a) takes metaphysicians to be wrong about whether they disagree with each other, and about whether they are saying something trivially true. (b) entails that they are discussing meaningless, pseudo-problems, or (on the pluralist interpretation) that they doubt a trivial inference for no reason. (Of course, something along the lines of (b) may still be a plausible view of metaphysicians who assign technical meanings to ontological questions, but this issue falls outside the scope of this paper.) Finally, (c) entails that they are discussing normative questions about language but have the false belief that they are only debating substantial, non-linguistic questions. As I have been arguing, (a)–(c) therefore entail that metaphysicians are wrong about what they mean, what they say, and similarly conceptual matters that people, and especially philosophers, can be expected to get right.

It may be thought, however, that (d) is in a worse position than I have admitted, since it takes metaphysicians not merely to have a false philosophical belief, but to take something trivial to be non-trivial, and, in the case of nominalists, to take something trivial to be false. How does this square with my claim that (d) allows us to give a reasonably charitable interpretation of said metaphysicians?

Let me first grant that EO indeed takes the answer to Q to be trivial, i.e., that it is trivial that there are numbers. This seems to me part and parcel of EO. Let me also simplify the discussion by discussing, more directly, the proposition that there are numbers, rather than the claim that Q can be truly answered in the affirmative. Let us also treat ‘it is easy to know that’, ‘it is trivial that’, and ‘it is obvious that’ as interchangeable. Then, it becomes clear that EO, as defined above (and given a few further innocuous assumptions) entails that it is trivial that there are numbers (and so on for properties, sets, etc.). These simplifications allow us to discuss the matter as one concerning the proposition that there are numbers and the “triviality operator”.

I will now first propose two possible answers to the question how expert philosophers can deny that there are numbers, even though this is trivial. This will then immediately yield a response to the objection that it is uncharitable to interpret them as denying something trivial. This line of reasoning can then be trivially applied to the case of realists who do not deny something trivial, but nevertheless take it to be non-trivial.

Carnap's solution to this problem would be to say that apparent rejections of what is trivial are only apparent, and that what nominalists actually reject (and realists take to be non-trivially true) is something else, i.e., an answer to an external question. The two alternative responses I will offer are very different. While the first one is not obviously untenable, the second one seems overall more plausible.

The first possibility is to reject S4 for triviality, and say that while it is trivial that there are numbers, it is not trivially trivial. The rejection of S4 for triviality could be motivated as follows. Triviality comes in degrees and is a threshold concept. Something is trivial if (and only if) it is easy enough to know it, i.e., if a relevant threshold is surpassed. Since the relevant threshold is plausibly not a precise boundary, the notion of triviality will be vague, but this is independently plausible.

Now, knowing that *p* may be easy enough for the proposition that *p* to be trivial, while knowing that trivially *p* is not easy enough to make the proposition that trivially *p* trivial. This is unsurprising, since the two propositions are typically rather different: whereas the proposition that trivially *p* always concerns epistemic matters, the proposition that *p* does not, for many *p*. Further, if one proposition is more complex than another, it can be more difficult to know, and 'Trivially *p*' is of course more complex than '*p*'.

Triviality is different from *a priori* knowability. Chalmers (2012, p. 154) claims that the latter obeys S4. Although we may stay neutral on that matter, we can note that this claim at least cannot be refuted by any argument analogous to the above argument against S4 for triviality. This is because apriority does not depend on how easy it is to know something in a certain way, but only on whether it is possible at all. Knowing something could be practically impossible because it is too difficult to know (e.g., because the proposition is too complex to grasp), but this does not affect *a priori* knowability, since the latter is meant to concern possibility in principle.

One might now go on to say that it is no mystery that experts deny that there are numbers, although this is trivial, since it is not trivial that it is trivial. And one can then say that it is not overly uncharitable to take experts to deny trivial truths, as long as they are not trivially trivial.

Denying S4 also allows us to shed some light on the dialectics surrounding EO, which some might find confusing. Consider the following instance of *modus ponens*:

(i) Simple Reasoning is sufficient for knowing that *p*
(ii) If Simple Reasoning is sufficient for knowing that *p*, then it is easy to know that *p*
(iii) It is easy to know that *p*,

We could say that (iii) is not easy to know, even though (ii) is, because (i) is not easy to know. This diagnosis is open to EO, for EO is not committed to the claim that it is itself easy to know.

Perhaps, finally, we could also propose that the reason Carnap thought that metaphysicians "must have some other question in mind" than the internal question is that he mistakenly accepted S4 for triviality. This would then have lead him to think that it is trivial that Q can be easily answered, and, hence, that metaphysicians cannot well have denied this.

While this response to the objection is possible, I believe there is a more compelling one, which is neutral on S4 for triviality. It begins with the observation that Q is not (of course) considered difficult because it involves advanced mathematics. Rather, it is held non-trivial due to certain philosophical considerations that seem compelling to many. But a proposition may be trivial even though there are considerations that experts take seriously, and which lead them to think the proposition in question is non-trivial. Thinking otherwise would involve a sloppy inference from a proposition being trivial to the conclusion that any consideration making it seem non-trivial must be clearly incorrect (and, hence something that experts would not be taken in by).

The proposition that there are numbers is trivial in that one can easily know its truth (namely, through Simple Reasoning), but this does not entail that an expert might not come to think of arguments against it that they find compelling. Carnapians are committed to taking these arguments to be unsound, of course, but they can easily do so, since being an expert does not entail never being moved by unsound arguments.

Thus, it is no mystery that experts can deny or doubt something trivial, and, therefore, it is not unduly uncharitable to sometimes take them to do so. In some cases, however, like those figuring in the so-called "Bad Company Problem", the considerations making certain existence questions seem non-trivial cannot as easily be dismissed as mistaken. This is a serious problem for EO, but it has no relevance for the question of charity discussed here, and hence of no relevance to the reform of EO advocated in this paper.

Funding The funding organization is the FCT (Fundação para a Ciência e a Tecnologia), and rthe grant no. is: DOI: https://doi.org/10.54499/2021.02647.CEECIND/CP1702/CT0014.

References

Balcerak Jackson, B. (2013). Metaphysics, verbal disputes, and the limits of charity. *Philosophy and Phenomenological Research, 86*, 412–434.

Balcerak Jackson, B. (2014). Verbal disputes and substantiveness. *Erkenntnis, 79*, 31–54.

Båve, A. (2015). A deflationist error theory of properties. *Dialectica, 69*, 23–59.

Båve, A. (2020). Semantic dispositionalism without exceptions. *Philosophical Studies, 177*, 1751–1771.

Belleri, D. (2017). Verbalism and metalinguistic negotiation in ontological disputes. *Philosophical Studies, 174*, 2211–2226.

Burgess, A., & Plunkett, D. (2013a). Conceptual ethics I. *Philosophy Compass, 8*, 1091–1101.

Burgess, A., & Plunkett, D. (2013b). Conceptual ethics II. *Philosophy Compass, 8*, 1102–1110.

Carnap, R. (1950/1956). *Empiricism, semantics, and ontology* (Reprinted in Meaning and Necessity, 2nd ed.). University of Chicago Press.
Chalmers, D. (2011). Verbal disputes. *Philosophical Review, 120*, 515–566.
Chalmers, D. (2012). *Constructing the world*. OUP.
Eklund, M. (2007). Meaning-constitutivity. *Inquiry, 50*, 559–574.
Eklund, M. (2013). Carnap's metaontology. *Noûs, 47*, 229–249.
Eklund, M. (2019). Fictionalism. In E. N. Zalta (Ed.), *The Stanford encyclopedia of philosophy* (Winter 2019 ed.). https://plato.stanford.edu/archives/win2019/entries/fictionalism/
Hirsch, E. (2007). Ontological arguments: Interpretive charity and quantifier variance. In T. Sider et al. (Eds.), *Contemporary debates in metaphysics* (pp. 367–381). Blackwell.
Hirsch, E. (2009). Ontology and alternative languages. In D. Chalmers, D. Manley, & R. Wasserman (Eds.), *Metametaphysics* (pp. 231–259). OUP.
Hirsch, E. (2011). *Quantifier variance and realism*. OUP.
Hofweber, T. (2016). *Ontology and the ambitions of metaphysics*. OUP.
Horden, J. (2014). Ontology in plain English. *The Philosophical Quarterly, 64*, 225–242.
Horwich, P. (1998). *Truth* (2nd edition), Clarendon Press.
Horwich, P. (2005). *Reflections on meaning*. OUP.
Korman, D. (2019). Easy ontology without deflationary metaontology. *Philosophy and Phenomenological Research, 99*, 236–243.
Ludlow, P. (2008). Cheap contextualism. *Philosophical Issues, 18*, 104–129.
McGrath, M. (2008). Conciliatory metaontology and the vindication of common sense. *Noûs, 42*, 482–508.
Plunkett, D. (2015). Which concepts should we use? Metalinguistic negotiations and the methodology of philosophy. *Inquiry, 58*, 828–874.
Plunkett, D., & Sundell, T. (2013). Disagreement and the semantics of normative and evaluative terms. *Philosopher's Imprint, 13*, 1–37.
Price, H. (2011). *Naturalism without mirrors*. OUP.
Priest, G. (2005). *Towards non-being: The logic and metaphysics of intentionality*. OUP.
Sider, T. (2011). *Writing the book of the world*. OUP.
Thomasson, A. (2015). *Ontology made easy*. OUP.
Thomasson, A. (2017a). Metaphysical disputes and metalinguistic negotiation. *Analytic Philosophy, 58*, 1–28.
Thomasson, A. (2017b). What can we do, when we do metaphysics? In G. D'Oro (Ed.), *Cambridge companion to philosophical methodology*. CUP.
Thomasson, A. (2020). A pragmatic method for normative conceptual work. In A. Burgess, H. Cappelen, & D. Plunkett (Eds.), *Conceptual engineering and conceptual ethics* (pp. 435–458). OUP.
Williamson, T. (2007). *The philosophy of philosophy*. OUP.
Yablo, S. (1998). Does ontology rest on a mistake? *Proceedings of the Aristotelean Society, 72*, 229–261.
Zwicky, A., & Sadock, J. (1975). Ambiguity tests and how to fail them. In J. Kimball (Ed.), *Syntax and semantics* (Vol. 4, pp. 1–36). Academic.

Chapter 6
Possible Worlds and Propositions

María de Ponte

Abstract Possible-world semantics has been widely and successfully applied both in philosophy and in linguistics, and some of these applications are, or seem to be, largely independent of metaphysics. But not all are. Kripke claimed that the "distant planets" picture of possible worlds was a mistake. I agree. This picture is particularly pervasive and problematic when it comes to its treatment of intensional entities, such as propositions. According to the most prevalent version of possible-world semantics, a proposition is a set of possible words where it is true. What this set of possible worlds might be varies greatly to modal realists (e.g., Lewis, 1986a), actualists (e.g., Stalnaker, 2012) or deflationists (Thomasson, 2009). I take possible worlds to be abstract entities, akin to those postulated in science to help us clarify and understand the world: formal, heuristic tools, not explanatory ones. Following Perry (De Gruyter, 1989), I argue that propositions are abstract entities used to encode conditions on objects and properties: truth-conditions (for utterances) satisfaction-conditions (for desires), etc. I take my proposal to be similar to deflationary views, but I complement it with a realist position with regards to modality: modal properties and facts.

Keywords Possible worlds · Deflationism · Propositions · Truth-conditions

6.1 Introduction

Possible world semantics has supplied philosophers, logicians, and linguists with a wide array of methods and tools to analyze intensional phenomena, such as modality, tense, reports of cognitive attitudes, or conditionals, to name but a few.

M. de Ponte (✉)
ILCLI and Philosophy Department, University of the Basque Country, Donostia, Spain
e-mail: maria.deponte@ehu.eus

X. de Donato-Rodríguez et al. (eds.), *Deflationist Conceptions of Abstract Objects*, Synthese Library 516, https://doi.org/10.1007/978-3-031-98717-5_6

There are those who claim that possible world semantics's contributions are, to a great extent, independent of metaphysics (Partee, 1989, p. 107; Chihara, 1998, p. 3). Even so, it seems that a semantic theory that makes use of certain entities in its explanations either must be committed to the reality of those entities or, at the very least, must be able to account for their ontological status.[1]

> … talk of possible worlds … cries out for further analysis. There *are* no possible worlds except the actual one; so what are we up to when we talk about them? (Mackie, 1973, p. 90)

In this paper, I defend a realist position about modal properties, and a deflationary view about possible worlds. Like Mackie, I believe that there are no possible worlds, other than ours. It might be that one day physics will prove us wrong. But, so far, I do not think there is evidence enough to support believing in the existence of possible worlds.

I first briefly present what I take to be a common-sense view on modality, and, mostly, on modal statements and beliefs. This view has come under attack from different fronts, starting with Quine, continuing with Lewis, and including recent deflationist accounts of modality. I will discuss two alternative accounts, one realist (Lewis's) and one normativist (Thomason's), and I will then come back to the common-sense view.

6.2 Common Sense View

Consider the following modal statements:

1. There could have been talking donkeys
2. I could be wearing a green t-shirt today

One natural or intuitive way to think of them is as being *about* reality; about donkeys, talking, me and coloured t-shirts. Take the first statement. Intuitively, it is true if there are no talking donkeys, but there could have been, i.e., with some relevant changes in donkeys's anatomy and abilities, donkeys could have talked.[2] Another way of putting this is that even if it is a fact that donkeys don't talk, it is a contingent fact; things could have been different than they are. So, the statement touches upon the contingent versus the necessary nature of a certain fact.

---

[1] See for instance Menzel (1990, p. 363). Barwise and Perry (1985, p. 116) make a similar claim, but propose a different semantic theory, committed to the existence of situations, rather than possible worlds. Chihara, on the other hand, accepts possible world semantics, but claims that "that one can apply systems of modal quantificational logic to evaluating modal principles and to assessing modal arguments, without incurring a commitment to the existence of possible worlds" (1998, p. 3).

[2] An alternative interpretation of this example would be considering talking donkeys as mythical or fictional entities, very much like unicorns. So what we are saying is that the existence of these fictional entities is possible. I consider briefly this reading below.

If the changes required for donkeys to talk are possible, statement (1) is true. If they are not, then it is false. The conditions for the truth of (1) are then conditions that concern real donkeys and the ability to talk (i.e., what is required for some noises to count as "talk," what is required for donkeys to be able to do that, etc.).

Consider the following everyday situation. I wake up, take a shower and put on a t-shirt. Let us suppose I choose a pink t-shirt. Me wearing a pink t-shirt today is a fact, if we understand facts in a simple way as being combinations of objects, properties and relations.

But it is not a necessary fact. I could have picked up a green t-shirt this morning, or a yellow one, or I could have decided not to wear a t-shirt at all. All these are what we might call possible states of affairs; i.e., different ways in which things might have happened, or in which things might have been. States of affairs can be merely possible or actual, in which case, they are facts. States of affairs can be actual, but contingent, i.e., they might not be necessary facts.

Following with the example, it is a fact that I am wearing a pink t-shirt today. But it also seems to be a fact that I could have been wearing a green t-shirt. Me wearing a green t-shirt is not a fact of reality; it is a possible state of affairs. But it is a possible state of affairs *of reality* or, rather, with regards to the facts of reality. It is a fact *of reality* that me wearing a green t-shirt is possible (a possible state of affairs); there is no law (of nature or of logic) that prevents this from happening. Possibilities, understood as possible states of affairs, are *real* facts of our world.

Accepting that things might have been different than they are, and that this is *a fact*, is important to account for many fundamental features of reality. It seems to be essential for any account of freedom and free will, for instance. It might also be important to understand laws of nature, causation and so on. I think, further, that it gives us a clearer view on the meaning and truth-conditions of modal statements such as (1)–(2) above.[3]

This common-sense view, however, has been widely challenged. It is nowadays common, for instance, to reduce metaphysical possibility to possible worlds, i.e., to interpret (1) as saying that there is a possible world in which there are talking donkeys. This interpretation can be traced, of course, to Lewis and his paraphrase argument (1973). Recently, some deflationist proposals have claimed that we should interpret modal statements not as describing features of reality, but rather as normative statements, with the function of making explicit the semantic rules of language (Thomasson, 2020).

These two proposals, Lewis's extreme form of realism about possible worlds and Thomasson's deflationist approach to modal ontology, are, in many senses, radically different. But they are also similar in an important respect: they both claim that modal statements do not describe modal features of reality. Lewis proposes a reduction of modal features to possible worlds. To say, for instance, that something is possible is to say that there is a world where it is the case. His is a form of extreme

[3] All these are modal statements that involve either physical or metaphysical modality. I will focus on metaphysical modality in this paper.

realism, in the sense that his view entails the existence of infinite concrete worlds. But it is also a form of modal anti-realism, in the sense that his view entails the reduction of modality features to possible worlds.

Thomasson does not propose the elimination or reduction of modality. She claims that modal properties and facts do exist but are not the truth-makers of modal statements. They are not what modal statements are *about* either.[4] Modal statements, she argues, are not descriptive statements but rather normative ones; their function is not to describe, or to give information about reality, but rather to convey the semantic rules governing the terms contained in them.

She combines this normativism with regards to modal statements with a deflationist meta-ontology, according to which, debates over ontological questions are pointless. She claims instead that *all* well-formed existence questions can be satisfactorily answered via empirical and/or via conceptual work. In most cases, existence questions are answerable with "easy arguments": a series of trivial inferences from an uncontroversial truth (Thomasson, 2015, p. 128).

In this paper, I argue that these proposals are not convincing, and that they do not give us good reasons to abandon the common-sense view briefly described above (or some similar view). Despite the apparent differences, both proposals lead to similar consequences: the loss of subject matter, an inflationary ontology and counterintuitive views about modality and existence.[5]

6.3 What Are Possible Worlds?

Mackie urges, in the quote above, to wonder what possible worlds are but, actually, we might need to start by wondering what philosophers mean by "world." In everyday talk, we mean different things. Most commonly, we mean something like "the earth and all the people, places, and things in it" or "a group of things such as countries or animals, or an area of human activity or understanding" (i.e., Christian world, subaquatic world, etc). But "world" can also mean either the whole physical universe, as in "The world contains many solar systems, not just ours," or a very small part of it, as in "he lives in his own little world" (all definitions from the Cambridge English Dictionary).

[4] On any philosophically relevant sense of "about" at least. I discuss this issue below.

[5] There are, of course, other options. With regards to possible worlds, the view that comes closer to the common sense view I defend in this paper is Plantinga's (1974), who takes them to be maximal states of affairs, made up of real object and properties (see also Armstrong, 1997). Stalnaker (1976, 2012) on the other hand takes them to be abstract objects, useful as tools for logicians and philosophers, but real in the same way the actual world is. With respect to modality, Arthur Prior (1968) defends a modal realist view, where modalities are primitives and only actual entities exist. His view has many similarities to the one I advocate. See also Fine (2005).

We don't have a clear or coherent view of what philosophers mean by "worlds." Far from it. Consider these two well-known quotes, made by two staunch defenders of possible world semantics:

> In the present monograph I *argued against those misuses of the concept that regard possible worlds as something like distant planets*, like our own surroundings but somehow existing in a different dimension, or that lead to spurious problems of 'transworld identification.' (Kripke, 1980, p. 15. Italics mine)

> There are countless other worlds, other very inclusive things … *The worlds are something like remote planets*; except that most of them are much bigger than mere planets, and are not remote. Neither are they nearby. They are not at any spatial distance whatever from here. (Lewis, 1986a, p. 2. Italics mine)

Lewis famously claimed that reality is populated with infinitely many possible worlds, and that these are concrete entities, just like our own world. Following the quote, there would be infinite planets just like ours, although most of them are "much bigger than mere planets" (whatever that means). But this is surely wrong. An explanation of modal statements and modal properties cannot be reduced to entities that are—somehow—similar to remote planets.

It seems that by "world" philosophers mean something closer to reality, that is, not planet earth, but the whole universe. But the universe is a very big place. One might say that it is all there is, including all that has ever been and all that will ever be. So, what Lewis seems to be saying is not only that there are infinite planets, but rather that there are infinite universes, all complete and all with spatiotemporal properties, but not spatiotemporally located in relation to us (or anything else). Lewis calls these infinite concrete universes "possible worlds," but, on his view, they are all existing universes, each of them actual for the people or entities living in them. The term "possible," in "possible worlds," is also used differently than we usually do. That is, our universe is not less, or more, possible than the others. Our universe is ours, just because it is the universe we happen to be on. "Actual" is, on his view, an indexical. Lewis claims further than something being "possible" means that it is the case in at least one universe (either in ours, or in some other one).

On Lewis's view, the fact that I, in the example above, might have worn different t-shirts on a given day basically means that, on that given day, there is a different universe (world) where my counterpart (basically, a different person, but similar "in all relevant aspects" to me) decided to put on a green t-shirt, and another one where a third counterpart decided not to wear a t-shirt for the day, and so on. Something being possible is defined as there being a universe where that thing happens (to counterparts of things in *our* world).

Equivalently, a necessary fact, such as the fact that Kepa and John are not identical, would be explained by saying that in no universe, Kepa's and John's counterparts are identical, or that none of Kepa's counterparts are identical with any of John's counterparts.

It is difficult to make sense of all this, and to think of a plausible justification of this view. Lewis's main argument for it, as far as I know, is based on the theoretical advantages of it, an indispensability argument of sorts. Lewis was a mathematical realist—or platonist—and the arguments he uses to defend the two kinds of realism are closely related.

> We have only to believe in the vast hierarchy of sets, and there we find entities suited to meet the needs of all the branches of mathematics... Set theory offers the mathematician great economy of primitives and premises, in return for accepting rather a lot of entities unknown to *Homo javamensis*.... It's an offer you can't refuse. The price is right; the benefits in theoretical unity and economy are well worth the entities. (Lewis, 1986a, pp. 3–4)

So, just as set theory, with all its rich ontology, is a paradise for mathematicians, so is possible worlds, with all its rich entities, a paradise for philosophers. Possible worlds, claims Lewis, allow us to greatly reduce the number of primitives needed, "and thereby to improve the unity and economy of theory that is our professional concern—total theory, the whole of what we take to be true" (Lewis, 1986a, p. 4).

This "total theory" Lewis talks about includes modality, which is now analyzed as quantification over possible worlds. Possibility is existential quantification over possible worlds; i.e., "*p* is possible" is true if there is a world *w* in which *p* is true. But not only that, with our total theory at hand we can define intensionality *extensionally*. Thus, a property is a set of things (actual and possible) and a proposition is a set of possible worlds (rather, a proposition *p* is the set of all worlds at which *p* is true).

Let us leave aside difficulties of believing, without further proof or argument, in the existence of infinite concrete universes and infinite counterparts. This is leaving aside a lot, but it is not this aspect of Lewis's extreme realism that I want to refute in this paper. Or at least not only. There is one fundamental issue, I think, that makes Lewis's proposal problematic and that poses questions for more "moderate" versions of possible worlds: The attempt of reduction (or definition) of possibility and necessity to possible worlds. This reduction follows from Lewis's so-called "paraphrase argument" and it entails reduction of both modal facts to happenings in possible worlds and explaining intensional concepts as sets or functions to possible worlds.

6.4 Ways Things Might Be and the Paraphrase Argument

Lewis's notion of possible world is problematic and rather counter-intuitive, to put it mildly. But the notion of possible world we nowadays use in philosophy is undeniably attached to his views and, mostly, to the way in which he framed the debate. Lewis's notion of possible worlds was not only intended to serve as a model or heuristic aid for logic or semantics. Far from it. The idea was to define, or reduce, all intensional entities to possible worlds, thus reducing the primitives needed for our "total theory" of reality, and solving the problem of intensionality and intensional entities, such as properties, propositions, etc.

Lewis, Stalnaker, Plantinga, and others use the phrase "a way the world might be" to explain what they mean by a possible world. Here is what Lewis says in *Counterfactuals*:

> I believe there are possible worlds other than the one we happen to inhabit. If an argument is wanted, it is this. It is uncontroversially true that things might have been otherwise than they are. I believe, and so do you, that things could have been different in countless ways. But what does that mean? Ordinary language permits the paraphrase: there are many ways things could have been besides the way that they actually are. On the face of it, this sentence is an existential quantification. It says that there exist many entities of a certain description, to wit, 'ways things could have been'. I believe things could have been different in countless ways. I believe permissible paraphrases of what I believe; taking the paraphrase at its face value, I therefore believe in the existence of entities which might be called 'ways things could have been'. I prefer to call them 'possible worlds'. (Lewis, 1973, p. 84)

More often than not, this argument has been taken as uncontroversial or even trivial. From it, of course, it does not follow that worlds are concrete entities, only that they exist (however we think of them). Van Inwagen, for instance, labelled it as "unexceptionable" (van Inwagen, 2001, p. 206) and Stalnaker claims that it is "only a presumption that the sentences of ordinary language be taken at face value" (Stalnaker, 1976, p. 66).[6]

Lewis's argument contains three premises and one conclusion:

(a) Things might be otherwise than they are (or, things could have been different in countless ways). (*Uncontroversial truth*)
(b) There are many ways things could have been besides the way they actually are. (*Paraphrase of a.*)
(c) There are many entities of a certain description: "ways things could have been", i.e., possible worlds. (*Taking (b.) at its face value*)
(d) Therefore, if one believes (a.), one believes (c.)

But I do not think that we should accept this argument. It seems to me that believing that things (in our world) could have been different is not the same as believing that there exist (countless) possible worlds where things are different. Something has gone amiss.

There is, I think, a tension between two steps in the argument. First, Lewis's claims that the *meaning* of (a.) is captured, or explained, by (b.). He is proposing that the meaning of a counterfactual statement such as (a.) *is* an existential quantification statement, such as (b.). Second, Lewis claims that the paraphrase has to be taken *at face value*; i.e., by accepting a common use of an everyday "paraphrase" in

[6]Lewis claims here that it is an uncontroversial truth that things might have been otherwise than they are. But is this so? It doesn't seem to me uncontroversial at all. Determinism, or at least certain forms of determinism seem to go against it. And determinism has proven to be a difficult doctrine to reject. What is more, this contradicts Lewis's own—later—view, according to which there are not different ways things might have been in this world, but rather infinite possible worlds whether things are (in a certain way). And, what is worse, things and people are word-bound in his theory, so, if on our world one thing is in a certain way, in all other worlds that thing's counterparts are either different or not.

ordinary language, we commit ourselves to the existence of certain entities: possible worlds.

But if we interpret the existential quantifier literally, and if by a paraphrase we mean "the same thing written or spoken using different words," (Cambridge Dictionary), then (b.) is not a paraphrase of (a.). (a.) is a statement about reality, concerning the contingent nature of (at least some) facts, and about how things could have been different; (b.) is a statement about the existence of ways or possible worlds and what is contained in them.

Let us go back to statement (1) above ("there could have been talking donkeys"). Suppose Arthur, a biology professor, is the speaker. It seems obvious that, assuming Arthur is being sincere and using words in their literal meaning, he is stating that there could have been talking donkeys, that is, expressing his belief that there could have been talking donkeys. I can see two possible readings of what Arthur states. First, that it is possible for donkeys to talk, or that it is possible for a certain kind of entity to be a donkey and to be able to talk. The first reading includes the belief that it is not impossible for donkeys to talk, or, in other words, that the fact that donkeys do not talk is contingent. On a second reading, the belief (1) expresses might be that the existence of certain entities, quite different from normal donkeys, and called "talking donkeys," is possible.

The second reading entails considering "talking donkeys" as a term for some sort of mythical, imaginary or fictional entity, substantially different from real donkeys. Something like unicorns, which might seem like horses with one horn, but are actually an entirely different kind of (mythical) beings. I will, for the most part, ignore this interpretation. I do not think it is the most intuitive one, and, in any case, is quite irrelevant for our purposes here.[7]

From the fact that Arthur believes that there could have been talking donkeys, it does not follow that Arthur believes that there are talking donkeys, in this or in any other possible world (regardless of whether they are normal donkeys that talk or imaginary entities called "talking donkeys"). It does not follow that Arthur believes that there are worlds where normal donkeys talk (or where beings called "talking donkeys" exist). In other words, that there are other worlds, abstract or concrete, where there are talking donkeys.

Conversely, believing that there is a possible world where donkeys talk is quite different from believing that donkeys could have talk (in our world). Believing that there is a possible world where there are talking donkeys is believing that there are talking donkeys (in that world). It is not believing that it is possible that there are talking donkeys, but rather that there are. By accepting the paraphrase argument, it seems, we modify the content of what we believe and of what we assert when we use modal statements to express our beliefs. We modify the subject matter of the statement.

[7] Also, because it raises a different set of issues about modal logic and about the proper account of name reference, such as "talking donkeys" or "unicorn." Kripke (1980, pp. 24, 156), Dummett (1993).

Statement (2) is even more problematic for Lewis. It is a fact, according to our example above, that I am wearing a pink t-shirt today. It is also a fact, presumably, that, even if there are many different worlds, a human being can only inhabit one world at a time. According to the paraphrase argument though, the meaning of (2) is something like: *that there are possible worlds (W) other than our world (W_0) where María is wearing a green t-shirt*. The truth-conditions of (2) would then be something like: *that there be (at least) a possible world (W_1) other than our world (W_0) where María is wearing a green t-shirt*. But I live in W_0 and, as far as we know, I cannot be at two worlds at the same time. The truth-conditions of (2), according to this, cannot be met. But then (2) would not only be false, it would be a contradiction.

But this makes no sense. (2) seems uncontroversially true.[8] To avoid this, as we mentioned above, Lewis introduced the idea of counterparts. According to this, statement (2) really means *that there is (at least) a possible world (W_1) other than our world (W_0) where my counterpart (María) is wearing a green t-shirt*.[9] But this has little, if anything, to do with (2). Why would (2) involve the existence of infinite possible worlds and infinite María-counterparts? And, even if there were infinite María counterparts, why would it be relevant for me (the *real* María), for what I can or cannot wear, what any of my counterparts are wearing in their respective worlds?

To repeat, it seems that Lewis, in an effort to explain modality and to avoid suspicious intensional entities, ended up changing the subject matter and thus offering an account of the meaning of modal statements that has little, if anything, to do with what we normally mean by them (both in ordinary life and in philosophy). The truth-conditions of a modal statement, like "I could be wearing a green t-shirt," are different from the truth-condition of a statement about possible worlds, like "there is (at least) a possible world where I (my counterpart) is wearing a green t-shirt."

There are, however, different ways to read the paraphrase argument. In particular, one can accept the possibility of *explaining*—not restating or rewording—possibility in terms of ways things might have been without accepting Lewis's reductionism or his ontological conclusions. Amie Thomasson, for example, accepts the paraphrase argument but not the reductionist element. She reads it as an "easy argument."

---

[8] Leaving aside, again, issues of determinism. Also, notice that there are different ways in which (2) might be true, or different possible meanings of (2). The statement may mean that I could have chosen a different t-shirt this morning, or it could mean that the t-shirt I chose could have been green, instead of pink. We assume the first reading.

[9] Or, rather, a counterpart of an actual green t-shirt (green t-shirt1). The idea of counterparts give raise to a whole set of new problems. Most notably, that of transworld identity. I believe, with Føllesdal (2004), that this, and related issues, are pseudoproblems.

6.5 Easy Arguments and Possible Worlds

Easy arguments consist of a series of trivial inferences starting with an uncontroversial claim and a conceptual claim, and reaching an ontological conclusion. The paraphrase argument, according to Thomasson, is an easy argument, involving only a series of "trivial inferences:"[10]

A1. There could have been talking donkeys.
A2. There is a way the world could have been: a way in which the world contained talking donkeys.
A3. There is a possible world in which there are talking donkeys (Thomasson, 2020, p. 134)

Lewis begins with an uncontroversial claim (A1) and a paraphrase, a "conceptual truth," in Thomasson terms. From these two claims, the truth of (A3) follows trivially, if we take the paraphrase *at face value*. As far as this goes, Lewis and Thomasson seem to agree. The difference comes in how to interpret what these possible worlds are and, more importantly for us, what is the function of both (A1) and (A2).

Thomasson does not think we need to take the inference from (A1) to (A2) as a reduction of modal talk into possible worlds talk, and thus, as an elimination of modal properties and facts. Modal properties and facts exist, she claims, and so do possible worlds. But neither are what make true (or false) statements like (A1) or (A2).

The reason is that by using these statements, we do not describe modal features or features of different worlds. What we do is showing or conveying the rules that govern the concepts involved. The main function of modal statements is normative, not descriptive.

Possible world semantics, and talk of possible world in general, has, according to Thomasson, the distinctive function of adding expressive power to the language:

> [P]ossible worlds talk serves the function not of tracking features of additional worlds and reporting on their features, but rather of adding these sorts of expressive power to our language. (Thomasson, 2020, p. 133)

Possible world semantics function then is not to describe or track worlds, just as the function of modal statements is not to describe or represent modal features of reality. Its function is to help us saying things we could not have said otherwise, and, presumably, to say things in a clearer way, that is, to help us *clarify* modal talk.

Easy arguments, are, according to her, the type of argument we need to derive the existence of entities like worlds, properties or propositions, but also of ordinary objects like tables, or trees. On both statements (A2) and (A3) the entity's existence is stated (ways and possible worlds) and a property is being attributed to it (there being talking donkeys in it). But, since Thomasson does not accept Lewis's

[10] For a detailed defense of easy arguments, see Thomasson (2015).

reductionist claim, she needs to explain how the introduction of a new entity in (A2), an entity not present in (A1), can be considered a trivial inference.

The key is in Thomasson's meta-ontological deflationism, encapsulated in her formal criterion for existence, according to which entities of kind K exist iff the application conditions associated with "K" are fulfilled. (Thomasson, 2015, pp. 116–7). There is nothing more in the notion of existence, according to this criterion, than what is captured by the rules of use of the relevant terms. There is nothing further, nothing substantial about existence, so no other criteria should be needed to claim that an object, a property or an event exist.

Consider everyday terms, like "tree." According to the formal criterion, trees exist if the application conditions associated with "tree" are fulfilled. It seems clear that any, or at least most competent speakers are capable of applying the term "tree" adequately. This is so, even if they are incapable of stating its precise conditions of use. Basically, any competent speaker should be able to know that the term "tree" is the appropriate term to use when confronted with a pine but not, say, when confronted with a geranium. This ability, combined with empirical data concerning trees (being confronted with and perceiving tree-like objects), is all we need to conclude that trees exist. (Thomasson, 2015, p. 118).

Similarly, in the case of possible worlds, they are said to exist if (and only if) the application conditions of the term "possible world" are met. In this case, of course, one cannot appeal to empirical data to justify the introduction and adequacy of the rules of use; to know when to use the term and when not, for instance. What we have, instead, are arguments containing conceptual truths and trivial inferences, "easy arguments," like (A1)–(A3) above.

Thomasson's way of solving existence questions has obvious advantages. It avoids the danger of falling into "pseudo-questions," as Carnap intended. It also avoids Benacerraf-style concerns: how to explain knowledge of whether or not abstract entities exist, given that we are concrete beings. According to Thomasson knowing the existence of abstract entities such as numbers, properties, propositions, or possible worlds, is just a matter of knowing the corresponding concepts (and their rules of applications). Finally, with her "simple realism," she answers to Quine concerns (location issues), while retaining the use of modal language.

6.6 Challenging Easy Arguments

Easy arguments are pretty easy to come by, once one understands how they work. Other than an uncontroversial first premise—there could have been talking donkeys—and a second, true, premise—there is a way the world could have been: a way in which the world contains talking donkeys—not much else is needed. The second premise doesn't have to be uncontroversial, but there must be good reasons to accept it, i.e., to consider it a conceptual truth.

It is the job of philosophy to study concepts, according to Thomasson, and doing so entails studying the conditions of application of the relevant concept terms.

Existence questions can only be discussed, and answered, through this conceptual analysis. So the reasons why we should accept the existence of a certain entity are not derived from any "ontologically-loaded" sort of reasoning, but rather from conceptual analysis:

> We should accept that there are properties, or numbers, or tables, because we have good reason to accept that the application conditions for the relevant terms are fulfilled. (Thomasson, 2019, p. 251)

Consider the following example, due to Daniel Korman (2019). Imagine an easy argument designed to prove the existence of free will. According to some compatibilists, if a person chooses to φ without being coerced or manipulated, then that person freely chose φ. Suppose this is correct: Lenny, who is unaware of the arguments of incompatibilism, can know that some of his choices are free simply by means of a trivial inference from the uncontroversial premise that some of his choices aren't coerced or manipulated:

B1. I chose to order dessert without being coerced or manipulated.
B2. If so, then I freely chose to order dessert.
B3. So, I freely chose to order dessert. (Korman, 2019, pp. 237–8)

(B2) is not uncontroversial. Just as (A2) above was not. Incompatibilists can and have developed several arguments against its truth. Regardless of whether or not these arguments are sound (one hopes not), the question is that the philosophical debate between compatibilists and incompatibilists is "perfectly sensible." But then, only if one accepts (B2), would (B3) follow trivially from (B2) and it would be easily known. And it is not clear that one should accept (B2). The existence of freedom is not settled, because we do not know if the conditions of application of the term "freely" are fulfilled.

I do not want to question here whether easy arguments are indeed appropriate to determine the existence of relevant entities, or whether or not philosophy's work is merely conceptual analysis. What I argue is that the existence of possible worlds cannot be determined by an easy argument, because neither the conditions of application of modal terms nor those of the term "possible world" are fulfilled. Or, more precisely, because it is not clear that they do or how they do.

In the case of the concept or term "possible world," for instance, Thomasson seems to be overlooking the fact that such arguments occur when there is some conception of what kind of possible worlds are being argued for—Plantinga's maximal states of affairs, Kripkean or Stalnakerian models, or Lewisian concrete universes. The easy argument won't get us to any of those. In fact, given that all these conceptions of possible worlds are total—in the sense that they resolve every issue—Thomasson's interpretation doesn't work even as the opening argument intended to get us to accept possible worlds so we can then explore the different conceptions.[11]

[11] Thanks to John Perry for pointing this out.

Consider another easy argument, designed to proof the existence of propositions:

C1. Snow is whiteC2. If P then that P is trueC3. That snow is white is true C4C4. So, there is a proposition(Schiffer (2003), mentioned in Thomasson (2015, p. 135))

(C1) is an uncontroversial claim, and (C2) is a "conceptual truth," according to Thomasson. (C3) and (C4) trivially follow from (C1) and (C2), and the existence of propositions —pleonastic propositions, Schiffer claims—would unproblematically follow.[12]

But, like in the freedom case, there is a "sensible debate" to be had about the application conditions of the term "proposition" and of the term "true." Schiffer's argument presupposes that a sentence S expresses a proposition, and that that proposition is the bearer of truth-value. But a philosopher can sensibly question these claims, and several have indeed done so. Take for instance the view defended by Korta and Perry, Critical Referentialism (CR, from now on).[13]

On this view, which I endorse (de Ponte et al., 2020, 2023, 2025), neither propositions nor sentences are the bearers of truth-value: utterances are. Assuming that snow is white, utterances of "snow is white" would be true. Also, CR rejects that believing or saying that *P consists* in a relation between an agent and the proposition *that P*. These are the core claims of the so called "doctrine of propositions," forcefully rejected by Perry (1977). Being a reasonable philosophical view, it shows that the truth of (C2) can be sensibly challenged, and because of this, that (C1)–(C4) are no longer trivial inferences, and the philosophical debate over existence questions of propositions remains open, and legitimate. We know of propositions, because we are participants in a linguistic and conceptual practice that makes use of the concept of "proposition," we understand the criteria of applicability of the concept proposition. Still, or maybe because of this, we reject (C2) and the ontological conclusion, (C4).[14]

According to CR, utterances of sentences like "snow is white" or "there could have been talking donkeys" are utterances about reality, and as such, they set some conditions for it to be true. More specifically, they set conditions on reality, presumably, something like "that snow is white" or "that donkeys could have talked" (or, maybe something along the lines of "that they do not talk, but that they meet all the requirements, so that they could have talk"). These truth-conditions are captured by what Korta and Perry have called the "referential content" of the utterance, which

[12] Schiffer gives an elaborate account of pleonastic entities and the criteria for their existence (2003). For our purposes here, suffice to say that, according to Schiffer, propositions exist but they are pleonastic in the sense that there is nothing else to them than what their concept establishes. Pleonastic propositions do not lead to epistemic problems, because knowing that there are propositions is just being part of a conceptual and linguistic practice; we use the concept of proposition.

[13] Also called Reflexive Referential theory or Critical Pragmatics. See Perry (2001/2012), and Korta and Perry (2011).

[14] Also, like in the possible world case, Schiffer's easy argument overlooks the fact that these arguments happen when there is a conception of proposition that is being argued for—Fregean propositions, Russellian, structural, etc.

normally corresponds to what is said by the utterance. But the utterance also puts some conditions on itself, or on some of its constituents: on the speaker, location, time or the names involved. So, for instance, for these statements to be true, it has to be the case "that the kind of stuff referred to by the speaker's use of 'snow' has the property of being white" (at t, l, w) or "that the objects referred to by the speaker's use of 'donkeys' do not have but could have had the property of being able to talk." These are conditions on the nouns used—"snow" and "donkeys"—or, rather, utterance-bound truth-conditions.

If these different kinds of truth-conditions are met, and, ultimately, if snow is white and donkeys could have talked, these utterances will be true. Otherwise, they will be false. CR does not make use of any substantial notion of proposition.

Propositions, on this approach, are tools used by philosophers and linguists to keep track of these truth-conditions. Just as possible worlds are tools used by philosophers and linguists to keep track of the truth-conditions of modal sentences. But saying that so-and-so does not *consist* in a relation to the proposition that so-and-so. And saying that something could have been the case does not *consist* in saying that there is a world where it is the case. We can, of course, use possible world talk as an expressive or theoretical tool to analyze modal statements (and, mostly, modal logic). But just as from an utterance being true it does not follow that a proposition is true and, much less, that a proposition exists; from a modal statement being true, it does not follow that possible worlds exist.

CR does not require or postulate the existence of propositions, particularly not as abstract entities with certain ontological features. But it does not preclude their use as theoretical tools. I believe all this is congenial to Thomasson's view and, in particular, to her defence of a minimal ontology (and a deflationist meta-ontology). The disagreement comes with the acceptance of easy arguments as giving sufficient reasons for the existence of entities; i.e., with the criteria of existence of abstract entities such as propositions and possible worlds. Consider a somehow simpler version of the easy argument concerning possible worlds, due to Steinberg (and endorsed by Thomasson):[15]

D1. It is possible that p.
D2. If it is possible that p, then there is a possible world at which p.
D3. There is a possible world at which p. (Steinberg, 2013, p. 770)

Steinberg uses this argument to develop a pleonastic view of possible worlds, in clear similarity with Schiffer's defense of pleonastic propositions. Again, premise (D2) is key and, once more, it seems to me that it can be sensibly challenged. Steinberg and Thomasson's proposal is epistemologically speaking much simpler than Lewis's: knowing what possible worlds are and that they exist is, simply, knowing (D2). But, admittedly, knowing (D2) is not knowing much. Certainly, it is not knowing what the concept of "possible world" stands for. A pleonastic possible world is not necessarily a Lewisian world, a concrete entity. But it might be. Ontology becomes less determined, and more awkward.

---

[15] Thomasson is a bit more cautious here. She does not claim that we can know the answer to existence questions concerning possible worlds via this argument, rather, she says that "one is entitled to say that there are possible worlds" (2020, p. 135).

Pleonastic entities, worlds or propositions, exist in the same sense as a table or Lewis's concrete worlds are supposed to exist. They also exist in the same sense as numbers or sets exist. Pleonastic entities are, on the one hand, an "ontologically free lunch" (in Armstrong's terms) and, on the other, ontologically underdetermined entities. The price to pay for epistemic simplicity of a deflationist metaontology of this kind is very high: an inflated ontology and underdetermination of certain concepts.

Defenders of easy arguments for possible worlds are fully aware of the fact that these arguments say very little about what a world is. A whole apparatus needs to be added, for example, to ensure that they are maximal and complete (that all questions can be answered in them). This apparatus, in Steinberg's case, involves proto-possibilities and refinements of proto-possibilities (Steinberg, 2013). This complicates greatly the theoretical apparatus of modal speech and logic.

Also, ontology is further inflated because (D2) is not a reduction of modality to (quantification over) possible worlds. It cannot be, because that would involve renouncing to the meta-ontological deflationism: it would involve existential claims about what exists (possible worlds) and what doesn't (modal properties). Easy arguments are easy because the existential claims follow from conceptual truths alone; without ontological presuppositions, reductions or requirements beyond what is contained in the uncontroversial claim.

But because it is not a reduction of the modal into the non-modal, we seem to end up with a very rich ontology, overpopulated with modal properties *and* with possible worlds (plus other entities we might require to give content to the concept of possible world, like proto-properties). Granted, on Thomasson's view one does not "begin by 'positing' modality as a fundamental feature of the world" (Thomasson, 2020, p. 137), keeping the ontology simple at the start and avoiding ontological assumptions. But we do end up accepting the existence of all entities designated by our concepts. Thomasson, being deflationist at a meta-ontological level, allegedly avoids complex and, possibly, unanswerable existence questions but, in doing so, she has to allow a particularly, and unnecessarily, rich ontology: she ends up defending an inflationionary ontology.

What I claim, in short, is that the inferences involved on (at least many) easy arguments are not trivial, and unnecessary. Using the notion of proposition or possible world as a theoretical tool is enough to claim that the concept of proposition/possible world exists. But it is one thing to claim that a concept or a notion of something exist. And another to say that the entity the concept or its corresponding term refers to exists.

To sustain this normativist view, and the deflationist meta-ontology that accompanies it, Thomasson requires "a fundamental distinction between language that is normative and language that is descriptive" (Thomasson, 2020, p. 137). This, it seems to me, is a very peculiar way of approaching language. A way that involves, as we will see, distinguishing between different types of descriptions, and quite a strict approach to statements which, in my opinion, do not have a main or fundamental function *by default*. I propose a rather different way to think of statements and their relation to reality.

6.7 Modal Properties and Facts

Central to Thomasson's view is the claim that statements have more than one function and that they can be used in different ways. So, even if the main function of modal statements is to convey semantic rules, they also have the function of describing the world, in an "everyday sense of describing." When we say something like, "'Margaret Mead necessarily comes from the genetic origin she actually has' there is a sense in which we are describing Margaret Mead and saying something true about her" (2020, p. 52).

Similarly, when we say that there could be talking donkeys, there is a sense in which we are describing donkeys and saying something about their talking (in)abilities. This means that speakers "describe the world when they make metaphysical modal utterances. And if they do so in an everyday (not theoretically loaded) sense of 'describing,' they are not making a mistake" (2020, p. 69).

Thomasson distinguishes two senses of the concept "describe:" an everyday, shallow sense, and a deep, philosophically loaded sense. This seems to reflect a distinction between an *ordinary* conception of reality and a *metaphysical* one or, even, between what really is the case and what is only apparently the case. It is one thing to say something true about Margaret Mead, it is another to describe some *essential* property of hers: that which *makes* the statement true.

> The real difference between the normativist view and traditional heavy-weight realist views of modality lies not in whether each says there are modal facts, properties, or even possible worlds … Instead, the difference lies … in the role the properties and facts play. In more traditional forms of realism about the modal, modal facts or properties (or other worlds) are "posited" as truthmakers for our modal claims: they are supposed to be what *explains* what *makes* our modal claims true… on the normativist view, modal facts and properties (and possible worlds) are not "posited" to "explain" what makes our modal claims true. (Thomasson, 2020, p. 139)

It is only the metaphysical conception of reality, Thomasson claims, that gives us the truth-makers for statements; i.e., the facts that make them true or false. In the case of modal statements, these truth-makers simply do not exist, because modal properties and facts are not part of the metaphysical conception of reality or, as it is usually put, modal statements do not carve reality at its joints. Therefore, modal statements are not descriptive and neither true nor false in any philosophically relevant sense.

Thomasson explains this with the adoption of a deflationist notion of truth. Modal facts and properties do exist, but not in "a deep, philosophical" sense: they are not the truth-makers of modal statements. She admits, however, that one can describe these modal facts and properties in an ordinary sense, and say true (or false) things about them. Going back to statement (2) (I could be wearing a green t-shirt), this statement is, according to her, *about* me and my t-shirts, but only in a shallow sense; neither me nor my t-shirts are what make it true (or false). We are not modal properties or facts, after all. Thomasson would say that this statement is true if and only if I could be wearing a green t-shirt. Saying that it is true does not add anything to it, to the information it conveys or to the veracity of what it describes. Saying that it is true is just "saying it." Describing reality with (2) is just stating that things are such and such, or, rather, that they appear to be such and such. Nothing *makes* this statement true (or false).

But this is problematic. First, because this claim seems to contravene her meta-ontological deflationism, appealing to reasons that go beyond the rules of application of terms (concepts) to determine what entities are posited and how. But also, even accepting normativism, it seems that some modal statements are clearly true—I could be wearing a green t-shirt—and some are clearly false—donkeys cannot be donkeys and whales at the same time. And it seems clear that what *makes* these statements true or false, are ordinary objects, like t-shirts, donkeys and whales, and their properties and relations.

Thomasson does not explain in detail what is the relation between the ordinary and philosophical conceptions of describing.[16] Or between the two conceptions of reality. She doesn't explain what an everyday notion of describing facts of reality is either, other than saying that utterances would be "world-oriented," i.e., about facts or features of reality.[17] But then again, being world-oriented is, it seems, all that can be asked for a statement to be descriptive, in any philosophically relevant sense of the term.

6.8 Back to the Common Sense View

Modal properties and facts are involved in all kinds of explanations, both everyday and philosophical explanations. A child has to learn what the world is like to learn to do things. Roughly, given a certain goal and a certain desire to achieve that goal, the child needs some sort of understanding of how the world is like and how it works—some beliefs—in order to know what to do to get what they want. This involves grasping the relevant possibilities: possible ways things could be and possible actions they could perform.

Most of the possibilities that we care about are situated or relational properties. The child need only be concerned with possibilities with regards to their situation and with regards to their desires. If they want a cookie, for instance, they do not

[16] She discusses several "heavyweight" realist options, like Sider's (2011) or Dorr's (2005). But these views involve claims that are neither necessary to claim that modal statements are descriptive nor widely shared, such as differentiating two meanings for "exist" or claiming that metaphysic's job is to "carve reality at its joints" (i.e. to search for essential features of reality).

[17] She mentions as a proposal, due to Huw Price, on how to combine the world-oriented nature of the semantic content of metaphysical modal claims and their non-(theoretically loaded) descriptiveness: "allow that an utterance of a metaphysical modal claim may make a (minimal) *statement* understood as an utterance we can sensibly say is (minimally) true—that is, that it is an utterance that 'provides a well-formed substitution into the context 'It is true that P''" (Thomasson, 2020, p. 70, quoting Price, 2011, p. 66). But I find it really difficult to make sense of this.

need to worry about what could happen if they turned the shower on, or about what is or could be in the roof of the house.

Adults usually need to worry about wider situations than children. But not much wider, at least in our everyday lives. Philosophers and scientists try to use possibilities in the widest and less constrained way possible. But philosophers and scientists still need to distinguish between things that are possible in limited situations, in not-so limited ones and, whenever is possible, in all situations.

Philosophers, it seems to me, are concerned with ways things could have been, or things that are possible in situations that are relevant for animals, or even plants; but also, or mostly, with things that are possible in situations relevant for humans, with things that are possible in much wider situations (planets or galaxies, say) and, finally, with things that are necessary in all situations (like, $2 + 2 = 4$) or not possible at all (like $2 + 2 = 5$).[18]

Modal properties, understood this way, fit into a spectrum of situated properties, which we use all the time. Philosophers' qualms and understanding is also about situated possibilities. We wonder about what we can know (given x, or in certain situations, or even in any situation), what we can do (given x, or in certain situations), what can be or happen (given x, or in certain situation), etc. The concern for metaphysical possibility (or necessity) can be seen as the concern for what is possible *at all*. Or, more precisely, with independence of any situation—and of physical possibility—but within the boundaries of what is logically possible.

Questions about what is metaphysically possible or metaphysically impossible might not be satisfactorily answerable. But they are not essentially different from everyday questions about what is possible or not in a situation. And they are sensible, legitimate questions when they are understood as part of the search for properties of the type: "these kinds of things *X* never/always/sometimes happen in situations of the kind *S*, because situations *S* have the property *P*." In other words, questions concerning modal properties and facts make sense not as normative or linguistic ones, but rather as searches for modal (situation-related) features of reality.

A proper explanation of modal statements should be able to accommodate this common sense view of modal properties. I believe this can be done using the tools of CR. CR's explanation, as we saw, does not pose implausible ontological demands and, I argue, it is able to accommodate the variety of functions modal statements play (including Thomasson's normative function).

Statements, modal or not, are particular types of utterances and, as such, have truth-conditions: conditions the world has to meet to make them true. Propositions are theoretical tools to keep track of the conditions utterances and other cognitive episodes, like beliefs or desires, put on facts. More precisely, utterances have a wide array of incremental conditions: some are conditions the utterance itself has to meet in order to be true or false, some are conditions on the speaker, on the time of utterance or on the objects we refer to.

---

[18] This way of seeing modal properties is mostly due to John Perry (personal communication).

An utterance of a sentence "I could be wearing a green t-shirt" has truth-conditions that involve the speaker and what t-shirt they could be wearing. There are different ways to spell out a counterfactual statement and different reasons to choose one or the other. The important points for us here are that, first, on all of these ways the statement is "world oriented," it describes reality, and, second, that the ontological commitments are minimal: utterances, facts, their properties and time.

On a very basic level, with only the fact that an utterance of an English sentence has been produced (2), and a few other elements set, the reflexive conditions for (2) to be true would be something like,

(2r) ***that the speaker of*** **(2)** *could be wearing* ***a green t-shirt*** *but they are not* ***at the time of*** **(2)**,

or,

(2r*) ***that the speaker of*** **(2)** wearing ***a green t-shirt at the time of*** **(2)** *is a state of affairs, but not a fact (of reality)*,

or,

(2r**) *that the fact that* **the speaker of** **(2)** *is not wearing* ***a green t-shirt at the time of*** **(2)** *is contingent.*[19]

Reflexive truth-conditions might give us something like the rules of applications of terms or concepts. The normative element of modal statements, so to speak. This is not a function originally given to the reflexive truth-conditions in CR. But Shannon Bain (unpublished) has convincingly argued that they can be understood along those lines. It is at this level where the conditions of use of the terms included in the sentence uttered are set. Given the usual meaning of the terms used (and the rules of syntax), the reflexive content captures the conditions the use of these terms have to meet in order for the utterance to be true. In other words, (2r) shows or encapsulates the norms of use the modal terms in (2).[20]

But the reflexive truth-conditions, and the normative aspect, is only one among many conditions. Building from these truth-conditions, and given that I am the speaker and the time of utterance is 2 pm, October 14, 2022, the referential truth-conditions would be something like,

(2f) *that* **María** *could be wearing* ***a green t-shirt at*** **2 pm, October 14, 2022**,

or,

(2f*) *that* **María** *wearing* ***a green t-shirt at*** **2 pm, October 14, 2022** *is a state of affairs, but not a fact (of reality)*,

or,

[19] Following Perry (2001/2012), I use italics to distinguish propositions from statements; roman boldface to mark that it is the referent, and not any of its identifying conditions, which is the constituent of the content; and italic boldface, to indicate that it is the identifying condition that enters into the truth-conditions and not the object it designates.

[20] Bain (unpublished) presents an elaborate interpretation of Thomasson's normativism using the tools of Critical Referentialism.

(2f**) *that the fact that* **María** *is not wearing* ***a green t-shirt at*** **2 pm, October 14, 2022** *is contingent.*

Referential truth-conditions give us the descriptive component of (2). They are conditions on reality, set by (2). These truth-conditions might be called propositions, if by that we mean contents that encodes the conditions for truth of statements. But statements express beliefs (sincere) speakers have, and those beliefs are about the world: about what is the case, what is not, and what can or cannot be. In other worlds, beliefs about facts and their properties, about whether they are contingent, necessary or impossible.

6.9 Conclusion

Both Lewis and Thomasson aim at explaining modal statements without appealing to modal facts or properties, and both accept the existence of possible worlds. What they each mean by this, however, is very different. Lewis is committed to the existence of infinitely many concrete worlds and, by doing so, he not only ends up with an overpopulated and unintuitive ontology, but he also changes the subject matter of the statements and propositions.

Thomasson's account has several important advantages. It avoids positing worlds as concrete entities, it greatly simplifies epistemology, and it acknowledges the fact that modal statements might and usually have different functions. Her view, however, has some important drawbacks. First, despite the deflationist meta-ontology approach, her view is a sort of "ontological inflationism," since she accepts the existence of all entities derived from conceptual truths, be them possible worlds, propositions or tables.

Second, her view seems to accept a very particular notion of propositions, according to which they are the bearers of truth-value, and the object of propositional attitudes. Third, in order to explain what makes modal statements true or false, without falling into modal realism (or "heavyweight" realism with respect to possible worlds), she requires the adoption of a deflationist notion of truth. Whatever our views on truth are, requiring a particular view on truth to sustain your account of modal statements and modality is general seems to be a clear disadvantage.

A normativist about modality, like Thomasson, and a modal realist, like me, agree on many issues. First, we both claim that by uttering sentences containing modal terms we describe reality, in an "ordinary" sense of "describing." Second, we both defend that there are modal facts and properties. Therefore, third, we both believe that modal statements are either true or false.

But there are important differences on our views. I do not think that a modal normativist like Thomasson would disagree with most of what a modal realist like me claims. But a modal realist like me cannot accept some of the key theses of Thomasson's modal normativism.

I have presented the basic elements of an alternative, simpler, explanation of the meaning and truth of modal statements; one that does not require accepting the reduction of modality to quantification over possible worlds (a' la Lewis) or rejecting that by using modal statements we are actually talking about (describing) our world (a' la Thomasson). What is more, the view sketched does not clash with our common sense intuitions about language or reality and, most importantly, is not committed to the existence of propositions, to the existence of possible worlds, or to a non-intuitive notion of existence altogether.

Talk about a "world" is not only confusing, but positively misguided. And so is talk about "propositions." We are not justified in believing in the existence of a world, understood as a totality of facts, or in a proposition, understood as the content of our utterances and propositional attitudes. Much less are we justified in believing in the existence of many worlds, or many propositions.

Acknowledgements Thanks to Shannon Bain, Kepa Korta and John Perry for long discussions about the topics discussed here and for detailed comments on earlier drafts. This work has been partially funded by grants from the Spanish Government (Grant PID2023-151480NB-I00 funded by MICIU/AEI/10.13039/501100011033 and by ERDF/EU) and the Basque Government (IT1612-22).

References

Armstrong, D. (1997). *A world of states of affairs*. Cambridge University Press.

Bain, S. (unpublished). *Bias, reflexive content, and full content*.

Barwise, J., & Perry, J. (1985). Shifting situations and shaken attitudes. *Linguistics and Philosophy, 8*, 105–161.

Chihara, C. S. (1998). *The worlds of possibility: Modal realism and the semantics of modal logic*. Oxford University Press.

de Ponte, M., Korta, K., & Perry, J. (2020). Utterance and context. In T. Ciecierski & P. Grabarczyk (Eds.), *The architecture of context and context-sensitivity* (pp. 15–29). Springer. Reprinted in de Ponte, Korta and Perry (2025) (pp.211–225).

de Ponte, M., Korta, K., & Perry, J. (2023). Language and luck. In A. Fairweather & C. Sotomayor (Eds.), *Linguistic luck. Essays in anti-luck semantics* (pp. 15–35). Oxford University Press. Reprinted in de Ponte, Korta, and Perry (2025) (pp. 245–265).

de Ponte, M, Korta, K, & Perry, J. (2025) Three demonstrations and a funeral and other essays . Stanford. CSLI Publications.

Dorr, C. (2005). What we disagree about when we disagree about ontology. In M. E. Kalderon (Ed.), *Fictionalism in metaphysics* (pp. 234–286). Oxford University Press.

Dummett, M. (1993). *The seas of language*. Oxford University Press.

Fine, K. (2005). *Modality and tense*. Philosophical Papers. Clarendon.

Føllesdal, D. (2004). *Referential opacity and modal logic*. Routledge.

Korman, D. Z. (2019). Easy ontology without deflationary ontology. *Philosophy and Phenomenological Research, XCIXX*(1), 236–243.

Korta, K., & Perry, J. (2011). *Critical pragmatics. An inquiry into reference and communication*. Cambridge University Press.

Kripke, S. (1980). *Naming and necessity*. Harvard University Press.

Lewis, D. (1973). *Counterfactuals*. Blackwell.

Lewis, D. (1986a). *On the plurality of worlds*. Blackwell.

Lewis, D. (1986b). Modal realism at work: Properties. In D.H. Mellor and A. Oliver (Eds.), (1997) *Properties*. Oxford University Press.
Mackie, J. L. (1973). *Truth, probability and paradox: Studies in philosophical logic*. Claredon Press.
Menzel, C. (1990). Actualism, ontological commitment, and possible world semantics. *Synthese, 85*(3), 355–389.
Partee, B. H. (1989). Possible worlds in model-theoretic semantics: A linguistic perspective. In S. Allen (Ed.), *Possible worlds in humanities, arts and sciences: proceedings of nobel symposium* (Vol. 65). Walter deGruyter.
Perry, J. (1977). Frege on demonstratives. *Philosophical Review 86* (4), 474–497. Reprinted in *The Problem of the essential indexical and other essays* (Expanded Ed.). CSLI Publications. 2000
Perry, J. (1989). Possible worlds and subject matter: Discussion of Barbara H. Partee's 'possible worlds in model-theoretic semantics: A linguistic perspective'." In S. Allen (ed.). *Possible worlds in humanities, arts and sciences: Proceedings of nobel symposium, August, 1986, 65. Walter deGruyter. Reprinted in The problem of the essential indexical and other essays,* Expanded Edition. CSLI Publications. 2000.
Perry, J. (2001 [2012]). *Reference and reflexivity* (2nd Revised ed.). CSLI Publications.
Plantinga, A. (1974). *The nature of necessity*. Clarendon Press.
Price, H. (2011). *Naturalism without mirrors*. Oxford University Press.
Prior, A. (1968). *Papers on time and tense*. Claredon.
Schiffer, S. (2003). *The things we mean*. Oxford University Press.
Sider, T. (2011). *Writing the book of the world*. Oxford University Press.
Stalnaker, R. (1976). Possible worlds. *Noûs, 66*, 65.
Stalnaker, R. (2012). *Mere possibilities: Metaphysical foundations of modal semantics*. Princeton University Press.
Steinberg, A. (2013). Pleonastic possible worlds. *Philosophical Studies, 164*, 767–789.
Thomasson, A. L. (2009). Non-descriptivism about modality: A brief history and revival. In *Baltic international yearbook of cognition, logic and communication* (Volume 4: 200 years of analytical philosophy, pp. 1–26).
Thomasson, A. L. (2015). *Ontology made easy*. Oxford University Press.
Thomasson, A. L. (2019). Replies to comments on ontology made easy. *Philosophy and Phenomenological Research* Vol. XCIXX, 1: 251–264.
Thomasson, A. L. (2020). *Norms and necessity*. Oxford University Press.
van Inwagen, P. (2001). *Ontology, identity, and modality: Essays in metaphysics*. Cambridge University Press.

Part III
Themes on Linnebo's Thin Realism

Chapter 7
Ontology and the Algebraic Conception of Logic

Øystein Linnebo

Abstract I articulate and defend the thesis that first-order logic is algebraic, in the sense that it does not have a single canonical application to reality. If correct, this thesis supports a flexible conception of ontology. Suppose we have a first-order language that purports to talk about certain objects, say *F*s, whose conditions for correct use are expressed in terms that are already assumed to be in good standing and thus, in particular, involve no talk about *F*s. Suppose further that these assertability conditions respect the laws of logic. Finally, suppose that the assertability conditions involve a criterion of identity that licenses the use of non-trivial identity statements. Then, according to the flexible conception, there really are *F*s. I also defend this algebraic conception of logic, and the conception of ontology to which it gives rise, against the objection that this approach involves a problematic form of anti-realism.

Keywords Flexible conception · Criterion of identity · Assertability · Algebraic conception of logic

7.1 Swimming and Ontology

When I was a boy, I loved to go swimming in lakes and the ocean. In the summer, I used to swim *a lot*, often several times a day. I also loved to count things. Unsurprisingly, these two passions intersected: I wanted to count how many times I went swimming. This posed a tricky conceptual problem. How to define—or, as I would now put it, *individuate*—a swim? The problem was pressing. On a warm day, I would dive into the water, swim, get out and mess around for a little while, and then dive in again. Was that one swim or two? In order to count the number of

---

Ø. Linnebo (✉)
University of Oslo, Oslo, Norway
e-mail: oystein.linnebo@ifikk.uio.no

X. de Donato-Rodríguez et al. (eds.), *Deflationist Conceptions of Abstract Objects*, Synthese Library 516, https://doi.org/10.1007/978-3-031-98717-5_7

swims, it was essential to answer that question. Without an answer, I would be unable to count how many times I went swimming on a given day.

I therefore formulated an answer that still strikes me as pretty good. If two episodes of being fully submerged in the water were separated by being out of the water and completely dry, then the two episodes belong to two different swims; but if not, then the two episodes count as merely different stages of one and the same swim. This criterion was natural, easily implemented, and sufficiently precise to enable me to get on with the business of counting of swims. Moreover, the criterion enabled me to predicate various "holistic" properties of swims, such as lasting 15 min, comprising five dives, involving play with three different friends, and so on.

In short, I seemed to have established a good way to refer to swims, to count them, and to predicate properties of them. This practice was logically coherent, sufficiently determinate to give a definite verdict in all of the relevant circumstances, and served the relevant purposes, which were important, if not exactly to science, then at least to my friends and me.

Nonetheless, these childhood adventures prompt a hard philosophical question. Do swims really exist, or had I merely established a linguistic practice that allowed me to speak *as if* there are swims?

7.2 The Flexible Versus the Rigid Conception of Reality

In my recent book *Thin Objects*, I describe two competing conceptions of reality that give very different answers to this question.[1] Let me to review these conceptions.

Suppose we want to expand our language so as to talk about a range of objects that we have not yet recognized. The situation in which I found myself when I wanted to talk about swims provides an example. What does it take to succeed? A minimal requirement is that we have assigned assertibility-conditions to all statements of the expanded language—for example, the language in which we talk about swims—and that these assignments respect all the laws of first-order logic.[2] Particularly important is what we may call *the logic of identity*, which describes the identity relation and its interaction with predication and in this way lays down the fundamental principles governing our conceptual apparatus for identification and predication. In brief: identity is an equivalence relation that interacts with predication as described by Leibniz's Law, namely, if x is identical with y, then x and y have precisely the same properties:

$$(\mathrm{LL})\, x = y \rightarrow (\phi(x) \leftrightarrow \phi(y))$$

[1] See Linnebo (2018, Sect. 2.4), on which the remainder of this section draws.

[2] As usual, a statement is a sentence used in some linguistic context.

As usual, Leibniz's Law must be restricted to contexts $\phi(\ldots)$ that are transparent, in the sense that the context is only concerned with the referents of the terms that fill the argument place, not with these terms themselves or any non-referential aspects of their meaning.

Suppose this minimal requirement is satisfied: assertibility-conditions have been assigned to all statements of the expanded language in a way that respects all the laws of first-order logic (including, of course, the logic of identity). Indeed, to strengthen the case further, we may suppose that these assertibility-conditions are expressed in terms that are already assumed to be in good standing. As noted above, as a boy who loved swimming and counting in equal measure, I had taken due care to ensure that my language satisfied this minimal requirement. The question is now whether satisfaction of the minimal requirement suffices to ensure that the expanded language refers—to swims or whatever other objects we wish our language to be capable of describing.

One conception of reality gives a negative answer.

> *The rigid conception* holds that reality is "carved up" into objects in a unique way that is independent of the concepts that we bring to bear. This conception introduces an element of risk into our proposed application of our conceptual apparatus for identification and predication. Although the attempted application is logically in good order, reality may fail to cooperate. Reality may simply not contain the sorts of objects we are trying to "carve out". (p. 31)

For example, either reality contains swims or it does not. If it does not, then my careful conventions governing the individuation of swims were all in vain: through no fault of my own, my expanded language would fail to refer. An uncooperative reality may simply fail to contain the objects that I sought to count and describe.

A different conception disagrees.

> *The flexible conception*, on the other hand, insists that reality is articulated into objects only through the concepts that we bring to bear. And we often have some choice in this matter. Of course, our choice is limited. It is not an option for a starving person to produce more bread by "carving up" reality differently. The choice we have is of a different character. As Frege observed, reality provides different answers to ontological questions about what there is depending on which concepts we bring to bear. If you bring to bear the concept of a loaf of bread, some part of reality (say, your kitchen) may answer that there are two such objects. If instead you bring to bear the concept of a molecule of organic material, you will receive a very different answer. This much should be uncontroversial. The flexible conception of ontology gets its bite by adding the controversial claim that there is no unique, privileged set of concepts in terms of which to "carve up" reality, namely the concepts that match some rigid concept-independent articulation of reality into objects. This means there is no risk that reality fails to contain objects answering to some coherent application of our apparatus for identification and predication. This coherent application "carves out" the appropriate objects. (p. 31; some footnotes have been omitted)

Clearly, the flexible conception entails a different view of whether as a boy I succeeded in referring to and quantifying over swims. Since I had taken proper care to ensure that my expanded language was well-defined and logically coherent, there was no *further* risk of failure. I had a coherent application of the language (not exactly of first-order logic, but of some natural language equivalent thereof). So this coherent application "carves out" the appropriate objects—in my case, swims—without being hostage to metaphysical fortune.

At heart, the disagreement between the two conceptions concerns the relative explanatory priority of the articulation of reality into objects and permissible applications of our conceptual apparatus for identification and predication. The rigid conception insists that *first* reality provides a range of objects, to which we *then* apply our conceptual apparatus. The flexible conception reverses the explanatory priority. We *begin* by providing a logically coherent interpretation of our language. *Only relative to such an interpretation* does it make sense to talk about what objects "reality provides".

Which conception is correct? My book provides a sustained—and often very detailed—defense of the flexible conception. Rather than rehearse this defense here, my plan in this article is to continue with a more abstract, or "big picture", exploration of the relation between logic, language, and the concept of an object.[3] In particular, I wish to develop and motivate what I will call an *algebraic conception* of first-order logic, which underlies and supports the flexible conception of reality. My hope is that this will shed some new light on the view developed in the book.

7.3 The Algebraic Conception of First-Order Logic

Let us say that a concept—or some representational device—is *algebraic* just in case it does not have a unique canonical application to reality but rather admits of a variety of different (although perhaps related) applications. For example, the notion of a partial order, typically written '$\leq$', is algebraic. There is no such thing as the *true* partial order of reality. It would be nonsense to say, without further qualification, that I bear $\leq$ to the Eiffel Tower. The notion of a partial order admits of countless different applications to reality. For example, I precede the Eiffel Tower in the order of mass, whereas the tower precedes me in the order of temporal origin. I do not think any philosopher would disagree.

Other cases are more interesting. For example, the concepts of geometry used to have a canonical interpretation, tied to physical space, that is, the space in which we live and move around. With the discovery of non-Euclidean geometries in the 1830s, however, this view became untenable. Both Euclidean and non-Euclidean geometries are good and legitimate mathematical theories, irrespective of the geometry of physical space. Mathematicians therefore started to distinguish abstract mathematical spaces from physical space. There is no need to choose between Euclidean and the various non-Euclidean geometries. All of these geometries are legitimate; they just describe different *mathematical* spaces. In this way, the geometrical concepts—such as *point* and *line*—went from having a canonical interpretation, tied to

[3] I am not always a fan of "big picture" philosophizing. But having given a detailed philosophical and logical exposition of my argument—some may even think *too detailed*—I hope to have earnt the right to have a go with a broader brush.

physical space, to being algebraic. Thus, as the great mathematician David Hilbert liked to emphasize, in a proper axiomatization of geometry, "one must always be able to say, instead of 'points, straight lines, and planes,' 'tables, chairs, and beer mugs' " (Hilbert, 1935, p. 403); that is, the mentioned geometrical primitives can be interpreted in any way whatsoever, provided that the axioms come out true.

Moving on to concepts dear to philosophers' hearts, we find considerable controversy. Consider the concepts of mereology, such as parthood, overlap, and sum. Do these concepts have a single canonical application or are they merely algebraic? Different philosophers give different answers. Many philosophers are happy to speak, without any quantification, about the *true* mereological structure of reality. Is there really a sum of me and the Eiffel Tower? Is it really the case that when x is not part of y, then y has a part that is disjoint from x (as asserted by the axiom of Strong Supplementation)? The literature abounds with discussions of such questions. Along with some other philosophers, however, I maintain that these questions rest on a fundamental confusion. The mentioned mereological concepts, just like that of a partial order, are algebraic; that is, these concepts do not have a single canonical application to reality but admit of countless different, equally legitimate applications. We can define perfectly legitimate mereological structures in which there is a sum of me and the Eiffel Tower, and others in which there is not. To ask, without further qualification, whether there really is a sum of these two objects makes no more sense than to ask whether I really bear the partial order $\leq$ to the Eiffel Tower.

Although this dispute about the concepts of mereology is interesting, this is not a battle I intend to fight today.[4] I mention the dispute simply to illustrate how it can be controversial whether certain concepts have a canonical interpretation or are merely algebraic.

My present concern is pure first-order logic—that is, first-order logic without non-logical predicates[5]—and the associated "logical" concept of an object—that is, something to which we can refer and over which we can quantify. Does pure first-order logic have a canonical application or is it merely algebraic? Advocates of the rigid conception of reality give the former answer. Since reality comes articulated into objects, independently of the concepts that we bring to bear, pure first-order logic has a canonical interpretation, namely, as talking about the totality of these objects. By contrast, advocates of the flexible conception of reality take pure first-order logic to be algebraic. By bringing to bear different concepts, we obtain different but equally legitimate ways to apply this logic to reality.

As should be obvious by now, in this case too, I favor the algebraic view. For a proper defense of this view, I refer the reader to my book, where I argue that

[4] When fighting that battle, I would build on Fine (2010), along with the theory of "liberal definitions" developed in Florio and Linnebo (2021, Sections 4.4, 5.8, and 12.3).

[5] In the literature, one sometimes finds talk about "the logical concept of object"; e.g. Parsons (2007, Ch. 1). While it is possible that Parsons and I are getting at the same thing, I find the adjective "algebraic" more informative. The fundamental point is not that our only general concept of an object has its provenance in logic, but that logic doesn't apply directly to reality, but only relative to a conceptualization of reality, which "carves" it into objects.

satisfaction of the minimal requirement—supplemented with appropriate criteria of identity, of which more shortly—suffices for the relevant language to refer. Suppose it is logically coherent to apply a first-order language in a certain way. What's to stop us from doing so? Moreover, doing so is conceptually stable. By using a meta-language with similar expressive resources, we can ascribe reference to our object language; indeed, we can even provide an account of how these relations of reference are constituted.[6]

To this positive case for the algebraic conception, one might add a challenge to defenders of a canonical application. What might possibly endow pure first-order logic with such an application? This seems like a magical philosophy of language.[7]

7.4 What the Algebraic Conception of First-Order Logic Does Not Say

To gain a better understanding of the algebraic conception, it will be useful to clarify what the conception does and does not say. I begin, in this section, with what the conception does *not* say, before examining some of its commitments in the next section.

First, the algebraic conception does not claim that the objects in question, such as the swims in my opening example, are indispensable to a scientific description of reality. There is a sense in which I give a complete description of the relevant aspects of reality when I report that at t_1, I was in the water; at t_2, I was out of the water and my skin was dry to such-and-such an extent; and so on. Facts about swims, individuated in the way I as a boy proposed, are implicit in such a report. Given such a report and an account of how we have decided to individuate swims, it is a routine matter to work out an alternative report in terms of swims and their properties. Thus, the proposed expanded interpretation of our language need not be obligatory for the purposes of science.[8] Although we *can* choose to apply first-order logic to reality in a certain way, it does not follow that we *must* so apply it. We may also leave the possibility of so applying it implicit in some more fundamental description of reality.

Second, the algebraic conception is not committed to denying that there may be other, equally natural objects that we might have chosen to single out. In the example concerning swimming, say, we might have chosen to count two episodes of being fully submerged as belonging to two distinct swims just in case the person in question is fully dressed (in some specified sense) between the two episodes. This

[6] See Linnebo (2018, especially, chapters 2 and 8).

[7] See Rayo (2013, Sect. 1.2).

[8] As Ismael Ordóñez reminds me, however, insofar as philosophy is concerned with *all* objects—even ones that are non-fundamental and highly parochial—philosophy should provide an account of all such expanded interpretations.

too would have been an acceptable way to individuate swims, although on a warm day, it would have yielded a different answer to the question of how many swims I had enjoyed. Thus, a way to use language to "carve out" objects can be acceptable even if it is not unique or even optimal. The objects that we "carve out" may well be of only parochial interest, as in our example of swims.

Third, had we not been around to "carve out" various objects, would these objects have existed? Would there have been objective facts about them? Some philosophers are doubtful.[9] However, I contend that the algebraic conception is not committed to the view that the objects we "carve out" depend on us for their existence and their properties.

To explain and defend this contention, let me begin by distinguishing the algebraic conception of first-order logic from the notion of social constitution. I am entirely happy to admit that some of the objects we "carve out" are socially constituted. Marriages and contracts, for example, are created by us through our practices. In the absence of various social and legal practices, these objects would not have existed. For a marriage to exist, two consenting individuals need to exchange a vow. Thus, by withholding your vow, you have the power to prevent a marriage from coming into existence. Contracts are not fundamentally different.

By contrast, although sticks and stones, numbers and sets, are "carved out" by us, they are not thereby created by us. It is one thing to single out an object for reference and quite another to create the object in question.[10] According to the algebraic conception, some work is required in order to single out an object—we cannot just rely on reality to do the job for us. This does not mean, however, that the object singled out is our creation. The object may be there, for anyone with the appropriate cognitive capacities to single out. Had *you* not singled out a certain stone or a set, say, someone else might have singled out that very object. They might also have found it to have the very same properties that you ascribed to it. By contrast, by withholding your vow, you could have caused your marriage not to exist; no one else would have had the power to make that very marriage exist.

More generally, realism about some domain has to do with reality providing an objective answer to every meaningful question about the domain. The algebraic conception does not restrict reality's ability to provide answers to meaningful questions. Rather, it is a thesis to the effect that more work than what one might have expected is required in order to formulate a meaningful question, which we can then put to reality. Consider the question "How many objects are there in my office?" As it stands, this is not a fully meaningful question: we additionally need to know what kind of object the questioner has in mind. On the algebraic conception, it takes more work than one might have expected to formulate a meaningful question. Yet once we have formulated a meaningful question, reality will provide an objective answer. In short, we must not confuse reality's inability to answer questions that are not even fully meaningful with anti-realism.

---

[9] See, e.g., MacBride (2016).

[10] Rayo (2013, pp. 24–25), makes similar remarks.

Let us apply this discussion to our standard example of swims. Are swims ontologically dependent on us? In a straightforward sense, the answer is "yes". Had I chosen not to dive into some particularly cold water, a certain swim would not have existed. In another and philosophically more interesting sense, however, the answer is "no". I did not bring swims into existence when as a boy I set up my language game for talking about swims. This language game enables us to count the swims of children who do not participate in the language game. We can even count the swims of children in a counterfactual scenario in which there is no language whatsoever. Once we have decided how to use the language, it is a language- and mind-independent matter whether a statement in the resulting language is true or false.

7.5 Some Consequences of the Algebraic Conception of First-Order Logic

The algebraic conception does have some important consequences, however.

First, the algebraic conception of first-order logic supports a flexible conception of reality. Suppose we have defined an application of some first-order language that purports to talk about certain objects. Suppose further that the conditions for the correct use of sentences of this language involve only objects that are accepted by all parties. Suppose finally that the laws of logic are respected and that we have clear criteria of identity for the relevant objects. Then these objects exist; we can make no sense of reality brutely failing to contain such objects. A brute existence failure would presuppose that pure first-order logic has a canonical interpretation: only this would provide a vantage point from which to make sense of such a brute existence failure.

By contrast, if pure first-order logic had a canonical interpretation, we would get a canonical and all-encompassing domain. Then every other attempted use of first-order logic would have to find a place within that domain. This would provide a way to make sense of some attempted application of first-order logic brutely failing to latch on to objects. In the absence of such an interpretation, however, the best we can do is to ensure that we have provided our first-order languages with clear and logically coherent assertibility-conditions.

Second, the algebraic conception does involve a form of conceptual relativity: what objects we find in reality will depend on what concepts we bring to bear. Yet there is nothing spooky here. As observed above, my view is fully compatible with realism, understood as reality providing a determinate answer to every meaningful question.

Third, the algebraic conception entails that it is open-ended what objects there are. Suppose we can talk about some domain of objects.[11] Based on this talk, we can

[11] I am tacitly restricting attention to *determinate* domains, that is, domains that can be specified extensionally, say as a plurality of objects. See Linnebo (2018, Sections 3.3 and 3.7), as well as

expand the interpretation of our language to talk also about sets of objects from the domain. On pain of paradox, some of these sets must lie beyond the domain with which we began. Thus, we cannot speak of a determinate totality of all objects that exist. For given any such a totality, we can define even more objects, namely those objects *and sets thereof*. This too is a permissible way to organize and convey information about pluralities of objects from the determinate totality with which we began.

7.6 Why Insist on Criteria of Identity?

I have suggested that in order to talk about certain objects, we must have provided clear criteria of identity for the objects. But why insist on criteria of identity? Might we not obtain an even more abstract and general notion of objecthood by relaxing this requirement? These questions are pressed with great force in recent work by Agustín Rayo.[12]

I have considerable sympathy for the ensuing "ultra-thin" conception of objecthood, which leaves out the requirement of criteria of identity. Indeed, I believe Rayo argues convincingly that this too is a legitimate and valuable notion of objecthood.[13] However, I will now argue that we obtain a more robust form of objecthood by additionally requiring criteria of identity. For many purposes, at least, this robustness is worth the slight loss of generality.

Once again, let us use the case of swims as an illustration. For me as a boy to talk about swims, it was imperative to provide a criterion of identity for swims. Without such a criterion, I would have been unable to count swims or to ascribe to them "holistic" properties such as lasting 15 min. My language would have been impoverished. I might perhaps have (truly) said that a swim took place and that it was cold, but not that I had precisely two swims or that each of these comprised three dives.[14]

There is a more general point here as well. Objects typically have a variety of properties that are not even implicit in how the objects are presented to us. While a

---

Florio and Linnebo (2021, Sections 12.2–12.4).

[12] See Rayo (Forthcoming).

[13] As Lev Lamberov reminds me, it is doubtful that quantum particles have sharp criteria of identity. So we must be open to the possibility that such particles qualify as objects only on the ultra-thin conception. A similar concession is made for mathematical objects, understood in a structuralist manner, in Linnebo (2018, pp. 24–26).

[14] As Agustín Rayo reminds me, individuation is a matter of degree. There could be a practice of talking about swimming entirely without criteria of identity for swims. Even the loose talk about swims that I encountered as a boy provided some criteria, e.g. if you go to sleep for 8 hours between two episodes of submersion, then these episodes certainly belong to different swims. While my youthful regimentation improved on this loose talk, it still fell short of strict scientific standards, e.g. by failing to stipulate precisely how dry one has to be between two distinct swims. How sharp the criteria need to be will of course depend on one's purposes. The more informal the purposes, the less sharpness will be required.

swim can last 15 min, for example, this property is not implicit in any single episode of being submerged in water. We additionally need a criterion of identity to connect a suitable sequence of such episodes into a single temporally extended swim.

Furthermore, there is a close connection between criteria of identity and objectivity. I insisted above that swims are objective, despite being "carved out" by us. To make this point, I argued that swims are available to be picked out by anyone (with appropriate conceptual capacities) who cares to do so, who can then find these swims to have the same properties that I ascribe to them. For this to be possible, we need to distinguish a swim from a mere representation of a swim. It must be possible for one and the same swim to be represented from different points of view, at different times, and by different individuals. This requires that we can make sense of non-trivial identity statements concerning swims, that is, to tell whether two episodes of being submerged in water belong to one and the same swim.

Once again, there is a more general point here. As Tyler Burge (2010) observes, "an objective subject matter is one that is constitutively non-perspectival" (p. 47). That is, the objectivity of some subject matter has to do with its independence from the particular perspective one may have on it—whether a spatial or temporal point of view, an occasion of representing the object, or the individual who represents it. Thus, for an object and its properties to be objective, both the object and its properties must be independent of the idiosyncrasies of how they are presented to us. This independence requires that we can make sense of representing one and the same object despite being presented with it in very different ways. Criteria of identity are required to make this possible.[15]

References

Burge, T. (2010). *Origins of objectivity*. Oxford University Press.

Fine, K. (2010). Towards a theory of part. *Journal of Philosophy, 107*(11), 559–589.

Florio, S., & Linnebo, Ø. (2021). *The many and the one: A philosophical study of plural logic*. Oxford University Press.

Hilbert, D. (1935). *Gesammelte Abhandlungen* (Vol. 3). Springer.

Linnebo, Ø. (2018). *Thin objects: An abstractionist account*. Oxford University Press.

MacBride, F. (2016). NeoFregean metaontology: Just don't ask too many questions. In P. Ebert & M. Rossberg (Eds.), *Abstractionism: Essays in philosophy of mathematics* (pp. 94–112). Oxford University Press.

Parsons, C. (2007). *Mathematical thought and its objects*. Cambridge University Press.

Rayo, A. (2013). *The construction of logical space*. Oxford University Press.

Rayo, A. (Forthcoming). The ultra-thin conception of objecthood. *Inquiry*. https://doi.org/10.1080/0020174X.2025.2466700

[15] Thanks to Lev Lamberov, Ismael Ordóñez, Agustín Rayo, Andrea Sereni, and Luca Zanetti for discussion.

Chapter 8
Quinean Lightweight Objects

Concha Martínez-Vidal

Abstract The chapter aims to clarify the metaphysical picture underlying the so-called Quine-Putnam Indispensability Argument (IA) by proposing an alternative reading that allows for minimal notions of object and existence (in Linnebo's 2013 sense). In such a framework, the conclusion of the revised IA is not that mathematical entities exist, as existence comes with reference and objecthood and is, therefore, quite easy to establish. Rather, this revised version of the argument claims that mathematical entities make a demand on the world, thus asserting that mathematical entities are heavyweight entities.

The chapter begins by introducing the debate over the metaphysical picture underlying the IA and goes on to present Linnebo's notions of minimal object, existence, and the distinction between thick versus thin objects. It then shows that it is possible to take some of Quine's theses on ontological commitment, existence, and reference to formulate a notion of existence that is minimal in nature. Furthermore, the chapter explores what is required to characterize light and heavyweight objects in a Quinean manner. Finally, the Quine-Putnam IA is reformulated in terms of the light versus heavy distinction, with the aim of uncovering the metaphysical assumptions underpinning the original argument and demonstrating that its conclusion is implausible by Quine's naturalistic standards.

Keywords Linnebo · Thin objects · Indispensability argument · Enhanced indispensability argument · Mathematical explanation · Explanatory virtues · Heavy-weight objects

C. Martínez-Vidal (✉)
Department of Philosophy and Anthropology, University of Santiago de Compostela, Santiago de Compostela, Spain
e-mail: mconcepcion.martinez@usc.es

X. de Donato-Rodríguez et al. (eds.), *Deflationist Conceptions of Abstract Objects*, Synthese Library 516, https://doi.org/10.1007/978-3-031-98717-5_8

8.1 Introduction

The so-called Quine-Putnam Indispensability Argument is intended to establish that, given confirmational holism, we are ontologically committed to mathematical entities since they are indispensable for science.[1] One primary objection to the argument rejects that using mathematics in our best scientific theory is indispensable. Thus, Hartry Field (2016) reconstructed Newtonian physics without using mathematics. Nevertheless, nothing similar has come up for the case of quantum physics, and the pervading idea in the literature is that it is not possible. Next, nominalists argued that, even if mathematics is *descriptively* indispensable, such indispensability does not entail a commitment to the existence of mathematical entities.[2] As a result, realists offered a revised version of the argument for platonism, which moves from indispensability to *explanatory* indispensability, and does not rely on confirmational holism; as we shall see below, this is why it is interesting for us. It aims to establish that, because mathematics itself explains certain physical phenomena, we are ontologically committed to mathematical entities just as we are to physical entities that play a *substantial explanatory role* in physics.[3] This version of the argument, known in the literature as the *Enhanced Indispensability Argument* (EIA, in what follows), claims that the specific explanatoriness of mathematics implies the commitment to mathematical entities. The debate turns into whether the explanatoriness at stake is or is not ontologically committing, with the nominalists claiming that the sort of explanatoriness involved is not ontologically committing and platonists contending that it is ontologically committing.[4] The debate about the metaphysical picture underlying the IA has already been addressed by Colyvan (2001, p. 62ff.; Idem 143ff.) when he discusses whether mathematical entities are causally inert or not.[5] More recently there seem to be two clear positions about how to understand the metaphysics underlying the EIA, that is, the relations between the mathematical and the physical: there are those who argue that the mathematical determines how the physical is, and those who defend that they are orthogonal (Liggins, 2016; Yablo, 2012). The success of this second version of the argument would also entail that mathematical entities make a demand on the world, meaning that they determine the physical.

[1] There are various reconstructions of the argument. The chapter uses Colyvan's reconstruction (2001; but see Colyvan [2024] for a detailed account) and does not address hermeneutic issues related to whether Quine or Putnam actually proposed the argument discussed in the paper.

[2] Bangu (2017); Colyvan (2019) or Marcus (2022) provide overviews of the debate.

[3] See the discussion below.

[4] Here, 'being ontologically committing' conveys making a demand on the world: that is, when the mathematical determines the physical.

[5] Colyvan presents different positions that try to formulate the relation of determinacy between the mathematical and the physical as causal in some sense. Several more present-day authors have addressed the issue. For instance, Yablo (2012) discusses whether the relation between the physical and the abstract is or is not orthogonal. At this point, considering mathematical entities as causally active seems to be out of the picture. See also Liggins (2016).

The purpose of this chapter is to advance an alternative theoretical framework that clarifies the metaphysical implications of the different versions of the Indispensability Argument (IA, from now on). Drawing on Linnebo's minimal notions of object and existence, as well as his distinction between thin and thick objects, this chapter formulates a revised version of the argument in which what is at stake is not the existence of mathematical objects, but whether they exert demand on the world.

Linnebo's notion of a thin object is an evolution of Frege's notion of a logical object. Brandom (1996) beautifully explains Frege's idea by spelling out the difference between the Kantian and the Fregean notions of an object. Kant sustained that for something to be an object it must be given in the a priori forms of sensibility: space and time. Hence, mathematics is synthetic a priori because mathematical entities are given to us as objects on these a priori forms of intuition. According to Kant, then, there are no analytic objects. In contrast, Frege's platonism puts forward that, in the case of arithmetic, formal logic (what Frege means by logic) gives us all its content, "including reference to objects." (Brandom, 1996, p. 294) Mathematical entities are logical, and therefore, analytic objects. One significant difference between analytic and thin objects is that, as we shall see below, there are thin objects that are not abstract.

Linnebo maintains that Frege's notion of analytic objects is still relevant, as objects that "make no demand on the world" (Linnebo, 2018, p. 102), which he refers to as lightweight objects. Presumably, the underlying idea is that although the full content of mathematical objects cannot be given in logical or analytical terms, it can be given without making demand on the world. Like Frege with is logical objects, Linnebo is restricting 'the world' to what is in space and time. Brandom (1996) contends that Frege's project is essentially Kant's project, but limited to abstract objects—specifically, how abstract objects are given to us. Linnebo's project is that of figuring out how thin objects are offered to us.

Since thin objects are inspired on analytic objects, conceiving of the possibility of lightweight objects (somehow *thin*) in a Quinean-style framework sounds, to say the least, problematic.[6] Firstly, because Quine (1951) rejected the analytic *versus* synthetic distinction; secondly, because, due to their lightness, the existence of mathematical objects would presumably be too easily achieved. This last point stands in contrast to Quine's position in the traditional meta-ontological debate with Carnap, where Quine argued that that ontological debates and ontological questions are substantial (Berto & Plebani, 2015). Moreover, the Quinean framework requires the existence of particular mathematical objects as a result of his IA (Colyvan, 2001, 2019). According to the IA, the existence of mathematical entities is justified for the same reasons as the existence of physical entities—namely, because our best

[6] Note that the chapter does not argue that Quine would agree to the proposal. Talk about Quinean lightweight objects appears in Linnebo (2018, p. 25) where he launches a challenge to defendants of a Quinean-oriented notion of object, one that relies on quantified variables instead of singular terms: "Philosophers who prefer the Quinean conception are invited to develop their own account of lightweight objects. If they succeed, their account would supplement mine, not threaten it."

scientific theory confronts experience as a whole and quantifies over them. Hence, mathematical objects make demand on the world just as physical entities do.

A closer look reveals that it is possible to unravel various Quinean theses in such a way that renouncing or modifying some of them results in a notion of object that allows for lightweight objects in a Quinean style. The label 'Quinean' is used here because the proposed notion of lightweight object remains faithful to the Quinean slogan "To be is to be the value of a quantified variable", in contrast to Frege's approach, which characterises reference in relation to singular terms. Furthermore, it aims to preserve the role Quine attached to identity criteria, as we shall see below.

The chapter starts with a discussion of Linnebo's characterization of thin objects and continues with a reconstruction of Quine's notion of object, identifying what needs to be excluded to get a Quinean-style notion of lightweight object. In doing so, the chapter puts forth the implications of this lightweight notion of an object for Quine's platonism. The contention is that the notion of lightweight Quinean objects helps illuminate the metaphysical picture underlying the Quine-Putnam IA, revealing that the argument is either trivial or implausible according to Quine's naturalistic standards.

8.2 Linnebo's Thin Versus Thick Objects

Linnebo shares Quine's view that, to determine the ontological commitments of our theories, we need to use a regimented language. In fact, he states that he rejects the idea that natural languages are the proper tools for semantics because their semantics tend to be "complex and quirky" (Linnebo, 2018, p. 23). Quine's reasons for employing first-order logic as the language of regimentation are well-known. While Quine goes for first-order logic, however, Linnebo allows both first- and higher-order formal languages.

Linnebo emphasizes the importance of relying on a course-grained notion of proposition because such conceptions allow for what he terms 'the flexible conception.' The flexible conception (Linnebo, 2018, pp. 31–33) fathoms that "the world is articulated one way or another pending on the set of concepts we use to describe it" and that "there is no privileged set of concepts in terms of which to 'carve up' reality."[7] This flexible view opposes what Linnebo calls the 'Lego block' conception (the Bucket view, according to Rayo [2020]), which is the "idea of a domain of objects—or 'Lego blocks'—that are available independently of any application of our conceptual apparatus for identification and predication" (Linnebo, 2018, p. 32).

Linnebo (2018, p. 32 and ff.) continues to clarify, that, under the flexible conception, reference is easy to achieve. Several approaches to achieve reference support this flexible view. Linnebo distinguishes between thin and ultra-thin conceptions of

[7] For instance, identifying a proposition with a set of worlds—instead of backing, say, the Russellian structured notion of proposition—is compatible with the flexible conception: it allows for various ways of articulating the world to characterize the same proposition.

reference. What Linnebo calls the 'ultra-thin' conception (Hale & Wright, 2001; Rayo, 2013) claims that it suffices to assign truth conditions to our sentences in a way that agrees with the laws of logic. Linnebo proposes a less radical view—the thin conception—which adds another condition: identity criteria for objects must be formulated in terms of abstraction principles.

Rayo (2020, p. 2) nicely synthesizes the ultra-thin view and Linnebo's conception of reference as follows:

> Let c be a singular term of L and let [[…]] be an interpretation of L. The following conditions are jointly sufficient for c to refer, as interpreted by [[…]], according to the ultra-thin conception:
>
> 1. [[…]] preserves logical entailments;
> 2. [[⌜∃x(x = c)⌝]] is a true (course-grained) proposition;

Condition (i) establishes that the interpretation must be compatible with the laws of logic. Condition (ii) establishes that a proposition existentially quantifying over an identity statement containing the singular term 'c' is true. However, being true does not entail succeeding in identifying the 'real' structure of the world, as it is in the sense of the Lego block conception. Instead, it demands that the statement gets the value 'true' in a set of possible worlds. And this can happen based on pure linguistic considerations for proponents of the ultra-thin conception (See Rayo, 2020).

Linnebo's thin conception is obtained by adding a further condition:

> 3. There is a partial equivalence relation, ~, such that:
>   (a) [[…]] interprets the language in such a way that c is governed by a predicative abstraction principle based on ~;
>   (b) A suitably positioned agent with an adequate grasp of ~ uses the language in accordance with [[…]].

Condition (iii) introduces the requisite (a) that the new term 'c' must have an identity criterion. Thus, being a referent of a singular term and having an identity criterion are sufficient to establish existence, where identity criteria (condition b) are to be specified in terms of abstraction principles.

Linnebo's conditions for reference, though more demanding than Rayo's, still offer a criterion that allows for the resulting notion of existence to remain 'minimal.' The reason is that there are abstraction principles that provide identity criteria for objects that make no demand on the world.

Linnebo distinguishes between objects that are thin *in an absolute sense*— those that make no demand on the world—and objects that are thin *in a relative sense*, those who are dependant on other objects where the latter do make demand on the world. In the case of the latter type Linnebo states that "given the existence of these other objects, the existence of the object in question makes no substantial further demand" (Linnebo, 2018, p. 4). He provides some examples: "… some may consider that an impure set is thin relative to the urelements (i.e., non-sets) that figure in its transitive closure. For example, the existence of the set of all the books in my office requires little or nothing beyond the existence of the books. Moreover, a mereological sum may be thin relative to its parts. For example, the existence of a mereological sum of all my books requires little or nothing beyond the existence of these books" (Idem). Analogously, consider the introduction of letter types, which are defined by abstraction from their tokens. Letter tokens do make demand on the world since they are concrete, while the existence of letter types demands little more than the existence of their tokens. Hence, letter types make no demand on the world beyond those made by their tokens, rendering them thin relative to their tokens. Furthermore, some objects that are thin relative to other objects are not thin in an absolute sense. For instance, you and I are thin relative to the material out of which we are composed. However, we are thick in an absolute sense. This would also be the case of mereological sums, which are thin relative to the objects in the sum, but if the latter are thick, so is the mereological sum. As a result, Linnebo concludes that some abstract objects are thinner than others.

Abstract objects are those objects that are thin and shallow. Thus, the difference between a body and a letter type—which are both thin relative to their parts—can be explained as follows: any truth about letter types can be established by looking at its tokens, which form the basis for the abstraction principle that defines the type. The abstraction principle uses a partial equivalence relation that is *intrinsic*. In contrast, truths about bodies cannot be reduced to truths about their parts. Given Linnebo's definition of bodies, the partial equivalence relation is *extrinsic*: how the term 'body' contributes to the semantic value of a sentence cannot be determined by looking at the properties of the parts. Instead, we have to look at the parcel of space-time it occupies, which is external to the parts that compose the body. As a result, though both bodies and letter types are thin, only letter types are shallow.[8]

Now, what is the relation between advocating for the flexible conception and allowing for thin objects? Is it a necessary condition, a sufficient one, or maybe neither? Rayo considers it consistent to claim that the world is articulated independently of our conceptual apparatus and that there are objects that make no demand on the world although, presumably, against the backdrop of the Bucket-view conception, holding both theses is against the 'spirit' of the distinction between thin and thick objects, as the former would no longer fulfill their intended function: allowing various conceptualizations involving different logical forms to express the same proposition.[9] Thus, even the Bucket View is consistent with a minimal notion of reference. It follows there is no doubt that the flexible view and a minimal notion of existence enable objects that make no demand on the world. In other words, subscribing to the flexible view suffices for thin objects. Moreover, any non-demanding

---

[8] See Linnebo (2018, p. 192 and ff.). Note that parts of a body are already in space-time. This is why the body does not make further demands and whence is thin relative to its parts. But, to determine whether a given part is indeed a part of a specific body, we need to look beyond the parts themselves to assess whether they are continuous with the other parts of the body.

[9] "There might be a way of making the Bucket View consistent with the letter of a thin or ultra-thin conception of reference. For one might postulate an "ultra-capacious" bucket: one that contains not only 0 but also referents for any singular term that might pass an ultra-thin sufficiency test for reference. But such a reconciliation would go against the spirit of a thin and ultra-thin conception of reference" (Rayo, 2020, p. 4)

view of reference seems to be compatible with the flexible view, licensing thin objects to fulfill the sort of role minimalists intend for them.

Both Linnebo and Rayo avow that subscribing to the flexible conception does not imply giving up on realism. Neither does the acceptance of thin objects. As mentioned above, Linnebo's thin objects amount to a reformulation of analytic objects in the hope that they avoid the difficulties of analytic ones. Linnebo explains these difficulties encountered as follows: agreeing that there are analytic objects implies agreeing that there are objects whose existence depends on terms that refer to them in statements that are true by virtue of the meanings of the expressions in those statements.[10] But, meanings are cognitively accessible to us, and thus appear dependent on us. Since meanings determine the existence of analytic objects, those objects are not "independent of us and our linguistic and cognitive activities" (Linnebo, 2018, p. 14). Linnebo means to explore whether abandoning analytic objects and going for *objects making no demand on the world* results in a notion of object that is "independent of us and our linguistic and cognitive activities." He hypothesizes that it does allow for a certain independence and, thus, that realism is still in place.[11]

Linnebo's characterization of objecthood relies on what he calls the Fregean triangle, which represents how Frege thought the notions of object, reference, and identity criteria were connected (See Linnebo, 2018, p. 21). The triangle connects:

- Reference to objecthood: explaining what it takes to be the referent of a singular term amounts to explaining what it takes for that referent to be an object. However, there are other ways of accounting for objecthood (it counts as a sufficient condition).
- Objecthood to identity criteria: the latter constitutes a necessary condition for an object to be considered as such.
- Identity criteria to reference: objects require identity criteria. Once new objects are established, it is possible to define further objects employing new abstraction principles.[12]

In what follows, our purpose will be to unpack Quine's characterization of objecthood to determine what needs to be left out to allow for a Quinean-style version of thin objects, or lightweight objects in the Quinean sense.

---

[10] Linnebo mentions that analytic here should be understood in its metaphysical sense (Bogoshossian, 1996)

[11] Linnebo's more demanding thin conception of reference seems to be aimed at saving objectivity. Rayo's ultra-thin conception has a different view on the matter (See Rayo, 2013, 2020) Nevertheless, it is far from clear that the Independence allowed for by the thin conception suffices for objectivity. See Ordóñez (2024) for an interesting account of the objectivity of mathematics.

[12] Abstraction has to be predicative, according to Linnebo (2018).

8.3 Quine's Notion of Object

As mentioned above, this chapter aims to identify the aspects in Quine's view that are incompatible with allowing for light objects—that is, those objects that make no demand on the world.

Since the intuition behind the idea of thin objects involves elaborating those objects whose existence follows from the ontological commitments of analytic truths, it seems essential to reconsider Quine's dismissal of the analytic-synthetic distinction. This implies, as we shall see below, that confirmational holism must go. After all, Quine's defence of confirmational holism is central to his rejection of the analytic-synthetic distinction (Quine, 1951). Nonetheless, before addressing this issue, we analyse other features that might need to be revised.

Linnebo's approach agrees with Quine's in several respects that are relevant to our endeavour. As mentioned above, both agree on the need to paraphrase natural language in a regimented language. Both also conceive objects as possible referents, offering a semantic answer to the question, "What is an object?" Lowe distinguishes the semantic and the metaphysical answers to the question. He places both the Fregean (Linnebo's approach is neo-Fregean) and the Quinean answer as semantic: "[a]ccording to the semantic answer, an object is to be conceived as a possible referent of a singular term (Gottlob Frege's view, as interpreted and defended by Michael Dummett) or as a possible value of a variable" (Lowe, 1995, p. 509). The difference lies in the fact that Quine coincides with Russell that some singular terms are eliminable. Hence, he advocates singular reference is achieved through quantifiers, and that ontological commitment comes with bounded variables.[13]

Two trendy slogans synthesize Quine's characterization of the notion of object: the one Lowe mentions above, "To be is to be the value of a quantified variable," and "No entity without identity." Both Quine and Linnebo concur that objects require identity criteria. In Linnebo's case, abstraction principles provide (two-level) identity criteria, while Quine advocates no particular methodology for identity criteria. Nevertheless, he rejected Frege's approach to identity criteria, which permitted for attributes and properties. Quine rejects this possibility because (Quine, 1995, pp. 40–41) he contended it is impossible to give adequate truth conditions for properties (understood as intentional entities) in terms of the physical objects that instantiate them. Some properties may be co-extensive, yet still not the same property (Quine, 1981, p. 100).

Counting on two-level identity criteria is crucial to the sort of metaphysical picture Linnebo has in mind: x being thin relative to y is intended to approximate the claim that the existence of x is grounded in the existence of y. Greimann (2001, 2018) interprets Quine as arguing for a physicalist and foundationalist ontology. Accordingly, Quine thinks we may introduce (presumably) abstract objects by defining sets of physical objects, sets of sets of physical objects, and so on.

---

[13] Linnebo's position differs from those of 'easy ontology'. He characterizes his position as 'meta-ontologial minimalism.' See Linnebo (2012, 2018, Chapter 1) for details.

For Quine, to be the value of a quantified variable and to have an identity criterion are necessary but not sufficient conditions for being an object. Sufficiency for objecthood comes with theory choice. We start by choosing our best theory about the world, paraphrasing the theory into a first-order (non-modal) logical system. Then, we furnish an interpretation for the language. This interpretation allows for quantifiers that range only over objects for which there are appropriate identity criteria.

Quine's proposal results in the inclusion of objects that are usually considered to be abstract, specifically those mathematical entities we quantify over in our best scientific theories, as well as classes. We mentioned above the Quinean argument leading to this result: the famous Quine-Putnam Indispensability Argument. According to this argument, mathematical entities exist just as physical entities do. Confirmational holism is a crucial premise in this version of the argument, which goes as follows:[14]

(P1) We ought to be committed to the existence of all and only the entities that are indispensable to our 'best' scientific theories.
(P2) Mathematical entities are indispensable to our 'best' scientific theories.
(C) We ought to be committed to the existence of mathematical objects.

As a result, Quine's proposal commits him to a notion of object that involves making demand on the world. Hence, to Quine, mathematical objects make demand on the world just as electrons do.

Quine's defence of the idea that we should rely on what we take to be our best theory of the world to determine what exists, combined with his adoption of confirmational holism, result in a view that is incompatible with the existence of thin objects (at least in an absolute sense). Given his confirmational holism, truths about mathematical objects are established together with truths about the rest of the entities featured in our best scientific theories.

In this context, there is another essential difference between the proposals considered here: Quine considers that the legitimacy of mathematical practice comes from scientific practice (which is understood as non-mathematic) (Marcus, 2022), while Linnebo thinks otherwise. Quine's view is linked to his naturalism and might prove relevant to the possibility of allowing for pure lightweight objects in a Quinean style. Nonetheless, the claim that we are committed to the existence of those entities quantified over by our best scientific theory, by itself (i.e., without confirmational holism) is compatible with allowing for light objects. Let us see how.

As mentioned earlier, Quine's conception of an object is metasemantic; it admits various ways of conceptualizing the world while defending one of them is best. This sort of view instantiates a label proposed by Rayo (2013, p. 11): moderate metaphysicalism. Moderate metaphysicalists claim that there is a correspondence between logical form and metaphysical structure only for assertions made in the ontology room. Rayo (2013, p. 11) points out that the moderate metaphysicalist and

[14] We use Colyvan's (2001) version of the argument.

the metaontological minimalist are on the same side because, although the former claims that a particular way of conceptualizing the world is better than the rest, they both accept that there are different ways of conceptualizing the world. In Quine's case, the 'real' entities are those quantified over by the regimented version of our *best* scientific theory.

8.4 A Lightweight Notion of Object Quinean Style

Allowing for lightweight objects in a Quinean-style seems bizarre because Quine (1951) rejects the analytic-synthetic distinction. According to Quine, our theories face experience globally. Therefore, any entity in our theory makes demand on the world. Given Linnebo's terminology, this means that all objects are thick in an absolute sense, as they all make substantial demand on the (concrete) world. To develop a lightweight notion of a Quinean-style object, we must abandon confirmational holism. As noted above, this comes as no surprise. Quine discards reductionism—the positivist view that every meaningful statement in science is reducible to statements about direct experience. For the reductionist, analytic statements are a limit case: those statements that get verified no matter what. Once the impossibility of reductionism is established, adopting confirmational holism leads to the rejection of the analytic-synthetic dogma. Consequently, there are no statements whose assertion has no impact on the world.

For Linnebo, 'thickness' (being heavyweight) comes from the kind of object in question. Thus, elementary particles are likely to be thick, in the sense that their existence makes some demand on reality: "Thickness derives from what it is to be an elementary particle, not from what it is to be an object" (Linnebo, 2018, p. 4). In contrast, given confirmational holism, being an object entails making demand on the world, so every object posited by our best scientific theory is thick. Making demand on the world comes with objecthood, not with the type of object involved. Given confirmational holism, there are no limit cases, that is, no statements that get verified no matter what. In other words, confirmational holism is incompatible with thin objects, not just with analytic objects.

Dummett (1973) noted that Quine, in a way, re-introduces (a version of) the analytic-synthetic distinction with his center-periphery opposition. Observation statements are in the periphery, and the evaluation of disciplines in the centre of the web is performed indirectly via their connection to empirical theories, from which contrastable statements follow. Hence, by Quinean standards, observable objects in observation statements do make demand on the world. In contrast, objects featured in statements mixing mathematics and physics make demand on the world, presumably only indirectly, through the connections of those statements to observational statements within our web of belief.

It is easy to see why this way of drawing the distinction will not do. The center-periphery distinction is gradual, whereas, in principle, the light versus heavyweight distinction is not. Statements in theoretical physics (which involve mathematical terms), and physical laws are not observable statements on their own. Instead, some auxiliary hypotheses need to be adopted for observable statements to follow from physical laws. In Quine's framework, given confirmational holism, it is our web of belief—our best scientific theory—that makes demand on the world. Mathematics and physics fall on the same side in this respect. This is precisely what Quine intended. Thus, we might characterize objects that are thin relative to other objects, for example, by taking objects featuring in inner parts of the web of belief to be thin relative to those featuring in directly observable statements. We would perhaps need to appeal to a physicalist criterion (palatable to the Quinean) to see whether there are any dependencies between those objects in the inner parts of the web. Deciding on dependencies between objects featuring in different statements in the web in physicalist terms is to be established case by case. The situation is the same for Linnebo: there is no general procedure that generates new objects in terms of previously existing ones by means of abstraction principles. Indeed, establishing abstraction principles for objects whose existence makes no further demand on the world beyond the existence of the objects in terms of which they are defined is not a trivial issue. The problem is that there is no way of characterizing objects that are thin in an absolute sense using the center versus periphery distinction. Again, given Quine's confirmational holism this does not seem to be possible. Hence, to formulate a notion of object that allows for objects that make no demand on the world, conformational holism must be abandoned.

8.5 Rejecting Holism: The Enhanced Indispensability Argument

The point of the Enhanced Indispensability Argument (EIA) is that there are uses of mathematics that are not purely instrumental. Take the example of the tiles (Steiner, 1995, p. 142). Suppose we need 120 tiles to tile a quadrangular floor. Why 120 and not 121? The floor is 10 units long and 12 units wide. The number of tiles needed is equivalent to m*n—where m is the length and n is the width; 10*12 is 120. Hence, we need 120 tiles. The mathematics involved is just instrumental, as there is a physical fact that explains why this is so. Numbers make no demand on the world beyond the ones the physical facts involved.

In 2009, however, Alan Baker argued that there are cases in which the use of mathematics is not instrumental. These are cases in which the existence of mathematical objects *does* make a difference—in an ontologically committing way—such as in the cicadas' case:[15]

1. Having a life-cycle period which minimizes with other (nearby/lower) periods is evolutionarily advantageous. [biological law]

[15] This is Baker's version of the argument with the exception of the numbering of the lines of argument that has been changed. Baker's version starts with number (4) (Baker, 2017, p. 2)

2. Prime periods minimize intersection (compared to non-prime periods). [number theoretic theorem]
3. Hence, organisms with periodic life cycles are likely to evolve periods that are prime. [mixed biological/mathematical law] (From 1, 2)
4. Cicadas in ecosystem-type E are limited by biological constraints to periods from 14 to 18 years. [ecological constraint].

Hence cicadas in ecosystem-type E are likely to evolve 17-year periods.

The underlying idea is that mathematics itself plays an explanatory role in cases such as the cicadas.[16] The platonist intends it to be the kind of explanatory role that is ontologically committing. Based on these sorts of cases, Baker advanced the EIA:

1. We ought rationally to believe in the existence of any entity that plays an indispensable *explanatory* role in our best scientific theories.
2. Mathematical objects play an indispensable explanatory role in science.
3. Hence, we ought rationally to believe in the existence of mathematical objects. (2009, p. 613)

Confirmational holism can be abandoned; the need to commit to the existence of mathematical entities arises because mathematical explanation plays an ineliminable role in scientific explanation.[17] Nonetheless, the idea that the legitimacy of mathematics depends on natural science is still here. It comes with the thesis that we are committed only to those entities that are explanatorily relevant to our best scientific theories. Thus, one issue is whether the primacy of scientific practice is compatible with allowing for the distinction between the light and heavyweight objects in a Quinean style. However, we have seen that by abandoning confirmational holism, and given Quine's views about reference and existence, the resulting notion of object allows for lightweight objects.

As mentioned above, the problem with the versions of the argument that rely on the explanatory role played by mathematical entities to conclude that they exist, is that their impact—their explanatoriness—can be understood differently. In the standard version, it is unclear whether this should be interpreted in an ontologically committing way (See Martínez-Vidal & Rivas-de-Castro, 2020) as the realist intends. Nominalists have objected that the explanatory indispensability involved is *not* ontologically committing (See Knowles & Saatsi, 2021).

Plebani (2018) provides an illuminating account of the metaphysical picture underlying the IA, claiming that the argument's success involves adopting a certain metaphysical view about the relations between the mathematical and the physical. As stated above, the success of the IA entails a commitment to the existence of mathematical objects because they make demand on the world. However, those

[16] The realist has put forward several cases such as the Königsberg Bridges, the honeycomb cells etc.

[17] As Baron (2016, ft. 5) claims "the case for realism about mathematics is supposed to 'piggyback' on the case for scientific realism. The case for scientific realism, however, is made through explanation. So, the case for mathematical realism should proceed via explanation as well."

demands do not seem to result from mathematical objects engaging in some causal relationship with the physical. Hence, the dependence relation must be metaphysical in nature: either a primitive metaphysical relation or a sort of grounding. According to Plebani, allowing this relation of metaphysical dependence to be primitive presumes encouraging a strong sort of platonism, referred to as 'heavy-duty platonism.' Take physical magnitudes like mass or temperature. According to the heavy-duty platonist (Field, 1989; Knowles, 2015), the relation 'x (a physical entity) having a certain mass (a number),' or the relation of 'a certain body x being in a certain temperature y,' are fundamental metaphysical dependence relations. In this view, the physical realm depends on the mathematical realm. If this is the metaphysical picture underlying the IA, then the IA does not allow for objects that make no demand on the world.

The alternative view, defended by those who reject the IA in any of its versions, supports the idea that the mathematical is grounded in the physical (Liggins, 2016). This relation exists not because the mathematical determines how the physical is, but because the mathematical allows us to say what goes on in the concrete world. Many authors subscribing to this second view are nominalists. Our contention here is that this second metaphysical picture is also compatible with platonism if mathematical objects are understood as thin objects or Quinean Lightweight objects. The idea is that, though the mathematical figuring in physical statements makes demand on the world, it does so indirectly, the mathematics is light relative to the physical.

As mentioned above, the reading of the argument proposed in this chapter—in terms of a Quinean notion of object that allows for lightweight objects—shifts the focus. The point is not whether the fact that our best scientific theories quantify over mathematical entities suffices for their existence. Instead, the problem is whether the explanatory role mathematics plays in our best scientific theories entails that mathematical entities are heavyweight objects, that is, objects that make demand on the world in an absolute sense or, on the contrary, are lightweight *relative to* concrete entities.

The opponents, in this case, are, on one side, the heavy-duty platonist and, on the other, the (moderate) platonist, who considers that the existence of mathematical entities is not a significant issue but rejects the idea that mathematical entities are thick or heavyweight objects in an absolute sense. The underlying metaphysical picture conveys that while objects involved in causal explanations do make demand on the world, those featuring in non-causal explanations, which include mathematical explanations, do so only indirectly (they make no demand on the world relative to physical entities).[18]

[18] Note that this version of the argument assumes that defenders of the heavyweight character of objects figured in mathematical explanations still allow for lightweight objects. As mentioned above, the combination of views sounds weird, but tenable. Weird because the very reason to permit objects that make no demand on the world was to have mathematical objects falling under this category. Tenable because we might intend objects other than mathematical to be those that make no demands on the world. This is not to say that some authors have gone this way but rather that some might.

A reading of the enhanced version of the argument in the proposed terms proceeds as follows:

1. We ought rationally to believe that any entity playing an indispensable *explanatory* role in our best scientific theories is a heavy entity (one that makes demand on the world).
2. Mathematical objects play an indispensable explanatory role in science.
3. Hence, we ought rationally to believe that mathematical objects make demand on the world.[19]

The problem with the original version of the EIA is that it provides no criterion for determining the commitments of an explanation. Similarly, the problem for this new version of the argument is that we need to determine whether the commitments of an explanation entail that the objects are lightweight relative to physical objects or heavyweight. In Marcus' (2022, p. 5) words: "EI[A] refers to the theoretical posits postulated by explanations but does not tell us how we are supposed to figure out what an explanation posits." Reading the argument in the terms we advance above does not solve this problem either. Nevertheless, reading the argument in terms of the lightweight relative to physical objects-heavyweight distinction offers a slightly different perspective that better illuminates what is at stake.

Participants in the debate have tried to settle the problem of 'figuring out what an explanation posits.' Yablo (2012, p. 1020) outlines levels of mathematical involvement in explanation. He starts by posing mathematical explanation in a "covering-law format: outcome E is explained as arising out of circumstances C by way of a generalization G that links them." (p. 1020). Then he distinguishes between (i) Descriptive mathematical explanation: where mathematics is needed to specify circumstances 'C'; (ii) Structural mathematical explanation: where mathematics allows us to run the explanation at the appropriate level of generality; and (iii) Substantive mathematical explanation: where mathematics provides the covering generalization 'G', as seen in the cicadas'case discussed above. This last type is the sort of mathematical explanation relevant to the debate around the EIA.

The debate has evolved (Baker, 2017; Colyvan, 2019; Knowles & Saatsi, 2021) to focus on the virtues of mathematical explanation, particularly in relation to substantial mathematical explanation. In the case of the new version of the argument our chapter advances, the issue is whether there are levels or virtues of mathematical explanation that entail the mathematical entities figuring in it are heavyweight.

Now, the notion of explanation allows us to bring in analogues to Linnebo's distinctions: (i) being thin versus being thick; (ii) being thin in a relative and absolute

[19] The traditional version of the argument can also be formulated in these terms. The problem is that it subscribes to confirmational holism, therefore not allowing for lightweight objects in an absolute sense:

(P1) We ought to be committed to the thickness of all and only the entities that are indispensable to our 'best' scientific theories.
(P2) Mathematical entities are indispensable to our 'best' scientific theories.
(C) We ought to be committed to the thickness of mathematical objects.

sense, and (iii) shallow versus non-shallow objects. Thus, (i) lightweight objects are those that make no demand on the world while heavyweight objects do; (ii) lightweight-objects-relative-to are those that feature in explanation while making no demand on the world *relative to* other objects; lightweight objects in-an-absolute-sense should be those that feature in explanation while making no demand on the world at all. There remains the problem of drawing the shallow/non shallow distinction since Quine's identity criteria do not rely on abstraction principles that involve partial equivalence relations on the objects from which the new objects are defined; as a result, neither the way to go, nor the reason to go for the distinction in terms of which Linnebo characterizes abstract objects (thin and shallow) are obvious.

In the following section, we analyze the virtues Baker (2017) links to substantial mathematical explanation to distinguish between shallow and non shallow objects and provide reasons for the need to allow for abstract objects. Then, we will argue, in line with the nominalist position, that although the explanatory role mathematical entities play on explanation is compatible with the idea of mathematical entities being heavyweight, it does not seem to imply it. Instead, it is sufficient to consider that they are light relative to the underlying facts, which are the ones that make demand on the world. Finally, the paper argues that the fact that mathematical entities feature on substantial mathematical explanation does not provide reasons to consider that those explanations are ontologically committing but to conclude that mathematical entities are shallow.

8.6 Explanatory Virtues and Heavyweight Objects

Consider the cicadas' case above. Baker takes it that "what is picked out by the mathematics are structural facts that go beyond any specific physical facts" (2017, p. 1). This follows from the virtues of mathematical explanation: scope generality and topic neutrality (unity).

Scope generality is a virtue of mathematical explanations which allows them "to collect together, generalize over, and make assertions about, various sets of physical facts in a given domain" (Baker, 2017, p. 7). Scope generality allows us to generalise by leaving out specific features of the physical facts. Thus, the mathematics in the explanation allows the *explanans* to operate independently "from the actual physical structure of the entities involved." (Jansson & Saatsi, 2019, p. 821).

According to Baker, it is topic generality (and explanatory depth) that provides evidence that mathematics relates to the structure of the world, as topic generality points to the fact that a mathematical theory applies across different domains.[20] Baker provides a case study in which the same piece of mathematics applies in two different domains. His point is that that sort of explanatoriness is scientifically relevant and *cannot be done in non-mathematical terms*. Mathematics abstracts on the

[20] See Baker (2017) for a detailed discussion.

properties of physical entities, unifying domains and unveiling the deep structural properties of reality (which he refers to this as 'explanatory depth').

Baker's point also amounts—though he does not explicitly say so—to arguing in favour of the idea that it makes sense to do mathematics independently of its applications precisely because science benefits from topic neutrality. Baker conveys that mathematics is about the structures in the world (given his realistic interpretation of the implications of topic neutrality). So, we could interpret Baker as claiming that we need mathematical objects because without them we cannot account for the structural features of reality. While physics uncovers electrons and so forth, mathematics reveals the structure of the world, and we need mathematics to account for the sort of structures used in science and their explanatory role.

These arguments are meant to establish that mathematical entities exist. In our case, the analogous arguments should entail that mathematical entities are heavyweight: they make (structural) demand on the world. This idea amounts to understanding the role of mathematics in applications in terms of structure. Bueno and Colyvan (2011) provide an inferential account of explanation that intends to be compatible with a realist and a non-realist reading of the role of mathematics in applications. Baker's realist way of interpreting these facts is, of course, not the only available one. All we need to distinguish between lightweight and heavyweight objects is a way of understanding the role of mathematics as providing structures that can be instantiated across different domains. This is compatible with an interpretation of the role of mathematics in explanation according to which mathematics is lightweight relative to the facts in the domains considered. A given structure that mathematicians have studied is either developed or found to be adequate to conceptualize and explain what is happening in the world. These structural mathematical facts are lightweight relative to the underlying (physical or biological) facts: they make no further demand on the world. The realist has not provided arguments that leave out this second way of accounting for the role of mathematical explanation and its virtues.

Given the flexible view—that is to say, the idea that "the world does not come equipped with a set of objects and a set of relations over them [...] rather," we talk about the structure of the world referring to the way in which our theories conceptualize it" (Bueno & Colyvan, 2011)—it seems quite natural to conclude that mathematics is thin (or lightweight) relative to other facts (such as the biological facts in the case of the cicadas).

It suffices for our purposes to recognise that there is a reading of the role of mathematics in explanation that is compatible with mathematical objects being lightweight,[21] with mathematical objects as not making a demand on the world in an absolute sense. This is so because, while conceiving of mathematical objects as lightweight relative to physical facts is palatable to the naturalist, conceiving of

[21] Knowles (2021) argues that the heavy-duty Platonist cannot provide a sensible account of mathematical explanation.

them as heavyweight is not (given the metaphysics underlying the lightweight and the heavyweight conceptions of mathematical objects).

Another issue mentioned above needs to be taken into consideration now. Natural science establishes that practising mathematics independently of its use in natural science is legitimate and advantageous. The possibility of mathematical explanations, such as the one in the cicadas' case, demands that the corresponding mathematical results are previously available. And, although mathematical results are sometimes motivated by the need to be used by a scientific theory, there are many cases in which mathematical results attained for independent reasons later come to be applicable in natural science.

Now, after all these reflections on the role of mathematical explanation, we can now advance a *more* satisfactory version of the distinction between lightweight and heavyweight objects in a Quinean style, and an analog of the distinction between shallow and non-shallow objects.

Heavyweight objects are those that figure in explanations that make direct demand on the world, i.e., causal explanations. The objects featuring in directly observable statements are heavyweight. Lightweight objects are those that figure in explanations but make no demand on the world. They can be lightweight relative to other objects in the explanation, or they can be lightweight in an absolute sense. To be thin in an absolute sense they need to be shallow; a sufficient condition for shallowness might be their allowing for scope generality and topic neutrality (unity). Without mathematical entities it is not possible to have explanations that have any of those properties.

8.7 Conclusion

The chapter reinterprets the IA in light of a the lightweight versus heavyweight objects in a Quinean style. The revised version of the argument results in a revised version of the classical debate about the IA which is no longer about whether mathematical entities exist but about whether mathematical entities make absolute demand on the world. Finally, a case is made against mathematical entities being lightweight in an absolute sense.

Acknowledgements Previous versions of this chapter were presented at the workshop 'Abstract Objects: Deflationist Approaches', and at the foundational conference for the Spanish Metaphysic Network (reM) that took place in Granada 23–25 November 2022. I would like to thank the audiences, and especially Marta Campdelacreu and an annonymous reviewer for their very helpful comments. Finally, financial support for my work was provided by the Grant number PID2020-115482GB-I00: Deflationist Views in Ontology and Metaontology (DEVIONME) of the Ministry of Science and Innovation.

References

Baker, A. (2009). Mathematical explanation in science. *The British Journal for the Philosophy of Science, 60*(3), 611–633.

Baker, A. (2017). Mathematics and explanatory generality. *Philosophia Mathematica, 25*(2), 194–209. https://doi.org/10.1093/philmat/nkw021

Bangu, S. (2017). *Indispensability of mathematics*. OBO. Retrieved 29 Sep. 2017, from http://www.oxfordbibliographies.com/view/document/obo-9780195396577/obo-9780195396577-0241.xml

Baron, S. (2016). Explaining mathematical explanation. *The Philosophical Quarterly, 66*(264), 458–480.

Berto, F., & Plebani, M. (2015). *Ontology and metaontology: A contemporary guide*. Bloomsbury Academic.

Brandom, R. (1996). The significance of complex numbers for Frege's philosophy of mathematics. *Proceedings of the Aristotelian Society, 96*(1), 293–315.

Boghossian, Paul Artin (1996). *Analyticity reconsidered. Noûs, 30*(3), 360–391.

Bueno, O., & Colyvan, M. (2011). An inferential conception of the application of mathematics. *Nous, 45*(2), 345–374. https://doi.org/10.1111/j.1468-0068.2010.00772.x

Colyvan, M. (2001). *The indispensability of mathematics*. Oxford University Press.

Colyvan, M. (2024). Indispensability arguments in the philosophy of mathematics. In E. N. Zalta (Ed.), *The Stanford encyclopedia of philosophy* (Spring 2019 ed.). https://plato.stanford.edu/archives/sum2024/entries/mathphil-indis/

Dummett, M. (1973). The significance of Quine's indeterminacy thesis. *In Truth and Other Enigmas, 1978*, 375–416.

Field, H. (1989). *Realism, mathematics & modality*. Blackwell.

Field, H. (2016). Science Without Numbers: *A Defence of Nominalism*, 2nd edition, Oxford: Oxford University Press (first edition 1980).

Greimann, D. (2001). Individuating abstract objects: The methodologies of Frege and Quine. *History of Philosophy & Logical Analysis, 4*(1), 121–142. https://doi.org/10.30965/26664275-00401008

Greimann, D. (2018). *Are Quine's criteria of adequacy for individuations unduly restrictive?* (Vol. 57, N°. 147, pp. 129–136). Revista de Filosofía de la Universidad de Costa Rica, ISSN 0034-8252.

Hale, B., & Wright, C. (2001). *Reason's proper study*. Clarendon Press.

Jansson, L., & Saatsi, J. (2019). Explanatory abstractions. *British Journal for the Philosophy of Science, 70*(3), 817–844.

Knowles, R. (2015). Heavy duty Platonism. *Erkenntnis, 80*(6), 1255–1270.

Knowles, R. (2021). Platonic relations and mathematical explanations. *The Philosophical Quarterly, 71*(3), 623–644.

Knowles, R., & Saatsi, J. (2021). Mathematics and explanatory generality: Nothing but cognitive salience. *Erkenn, 86*, 1119–1137. https://doi.org/10.1007/s10670-019-00146-x

Liggins, D. (2016). Grounding and the indispensability argument. *Synthese, 193*, 531–548.

Linnebo, Ø. (2012). Metaontological minimalism. *Philosophy Compass, 7*, 139–151.

Linnebo, Ø. (2018). *Thin objects. An abstractionist account*. UK.

Lowe, E. J. (1995). The metaphysics of abstract objects. *The Journal of Philosophy, 92*(10), 509–524. https://doi.org/10.2307/2940785

Marcus, R. (2022). *The indispensability argument in the philosophy of mathematics*. [Online] https://iep.utm.edu/indimath/. Accessed 12 Oct 2022.

Martínez-Vidal, C., & Rivas-de-Castro, N. (2020). Description, explanation and ontological commitment. In J. L. Falguera & C. Martínez-Vidal (Eds.), *Abstract objects: For and against* (pp. 35–57). Springer.

Ordóñez Miguéns, I. (2024). Objectivity with Objects. A modal explanation of the infinite (Ph.D., Universidade de Santiago de Compostela). Repositorio USC. http://hdl.handle.net/10347/33900.

Plebani, M. (2018). The indispensability argument and the nature of mathematical object. *Theoria: An International Journal for Theory, History and Foundations of Science, 33*(2), 249–263. https://doi.org/10.1387/THEORIA.17613
Quine, W. V. O. (1951). Two dogmas of empiricism. *Philosophical Review, 60*(1), 20–43.
Quine, W. V. O. (1981). *On the individuation of attributes*. Reprinted in Quine, *Theories and things* (pp. 100–112). Harvard University Press.
Quine, W. V. O. (1995). From Stimulus to Science, Cambridge, MA: Harvard University Press.
Rayo, A. (2013). *The construction of logical space*. UK.
Rayo, A. (2020) *The ultra-thin conception of objecthood*. Unpublished manuscript. http://web.mit.edu/arayo/www/ut.pdf
Steiner, M. (1995). The applicabilities of mathematics. *Philosophia Mathematica, 3*(2), 129–156.
Yablo, S. (2012). Explanation, extrapolation, and existence. *Mind, 121*(484), 1007–1029. http://www.jstor.org/stable/23407318

Chapter 9
Pragmatic Accounts of Justification, Epistemic Analyticity, and Other Routes to Easy Knowledge of Abstracta

Brett Topey

Abstract One common attitude toward abstract objects is a kind of platonism: a view on which those objects are mind-independent and causally inert. But there's an epistemological problem here: given any naturalistically respectable understanding of how our minds work, we can't be in any sort of contact with mind-independent, causally inert objects. So platonists, in order to avoid skepticism, tend to endorse epistemological theories on which knowledge is easy, in the sense that it requires no such contact—appeals to Boghossian's notion of epistemic analyticity are particularly common here, as are appeals to some broadly pragmatic account of the good standing of basic beliefs. I argue, though, that these appeals are hopeless: an argument adapted from the Benacerraf–Field challenge shows that, even if some such theory can deliver the verdict that our beliefs about abstract objects have some prima facie good standing, this good standing will inevitably be defeated.

Keywords Benacerraf–Field challenge · Entitlement · Epistemic analyticity · Epistemic luck · Platonism · Reliabilism · Safety · Sensitivity

9.1 Introduction

My concern here is with theories of abstract objects that are platonistic in the following minimal sense: they entail that there are such objects and that those objects are both mind-independent and causally inert. Theories meeting this description are widely endorsed, but they're also epistemologically puzzling. It's clear enough that our beliefs can't, on any naturalistically respectable story of how our cognitive faculties work, be a result of any sort of contact with a realm of mind-independent,

---

B. Topey (✉)
Department of Philosophy (GW), University of Salzburg, Salzburg, Austria
e-mail: brett.topey@plus.ac.at

X. de Donato-Rodríguez et al. (eds.), *Deflationist Conceptions of Abstract Objects*, Synthese Library 516, https://doi.org/10.1007/978-3-031-98717-5_9

causally inert objects. So how can we be in good epistemic shape with respect to such objects?

Platonists, if they're to have any hope of avoiding skepticism, must provide a satisfying answer to this question. This is why platonism is so often paired with an epistemological theory on which it's *easy* for us to be in good epistemic shape with respect to such objects, in the sense that we need not be in cognitive contact with those objects in order to know things about them (or to have justified beliefs about them, or to be entitled to our beliefs about them—different epistemological theories will attribute different sorts of good standing to the beliefs in question, but these differences won't matter much for our purposes).

It bears noting, though, that epistemological theories on which good standing doesn't require cognitive contact—we can call them *easy knowledge theories*—are themselves puzzling.[1] The reason there's an epistemological puzzle here in the first place is that, if we're not in any sort of cognitive contact with (say) a realm of abstract mathematical objects, there doesn't seem to be any mechanism by which we can determine what those objects are like, or even whether they exist. So when we ask how we can be in good epistemic shape with respect to those objects, the underlying worry is that it's utterly mysterious how it is that we're supposed to have arrived at true beliefs about them. And easy knowledge theories can be of no help whatsoever in resolving that mystery—to endorse such a theory is to attempt to finesse our epistemological notions so that we can deliver nonskeptical verdicts *without* resolving the mystery, so that we may simply marvel at how unaccountably lucky it is that, despite our lack of cognitive contact with the objects in question, we've managed to arrive at the truth about them. Or, to put the point another way: to respond to our worry by invoking an easy knowledge theory is not to answer the worry but to dismiss it as epistemologically irrelevant. Insofar as we're inclined to take the worry at all seriously, then, we might suspect that the alleged epistemological benefits of easy knowledge theories are, in the end, just the familiar old advantages of theft over honest toil.

Here I argue that this suspicion is correct. In particular, I show that, if our goal is to make a genuine case for the claim that our beliefs about abstract objects are in good epistemic standing, appeals to easy knowledge theories are hopeless. For even if we can make plausible, by appeal to some such theory, that those beliefs do indeed have some prima facie good standing, it turns out, given the mystery described above, that this good standing will inevitably be defeated. And the argument I give for this claim—an argument adapted from the Benacerraf–Field challenge—doesn't depend on any particular epistemological theory. Indeed, its only substantive epistemological premise is a claim that *all* reasonable epistemological theories, including easy knowledge theories, must accommodate: that one can't be in good

[1] Note that the sense in which knowledge (or justification, or entitlement) is "easy", on the theories I have in mind, is not the same as the sense that's in play in discussions of what Cohen (2002) has called the "problem of easy knowledge". What's relevant here is whether a theory implies that good epistemic standing doesn't require cognitive contact, not whether a theory implies that an agent can (say) bootstrap her way to knowledge that her color vision is reliable.

standing in believing something while at the same time accepting that, insofar as that belief is true, that it's true is merely a lucky coincidence.

What I show, in short, is that, no matter what *epistemological* theory we endorse, it's just inevitable that platonism will lead to skepticism. And if that's right, finessing our epistemological theory isn't a viable strategy here; if we want to deliver the verdict that we're in good epistemic shape with respect to abstract objects, our only option is to insist on a nonplatonistic *metaphysical* picture—i.e., a picture on which abstract objects aren't both mind-independent and causally inert.

9.2 Easy Knowledge Theories

We can begin with the question of how, exactly, easy knowledge theories are supposed to work. So: suppose we endorse a platonistic theory of (say) arithmetic. How might we nevertheless deliver the verdict that our arithmetical beliefs are in good epistemic standing?

One standard approach is to endorse some form of *reliabilism*, some epistemological view on which the main condition a belief must satisfy in order to be in good standing is for it to have been reliably formed—i.e., formed in such a way that, at some particular class of (metaphysically) possible worlds (where this may be the class containing just the actual world, the class containing all worlds sufficiently similar to the actual world, or the class containing all worlds that are "normal" in Goldman's (1986) sense), a sufficient proportion of beliefs formed in that way are in fact true.[2] It's clear enough that our arithmetical beliefs meet this reliability condition: we've formed those beliefs by standard mathematical methods such as calculation and proof, and these ways of forming beliefs are *extremely* reliable despite the fact that they don't work by putting us into cognitive contact with natural numbers.[3] (And this is so, notice, irrespective of what class of worlds we take to be relevant—the arithmetical facts don't vary across worlds.) So reliabilism does seem

[2] The lack of specificity in this reliability condition is intentional—different ways of filling in the details will be associated with different reliabilist theories. The condition is consistent, for instance, with (standard and modalized versions of) both process reliabilism (see, e.g., Goldman, 1979, 1986) and virtue reliabilism (see, e.g., Sosa, 1991; Greco, 1999). It's also consistent with safety theories (see, e.g., Sosa, 1999; Williamson, 2000; Pritchard, 2005)—what's needed in this case is a precisification on which the relevant class of worlds is the class containing those sufficiently similar to the actual world and on which a way of forming beliefs is narrow enough that it produces only beliefs with the same content as the belief in question (or perhaps only beliefs with sufficiently similar contents—see, e.g., Williamson, 2009). The differences between these theories won't be relevant for our purposes here.

[3] Of course, in insisting that standard mathematical methods are reliable, we're taking for granted that our platonistic theory of arithmetic is *correct*—i.e., that mind-independent natural numbers exist and have the basic features we take them to have. But that's not a problem, by reliabilist lights. For discussion, see, e.g., Kornblith, 2004.

to make available a story on which our arithmetical beliefs have some prima facie good standing.

Another approach is to insist that true arithmetical sentences are analytic in the epistemic sense introduced by Boghossian (1996)—i.e., are such that grasp of their contents (where this consists in being disposed to infer in accordance with the right axioms and rules) suffices for the prima facie good standing of beliefs in those contents. The thought here is that sentences can be epistemically analytic, can have the epistemological features traditionally associated with analyticity, without being analytic in the traditional metaphysical sense of owing their truth wholly to their (conventional) linguistic meanings.[4] If true arithmetical sentences are indeed analytic in this merely epistemic sense, then the prima facie good standing of our arithmetical beliefs requires only grasp of those sentences' conceptual contents and so doesn't require cognitive contact with the natural numbers: grasp of those contents is, once again, just a matter of being disposed to infer in accordance with the right axioms and rules.

Of course, an explanation is still needed of *how* a sentence can have the epistemological features associated with analyticity without owing its truth wholly to its meaning—in the absence of such an explanation, to appeal to the epistemic analyticity of true arithmetical sentences is not to give a full account of the good standing of our arithmetical beliefs but is merely to issue a promissory note. And though Boghossian has attempted various sorts of explanation here (see his 1996, 1997, 2000), the only one that turns out to have any hope of succeeding (see his 2003a, b) is one that goes roughly as follows: certain of our concepts are such that, for one purpose or another, we *need* them, and so there are sentences—namely, those that we must accept in order to grasp the concepts in question—that we aren't blameworthy for accepting, even in the absence of any positive reason for taking those sentences to be true. As Schechter and Enoch (2006) have pointed out, though, the notion of concept possession appears to be an unnecessary detour in this explanation; what's essential is just the thought that there are purposes for which we *must* accept certain sentences, in which case we aren't blameworthy for accepting them. So it's not clear that the approach here is genuinely analyticity-based after all.

Instead, what we appear to have is an instance of a third sort of approach, one on which the good standing of our arithmetical beliefs is ultimately to be accounted for in *pragmatic* terms. The thought is just that there are certain cognitive projects, certain forms of inquiry, that we're prima facie rational in engaging in, and the fact that we can't be faulted for engaging in these projects is supposed to be what explains why, when a claim is a presupposition of one of these projects—i.e., is such that we must accept it in order to engage in the project successfully—we can't be

[4] The claim that arithmetical sentences are *metaphysically* analytic isn't consistent with platonism: if arithmetical sentences owed their truth wholly to their conventional linguistic meanings, the objects those sentences are about couldn't be mind-independent. This, perhaps, explains why philosophers appealing to analyticity tend to insist that they're committed only to the epistemic variety (see, e.g., Chalmers, 2012; Thomasson, 2015; Nyseth, 2021).

faulted for taking that claim for granted.[5] And notice: whether a claim is a presupposition of a given cognitive project has nothing whatsoever to do with whether we're in cognitive contact with the part of the world that claim is about. So, insofar as a case can be made that foundational arithmetical claims are presuppositions of some cognitive project of the right sort, this pragmatic approach makes available another kind of story on which our arithmetical beliefs have some prima facie good standing despite our lack of contact with the natural numbers.[6]

There's certainly much more to be said about the epistemological theories underlying each of these approaches. But most of the fine details aren't going to matter for our purposes here—we aren't going to be criticizing the theories themselves. Indeed, we can grant for the sake of argument that one of these theories is correct, so that our arithmetical beliefs *do* have some sort of prima facie good standing. My claim, once again, is that even if this is so, it turns out—given platonism—that those beliefs will inevitably be subject to defeat. So what's crucial to notice here is just that, according to (plausible versions of) all of the approaches I've discussed, the sort of good standing enjoyed by our arithmetical beliefs is *defeasible*.

To see why, consider that, for any of the above proposed conditions for prima facie good standing, there are cases in which a belief meets that condition while at the same time the agent has very good reason to think that it's nevertheless not likely to be true; in such cases, whatever good standing the belief has is surely going to be defeated. Suppose, for instance, that I've formed some belief about my environment by relying on my vision. Presumably, this way of forming beliefs is reliable, in which case, according to reliabilist views, my belief has some prima facie good standing. But if I then gain strong evidence that I've been hallucinating, it's clear enough that I shouldn't hold on to my belief: it's no longer in good standing.[7] And there are going to be analogous cases of defeat for beliefs that plausibly are epistemically analytic and for beliefs that plausibly are presuppositions of some cognitive project that we're prima facie rational in engaging in.[8]

---

[5] Versions of this sort of pragmatic account have been developed by Enoch and Schechter (2008) and by Wright (2004a, b, 2014, 2016). Related (though in some respects more permissive) accounts have been developed by Putnam (see, e.g., his 1976/1983), Field (see, e.g., his 2000), Harman (see, e.g., his 2003), and Ebbs (see, e.g., his 2022).

[6] It's worth emphasizing that, on this story, what is accounted for by the relevant pragmatic considerations is the good *epistemic* standing of our beliefs. One might instead set epistemology aside altogether and insist only on a story on which we're justified in some *pragmatic* sense in holding on to the beliefs in question, but this isn't the sort of story that's under discussion here. (It also isn't the sort of story that Boghossian, Enoch and Schechter, and Wright are trying to tell.) The claims I'm targeting in this paper are claims about our beliefs' *epistemic* status.

[7] Quite a few different strategies have been tried for incorporating the phenomenon of defeat into a reliabilist epistemology (see, e.g., Goldman, 1979; Alston, 1988; Bergmann, 2006; Grundmann, 2009; Constantin, 2020; Beddor, 2021), but it's widely accepted that a reliabilist theory, in order to be adequate, must have *some* way of allowing for the possibility of defeat.

[8] Note also that those who defend the epistemological views that are relevant here are themselves committed to the possibility of defeat. Boghossian (2003b, p. 27), for his part, says that "no one should expect more than a defeasible entitlement, even from concept possession". Enoch and Schechter (2008) are explicit about their commitment to defeasibility as well, as is Wright (2016).

The point: platonists can do no better, by appeal to an easy knowledge theory, than to deliver the verdict that their beliefs about abstract objects are *initially* and *defeasibly* in good standing. Now to explain why it's inevitable that this good standing will in fact be defeated.

9.3 Epistemic Luck and the Benacerraf–Field Challenge

Suppose I've formed some empirical belief—say, a belief that Shirley won yesterday's lottery—on the basis of a report in what I, with very good reason, take to be today's newspaper. Presumably, this belief is (defeasibly) in good standing, on any reasonable account of what it takes for a belief to be in good standing. But suppose I then gain strong evidence that the newspaper in question isn't today's newspaper at all but is instead a novelty newspaper that was printed several years ago. It's clear enough that, on gaining this new evidence, I should give up my belief that Shirley won the lottery: that belief's good standing has been defeated. What I want to discuss is how, exactly, this defeat works.

Here, I take it, is one good way of describing what's going on in this case. There are, from my perspective, many ways the world might be with respect to who won the lottery. After I've read the newspaper report, though, I should be confident that one of those possibilities in particular—the one in which Shirley won—is the actual one. This is because I ought to take the newspaper report to be a good guide to the lottery results: given what I know about how newspapers work, I should take the newspaper's reporting to be (causally) related to what the world is really like in such a way that, if I form beliefs on the basis of that report, those beliefs will coincide with the facts. But then, when I gain the new evidence against the newspaper's authenticity, I should become confident that there's no such relationship after all.

From my own perspective, then, my situation is as follows: of the many ways the world might be with respect to who won the lottery, I've formed the belief that one possibility in particular is the actual one, but given the way I've done so, there just isn't any connection between my having this belief and the lottery results being what they are. So what I should think is that, if the belief coincides with the fact of the matter about who won the lottery—i.e., if the belief is true—this is just a matter of luck. This, I take it, explains why the good standing of the belief has been defeated: it would be perverse of me to continue to insist that Shirley won the lottery even after I've become convinced that, insofar as my belief that she did is true, its truth is merely accidental, nothing more than a lucky coincidence. And notice that, in order for my belief to be defeated in this way, it doesn't need to be *true* that the belief is merely accidentally true if true at all; what's doing the work is just that I'm required to come to *believe* that it's accidentally true if true at all—even if my evidence for this is in fact misleading. The problem, in other words, is that there's a kind of incoherence involved in believing *both* that Shirley won the lottery *and* that insofar as that belief is true, its truth is just a lucky coincidence.

To make this vivid, consider that, if I hold on to both of these beliefs, I thereby commit myself to concluding, without corroboration, that I have indeed gotten unimaginably lucky—i.e., that I've managed to hit on the truth about who won the lottery despite the fact that my belief about this matter is, by my own lights, no better than a wild guess. I hope the reader will agree that accepting this conclusion would be an absurd response to my situation.[9]

This suggests a general epistemological principle, something like a coherence constraint, along the following lines:

Lucky Coincidence Thesis An agent can't be in good epistemic standing in holding on to a belief while at the same time accepting that, insofar as that belief is true, its truth is merely a lucky coincidence.

And this principle is indeed plausible. Cases of defeat that have the general structure of our lottery case—i.e., cases in which the good standing of an agent's belief that *p* plausibly is defeated by evidence indicating that, given the way the belief was formed, there's no connection between the agent's having that belief and the part of the world relevant to whether it's true that *p* being the way it is, so that if the belief is true, its truth is a lucky coincidence—can be multiplied without end. And part of the explanation for this sort of defeat does seem to be that there would be something absurd about insisting both that *p* and that one's belief that *p*, if true, is merely accidentally true. Furthermore, it bears emphasizing that there isn't anything particularly subtle going on here, nor are there any sectarian commitments in play; one doesn't need to be in the grip of any particular epistemological theory, or even to have thought very much about epistemology, in order to recognize that in our lottery case, it would indeed be absurd for me to conclude, without corroboration, that, by relying on a way of forming beliefs that's no better from my perspective than wildly guessing, I've managed to arrive at the truth about who won the lottery.[10]

---

[9] Note also that if I do have corroboration, I *still* can't reasonably conclude that the truth of my belief is unimaginably lucky: if I have corroboration that Shirley won the lottery, I shouldn't accept that insofar as my belief that she did is true, its truth is just a lucky coincidence. (This isn't to say that I can't reasonably conclude, on the basis of some newly acquired corroboration, that the truth of my *past* belief was unimaginably lucky. But given that I now have corroboration, I shouldn't regard the truth of my *present* belief as lucky.)

[10] Notice: the thesis that self-ascriptions of epistemic luck are never rational—a thesis Skipper calls "No Luck" and suggests is so intuitively plausible that denying it without explaining away this plausibility would involve "an egregious amount of bullet-biting" (2023, p. 390)—is closely related to the Lucky Coincidence Thesis. But this latter thesis is weaker than No Luck in that it disallows only self-ascriptions of *veritic* luck (i.e., judgments to the effect that, given the way one's belief was formed, it's merely a lucky coincidence that the belief is true); it says nothing about self-ascriptions of other sorts of epistemic luck, such as judgments to the effect that it's a matter of luck that one is disposed to form beliefs in the way one does rather than in some less truth-conducive way. As a result, the sorts of putative counterexamples to No Luck discussed by Skipper—cases in which an agent learns that it's a matter of luck that she has the particular fundamental standards of reasoning she does but in which she's nevertheless purportedly rational in holding on to beliefs formed in accordance with those standards (see Elga, 2008; White, 2010; Schoenfield,

What I want to suggest, then, is that—even if we grant for the sake of argument that some easy knowledge theory is correct—we're in a position, now that the Lucky Coincidence Thesis is on the table, to see that platonists' beliefs about abstract objects just *cannot* be in good standing. Here's a statement of the argument—call it the *simple argument*:

1. Insofar as we accept a platonistic theory of abstract objects of a particular sort, we're committed to the claim that our beliefs about those objects are accidentally true if true at all.
2. We can't be in good standing in holding on to a belief while at the same time accepting a theory that commits us to the claim that this belief is accidentally true if true at all. (From the Lucky Coincidence Thesis.)

(C) We can't be in good standing in being nonskeptical platonists about abstract objects of any sort. (From (1) and (2).)

And from (C) it of course follows that, insofar as we can be in good standing in holding on to a platonistic theory of (say) arithmetic at all, we can do so *only* by giving up our first-order arithmetical beliefs. So even if platonists can, by appeal to some easy knowledge theory, deliver the verdict that their arithmetical beliefs have some prima facie good standing, (C) entails that this good standing will inevitably be defeated.

Now: the simple argument is valid, and (2) is an obvious consequence of the Lucky Coincidence Thesis. So, insofar as the Lucky Coincidence Thesis is correct, as I've suggested it is, all that's left for us to do, in order to establish our conclusion, is to show that (1) is true as well—to show, in particular, that platonists, as a result of their commitment to a view according to which they're not in any sort of cognitive contact with abstract objects, are committed also to taking any beliefs they've formed about those objects to be accidentally true if true at all.

Before we turn to that task, though, some clarificatory remarks about how the simple argument works are in order. Note, first of all, that it's *not* a radical skeptical argument. It doesn't present a challenge that platonists must answer without relying on their own beliefs—the question for platonists is instead whether, *according to their own theory of the world*, it turns out to be no more than a lucky coincidence that their beliefs about abstract objects are true. Nor does the argument depend on any controversial premises about the epistemic status of any particular belief or about what sorts of grounds a belief must have in order to be in good standing—the only substantive epistemological principle that's in play here is the Lucky Coincidence Thesis, which, as I've already suggested, is a fairly minimal coherence constraint. So we aren't going to be able to use an argument with the structure of the simple argument to call into question our beliefs in areas the epistemology of which isn't mysterious in the way the epistemology of abstracta is.

2014)—are, regardless of whether they succeed in showing that No Luck is false, no threat to the Lucky Coincidence Thesis.

In our lottery case, for example, the theory of the world to which I initially subscribe—a theory that includes beliefs about what sorts of causal relationships there are between the newspaper's reporting and what the world is really like—is *not* a theory according to which my straightforward empirical belief that Shirley won the lottery is merely accidentally true, and so the Lucky Coincidence Thesis doesn't suggest that this belief fails to be in good standing. It's only when I gain the evidence against the newspaper's authenticity that the belief is called into question. There just doesn't seem to be any real danger that the argumentative strategy here will lead us into anything like global skepticism.

In this respect, the simple argument bears a striking resemblance to the Benacerraf-inspired argument against mathematical platonism that appears in Field's *Realism, Mathematics and Modality* (1989), an argument known variously as the Benacerraf–Field challenge, the Benacerraf problem, the access problem, or the reliability challenge. And the resemblance is no accident—what I'm offering here is one way of construing (and generalizing) that challenge.[11] The main epistemological premise of Field's argument is that, insofar as it "*appears in principle impossible to explain*" how our beliefs about mathematical objects "can so well reflect the facts about them", this "tends to *undermine* the belief in mathematical entities, *despite* whatever reason we might have for believing in them" (1989, p. 26). And this claim, I take it, is plausible. But I also suspect that what *makes* it plausible is the thought that, if it is indeed impossible to explain the coincidence between our mathematical beliefs and the facts about mathematical objects—i.e., if there's no explanation available here, even in principle—then this coincidence is nothing more than an accident, just a matter of luck. If I'm right about this, Field's argument is best understood as working by (implicit) appeal to something like the Lucky Coincidence Thesis. And in that case it's more or less equivalent to the simple argument.[12]

It bears emphasizing, though, that in the literature on the Benacerraf–Field challenge, this construal isn't often endorsed (or even discussed).[13] In particular, this construal is *not* equivalent (though it is superficially similar) to a sort of interpretation that's quite common: one on which the primary epistemological principle that's in play is some general principle of coincidence avoidance. The basic thought behind this latter interpretation is that, all else being equal, a theory of the world according to which fewer inexplicable coincidences have occurred is better than one

[11] This generalization isn't novel: it's well known already that, though Field's discussion is about mathematical platonism in particular, the challenge applies in just the same way to platonism about abstract objects of other sorts (for discussion, see, e.g., Liggins, 2010). The challenge plausibly generalizes even further, to certain sorts of realism about logic (see, e.g., Schechter, 2010, 2013, 2018) and morality (see, e.g., Enoch, 2010), but our focus here is on theories of abstract objects in particular.

[12] This isn't an exegetical claim—I'm suggesting that an argument with the structure of the one given by Field *ought* to be understood as deriving its force from the plausibility of the Lucky Coincidence Thesis, not necessarily that Field himself understands his argument this way.

[13] Though I've endorsed it elsewhere (see my 2021, 2022).

according to which more of them have occurred, in which case nonskeptical platonism has theoretical costs that its alternatives don't have.[14] And maybe this thought is correct. But it isn't what's in play in the simple argument. What's relevant there is not how many inexplicable coincidences an agent's theory commits her to positing but only whether it commits her to self-ascribing a particular sort of epistemic luck. This is significant because it means that many of the responses to the Benacerraf–Field challenge that have appeared in the literature—in particular, those responses that target the coincidence avoidance interpretation, aiming to show either that platonism's benefits outweigh its costs or simply that these costs are more minimal than we might have thought (see, e.g., Burgess & Rosen, 1997; Enoch, 2010; Setiya 2012)—are just irrelevant to the simple argument.[15]

With these clarifications out of the way, we can now move on to the question of whether the simple argument's first premise is true—i.e., the question of whether platonists are indeed committed to taking their beliefs about abstract objects to be accidentally true if true at all.

9.4 Lucky Coincidence, Responsiveness, and Sensitivity

We can begin here by returning once again to our lottery case. In virtue of what am I committed to the claim that insofar as my belief that Shirley won the lottery is true, its truth is merely a lucky coincidence?

As suggested above, the explanation is roughly as follows. My perspective is one on which there are many ways the world might be with respect to the lottery results, one of which—the one in which Shirley won—I believe to be actual. But after I've gained the new evidence against the newspaper's authenticity, my perspective is *also* one on which, since there's no connection between my having that belief and the lottery results being what they in fact are, the belief isn't the product of any

---

[14] Those who construe the Benacerraf–Field challenge in roughly this way include, among others, Enoch (2010), Liggins (2010), Schechter (2010, 2013), Setiya (2012), Baras (2017), Berry (2020), and Bhogal (2023).

[15] Another standard response to challenges like Field's is that they go by way of principles that, while suitable to the epistemology of the empirical, aren't appropriately applied in the context of the epistemology of abstracta (see, e.g., Hale, 1994; Horwich, 2006; Linnebo, 2006). If the principle in question is some general principle of coincidence avoidance, this claim may be correct: perhaps the fact that inexplicable coincidences are improbable is something we've discovered empirically, not something we can assume to hold in abstract domains (though Schechter (2018) offers some reasons to think otherwise). But the claim is not at all plausible if the relevant principle is instead the Lucky Coincidence Thesis. It's clear enough that, irrespective of whether the content of a given belief is that Shirley won the lottery or that the continuum hypothesis is true, there's a kind of incoherence involved in holding on to that belief while at the same time accepting that insofar as the belief is true, its truth is just a lucky coincidence. (Of course, there remains the question of what the conditions are in which we should accept, in the case of a belief about something nonempirical, that insofar as the belief is true, its truth is just a lucky coincidence. I return to this question in the next section.)

genuine ability on my part to discriminate among the various possible results: there's no sense in which the belief is *responsive* to what the results in fact are. These two features of my perspective, it seems, suffice to commit me to taking the belief to be accidentally true if true at all.

The general thought behind this explanation is as follows: if an agent is such that

(i) her perspective is one on which there are multiple ways a certain part of the world might be, one of which she believes to be actual, and
(ii) her perspective is also one on which the procedure by which she's chosen that possibility is in no sense responsive to the way that part of the world really is,

then the agent should take it to be no more than a lucky coincidence if the possibility she's chosen happens to be the actual one. So, if the explanation is indeed a good one, what it suggests is that this general thought (a thought that, I take it, is independently plausible) is correct—i.e., that (i) and (ii) amount to a sufficient condition for an agent to be committed to taking her belief to be accidentally true if true at all.[16]

This sufficient condition, notice, is incompatible with at least one influential account of epistemic luck: the account developed and defended by Pritchard (see, e.g., his 2005, 2007), on which the truth of a belief is nonaccidental just in case the belief is *safe*—i.e., just in case, roughly, there are very few nearby possible worlds in which the agent falsely maintains that belief. To see why, suppose I both believe that p and take the nearest possible worlds where it's not the case that p to be far from the actual world. Then my belief that p is, by my lights, safe, since, if there are no nearby worlds where it's not the case that p, there certainly are no nearby worlds where I falsely believe that p.[17] But clauses (i) and (ii) may nevertheless be satisfied:

[16] There's a potential worry here. Suppose I (having seen no newspaper report) form the belief that Shirley *didn't* win the lottery, just on the basis of my knowledge that her ticket is one of an extremely large number only one of which is the winning ticket. (This is a version of a case discussed in Donaldson, 2014.) Presumably, I'm not committed to taking this belief to be accidentally true if true at all, but the belief isn't in any straightforward sense responsive to what the lottery results are. So we might think this case is a counterexample to our sufficient condition. There isn't space to do full justice to this worry, but note that the case here does *not* seem to be one in which, in forming the belief that Shirley didn't win the lottery, I've chosen one possibility among many: though my perspective is indeed one on which there are mulitple ways the relevant part of the world might be, the *vast majority* of those possibilities are ones in which Shirley didn't win. (Indeed, this is what led me to form the belief that she didn't win in the first place.) So clause (i) isn't satisfied, which means we can, consistently with our sufficient condition, insist that I'm not committed to taking my belief to be accidentally true if true at all. (I suspect, incidentally, that beliefs formed by enumerative induction, insofar as they fail to be responsive to their corresponding facts, are going to be amenable to a broadly similar treatment—see fn. 28 below.)

[17] Note that Pritchard's account is an account of the conditions under which a belief is *in fact* veritically lucky, while we're interested instead in whether my *perspective* is one on which my belief that p is veritically lucky. In order to apply Pritchard's theory in the present context, then, what we must determine is whether my belief is safe *by my lights*—i.e., whether, according to my own theory of the world, Pritchard's safety condition is met. Note also that what we're interested in is *not* whether my belief suffers from what Pritchard calls "reflective" luck—the conditions a belief must meet in order to avoid this sort of luck (conditions that, incidentally, are far more demanding than is justified by the motivations Pritchard offers for introducing the notion of reflective luck—

whether I take the nearest possible worlds where it's not the case that p to be far from the actual world tells us nothing at all about whether I take my belief to be a result of choosing one possibility among many in a way that fails to be responsive to what the relevant part of the world is really like.

Why, then, should we prefer our sufficient condition to a safety condition in the spirit of Pritchard's? The answer is that cases in which safety and responsiveness come apart in the way just described are precisely cases in which the verdicts delivered by safety-based accounts seem obviously incorrect. Consider, for example, a variant of our lottery case in which, even before reading the newspaper report indicating that Shirley was the winner, I was aware that there was an extremely well-coordinated conspiracy to rig yesterday's lottery, so that, whatever the lottery results in fact are, there aren't any nearby worlds in which they're different. (We can fill in the details so as to make the results as modally robust as we like.) It's quite clear that in this case, gaining the new evidence against the newspaper's authenticity should still lead me to think that, insofar as my belief that Shirley won is true, its truth is merely a lucky coincidence—the fact that I take the lottery results to be modally robust is just irrelevant here. But according to a safety-based account of epistemic luck, this is *not* what I should think, since my perspective is one on which, insofar as my belief is true, there aren't any nearby worlds where it's false.[18] So this is a case in which safety-based accounts go wrong, and do so precisely by neglecting the close connection between responsiveness and nonaccidental truth that our sufficient condition is designed to capture.[19]

Now: with our sufficient condition in hand, we can see that platonists are indeed committed to the claim that insofar as their beliefs about abstract objects are true, this is just a lucky coincidence. Clause (ii) seems clearly to be satisfied in the case of these beliefs; platonists are, again, committed to a view on which they're not in any sort of cognitive contact with abstract objects, which means their view is *not* one on which there's any mechanism by which their beliefs about those objects can possibly be responsive to what those objects are like. (Indeed, this is what makes

see Leite, 2006) have to do not with what follows from the agent's own theory of the world but with what follows from "what the agent is able to know by reflection alone" (Pritchard, 2005, p. 175).

[18] Lackey (2006) takes cases with roughly this structure to show that Pritchard's account of epistemic luck is fundamentally on the wrong track.

[19] Nor would it help to adopt a stricter safety condition according to which, for a belief that p to be safe, there must be very few nearby worlds where the agent is mistaken about whether p, so that the belief, in addition to failing to be safe if there are nearby worlds where the agent falsely believes that p, also fails to be safe if there are nearby worlds where the agent falsely *disbelieves* that p (see, e.g., Miščević, 2007; Pritchard, 2009). For even if some principled motivation can be offered for adopting this stricter safety-based account of epistemic luck (which is doubtful; see, e.g., Luper's (2012) discussion of the "false negatives avoidance" condition), we can again adjust the details of our lottery case, this time in such a way that my perspective is one on which there are no nearby worlds in which I fail to believe that Shirley won the lottery—say, because my new evidence, in addition to indicating that the report is from a novelty newspaper, also indicates that I was somehow *bound* to form my belief on the basis of that newspaper—and it remains entirely clear that, on gaining this new evidence, I should think that, insofar as my belief that she won is true, its truth is just a lucky coincidence.

platonistic theories epistemologically puzzling in the first place.) And as for clause (i): easy knowledge theories, recall, can do no more than to make available the verdict that platonists' beliefs about abstract objects initially enjoy some sort of *defeasible* good standing, and accepting that one's own belief is merely defeasibly in good standing involves acknowledging that the belief might turn out not to be true—i.e., that there are other ways the relevant part of the world might turn out to be.[20] So it's clear enough that (i) is satisfied as well.

I take it that these considerations are conclusive. Since (i) and (ii) are both satisfied, the first premise of the simple argument is true: platonists are committed to taking their beliefs about abstract objects to be accidentally true if true at all. And as we've already noted, the second premise is true as well, since that premise is just a consequence of the Lucky Coincidence Thesis. So the conclusion of that argument is true: it's simply not possible to be in good standing in being a nonskeptical platonist about abstract objects.

But this, we might think, is suspiciously easy. If things are as straightforward as all that, why is there so much controversy about whether the Benacerraf–Field challenge is a cogent argument against nonskeptical platonism?

I suggest that there are two main reasons for this. The first of these is that there's some temptation to insist that, because abstract objects are the way they are as a matter of metaphysical necessity, the kind of responsiveness that's in question here is irrelevant to whether we should take our beliefs about those objects to be, if true, merely accidentally true.[21] The world, the thought goes, can't *help* but be such that the objects have the features they do, and so there's no need to be responsive to what the world is in fact like in order to (nonaccidentally) form true beliefs about those objects.[22]

---

[20] Ebbs appears to think otherwise—he suggests that, if an agent finds herself unable to "specify a way in which a belief…may actually be false", it follows that it's not epistemically possible for her that the belief is false, even if she herself acknowledges that the belief isn't immune to revision (2022, p. 137), and he takes this sort of inability to specify a way in which a belief may be false to have a role to play in explaining why we're reasonable in holding on to (e.g.) certain of our mathematical beliefs. I doubt that this sort of inability to specify is really what's epistemologically relevant here—I take it that what's important, as I discuss below, is whether the agent takes there to be a *risk* that she's gone wrong, and whether this is so isn't determined by whether she can specify exactly *how* she might have gone wrong. But even if I'm mistaken about this, Ebbs's strategy is, it seems to me, going to be of no help to platonists. After all, it's straightforward to specify a way in which their beliefs about abstract objects may be false: the world may simply fail to contain any objects that are both mind-independent and causally inert.

[21] It's sometimes suggested (e.g., by Field himself) that truths about abstract objects aren't necessary in any genuine sense, in which case this sort of response to the Benacerraf–Field challenge is just based on a mistake. But this move isn't available in the present context. The simple argument, remember, isn't a radical skeptical argument: our question is whether *platonists' own theory of the world*—where this includes their theory of what the metaphysically necessary truths are—is one on which their beliefs about abstract objects are merely accidentally true.

[22] Lewis (1986, sect. 2.4), Hale (1994), Pust (2004), and Baras (2017) all express something in the neighborhood of this thought. See also fn. 24 below.

The second reason stems from the fact that, although platonistic theories' implications for the responsiveness of our beliefs are at the heart of what's epistemologically puzzling about platonism, the Benacerraf–Field challenge is almost never formulated in an explicitly responsiveness-based way. What tends to happen instead is that the challenge is formulated in terms of a modal property that's supposed to serve as a *proxy* for responsiveness—namely, sensitivity, where an agent's belief that p counts as sensitive just in case, roughly, in the nearest worlds where it's not the case that p, the agent refrains from believing that p (or, equivalently: just in case, if it weren't the case that p, the agent wouldn't believe that p).[23] Perhaps the thought is that we have a better grip on the modal notions relevant to sensitivity than we do on the notion of responsiveness itself. The problem, though, is that the usual formulations of the challenge in terms of sensitivity turn out *not* to be cogent: even setting aside cases of metaphysical necessity, there are lots of beliefs that clearly aren't undermined despite the fact that they very obviously fail to be sensitive. The result is that there's some temptation to retreat from the project of advancing a critique of platonism in the spirit of the simple argument, where this retreat involves either adopting some entirely different way of understanding of the Benacerraf–Field challenge (such as the coincidence avoidance interpretation mentioned above) or concluding that the challenge isn't cogent after all.

In fact, though, neither of these reasons turns out to reveal any genuine problem for the simple argument; further reflection makes clear that both of the temptations just mentioned are to be resisted. The first is just based on a misunderstanding of what the problem for platonism is supposed to be. And as for the second, the correct diagnosis here is not that our responsiveness-based critique of platonism fails to be cogent but simply that sensitivity turns out to be an imperfect proxy for responsiveness. We can see why these things are so by examining the temptations in turn.

9.4.1 Responsiveness and Metaphysical Necessity

Once again, the first of these temptations is to claim that, if our perspective is one on which abstract objects have the features they do as a matter of metaphysical necessity, responsiveness isn't relevant to whether we should take it to be a matter of luck whether our beliefs about those objects are true: since there's only one possible way for the world to be with respect to what those objects are like, we need not be able to discriminate among possibilities in order to get at the truth about those objects. Sometimes—in particular, when what's under discussion is a

[23] Roush (2005) is explicit about the fact that sensitivity is intended to capture a kind of responsiveness to what the world is like, though her concern is with accounts of knowledge in the spirit of Nozick's (see his 1981, chap. 3) rather than with the Benacerraf–Field challenge. Sensitivity-based formulations of the latter challenge have been discussed by Field himself as well as by Linnebo (2006), Liggins (2010), Donaldson (2014), Clarke-Doane (2016a, b, 2020a, b, 2022), Schechter (2018), and Berry (2020), among others.

sensitivity-based formulation of the Benacerraf–Field challenge—this takes the form of the claim that when it's necessarily true that *p*, there aren't any possible worlds where it isn't the case that *p*, in which case, if we believe that *p*, this belief is trivially sensitive (since, on the standard semantics for counterfactuals, a counterfactual whose antecedent isn't true in any possible worlds—i.e., a counterpossible—is vacuously true), or perhaps isn't evaluable for sensitivity at all (since, if there aren't any possible worlds where it's not the case that *p*, there's no answer to the question of what agent believes in the nearest possible worlds where it's not the case that *p*).[24] The underlying thought, though, remains the same: according to a view on which the realm of abstracta is the way it is as a matter of metaphysical necessity, there aren't multiple possibilities among which to discriminate, which means asking whether our beliefs about abstract objects result from an ability to discriminate among possibilities isn't going to be of any help in determining whether those beliefs are, if true, merely accidentally true.

What this amounts to, in the context of our sufficient condition, is the thought that if an agent's perspective is one on which her beliefs about abstract objects are true as a matter of metaphysical necessity, then clause (i) fails to be satisfied: the agent's perspective isn't one on which there are multiple possibilities one of which she takes to be actual. It of course follows that our sufficient condition isn't met, irrespective of whether (ii) is satisfied (i.e., irrespective of whether the agent's perspective is one on which she's formed her beliefs about abstract objects in a way that fails to be responsive to what those objects are like), in which case we can't show, by appeal to that sufficient condition, that the agent is committed to taking those beliefs to be accidentally true if true at all.

But when we put things this way, it's immediately evident that the thought here is based on a misunderstanding of what our sufficient condition is supposed to be doing. The reason failures of responsiveness are supposed to be worrying in the first place is that, from the agent's own perspective, if her beliefs fail to be responsive to the part of the world they're about, this indicates that the beliefs don't result from any genuine ability to pick out, from all the candidates for being her actual situation, the situation she's in fact in. Clause (i), then, is just intended to leave room for us to allow that there are cases in which getting at the truth doesn't *require* any such ability (e.g., cases in which most or all of the candidates are situations where the beliefs in question are true). So to insist that this clause fails to be satisfied in cases in which the agent takes her beliefs to be true as a matter of metaphysical necessity is to insist that, from the agent's perspective, the candidates for being her actual situation are worlds that, according to her theory, are metaphysically possible. But this is a mistake. The candidates, as previously mentioned, are simply ways things might, from the agent's perspective, turn out to be—i.e., situations the actuality of which the

---

[24] Linnebo (2006), Clarke-Doane (2016a, b), Schechter (2018), and Berry (2020) all express some sympathy with this sort of response to sensitivity-based formulations of the Benacerraf–Field challenge (though Clarke-Doane has more recently suggested (see his 2020a, sect. 5.7) that this isn't a good response, on the grounds that truths about abstracta aren't genuinely necessary in any "maximal" sense).

agent hasn't definitively ruled out from the start—and so, unless she takes herself to be infallible about what the metaphysical possibilities are, whether she takes her beliefs to be true as a matter of metaphysical necessity is irrelevant to whether clause (i), properly understood, is satisfied.

We can make this vivid by considering variants of our lottery case in which the content of the newspaper report is not a claim about lottery results but is instead some claim that, according to my own theory of the world, is necessarily true if true at all—e.g., that Shirley's biological parents (insofar as Shirley exists at all) are Geraldine and Leslie, or that every glucose molecule contains twelve hydrogen atoms, or that the continuum hypothesis is true. In these variants, it's quite clear that, despite my commitment to the metaphysical necessity of the facts in the relevant domains, gaining the new evidence against the newspaper's authenticity should still lead me to give up my belief in the claim in question, and for the same reason as in the original case: my perspective is one on which I've chosen among the various ways the relevant part of the world might be using a procedure that's no better than wildly guessing. And the reason this is worrying is precisely that, from my perspective, there's a risk that it will turn out that I've guessed incorrectly, that (say) Shirley's parents will turn out *not* to be Geraldine and Leslie; unless I take there to be some risk that I've gone wrong, even coming to accept that I've formed my belief by wildly guessing won't lead me to judge that, insofar as the belief is true, its truth is a lucky coincidence. But whether I take there to be any such risk is just determined by whether I take there to be *candidates for being my actual situation* where Shirley has other parents, and so, since I don't take myself to be infallible about what the metaphysical possibilities are, whether my theory of the world entails that it's metaphysically possible that she has other parents is irrelevant.

If what's under discussion is a sensitivity-based formulation of the Benacerraf–Field challenge in particular, our conclusion here can be formulated as the claim that when an agent is evaluating her own beliefs in the context of this challenge, the worlds she should take to be relevant to whether those beliefs are sensitive are not the worlds she takes to be metaphysically possible but the worlds that are *epistemically* possible for her, where what it takes for there to exist an epistemically possible world where *p*, for an agent, is just for the agent not to have definitively ruled out from the start that her actual situation is a situation where *p*.[25] But the larger point

[25] For a somewhat different argument for this same conclusion, see my "Realism, Reliability, and Epistemic Possibility" (2021), and for an explanation of how to construct a semantics that includes worlds that are epistemically possible in the relevant sense, see Collin, 2018, which refines some ideas presented in Chalmers, 2011. Incidentally, some of our epistemically possible worlds are presumably going to turn out to be metaphysically impossible, and it's sometimes suggested (see, e.g., Berry, 2020) that appealing to metaphysically impossible worlds in this context is problematic, on the grounds that we haven't fully worked out how to think about similarity in a semantics that includes such worlds. But the utility of this semantics doesn't require our having fully specified a similarity metric, any more than it does in the case of a more traditional semantics that includes only metaphysically possible worlds (a semantics for which, notoriously, we *also* haven't fully specified a similarity metric). What's important for our purposes is just that platonism, since it's a view on which there's no contact between the causal realm and the realm of abstracta, is

remains the same: whether platonists take the realm of abstracta to be the way it is as a matter of metaphysical necessity has no bearing whatsoever on whether there are multiple possibilities here, in the sense relevant to whether platonists are committed to taking their beliefs about abstract objects to be accidentally true if true at all. What guarantees that there are indeed multiple such possibilities—i.e., that clause (i) of our sufficient condition is satisfied—is, as suggested above, simply that platonists are committed to accepting that the beliefs in question are merely defeasibly in good standing.

9.4.2 *Insensitive But Nonaccidentally True Beliefs*

As for the second temptation, it arises, again, from the fact that the usual sensitivity-based formulations of the Benacerraf–Field challenge fail to be cogent. The problem with these formulations is that one of the principles on which they depend—namely, that an agent's coming to accept that one of her beliefs is insensitive suffices to undermine the good standing of that belief—is false: there are cases in which an agent's belief clearly remains in good standing despite clearly failing to be sensitive (even when it's uncontroversial that there are multiple possibilities here—i.e., that clause (i) is satisfied). Typically these are cases in which some insensitive belief is such that its truth is obviously entailed by the truth of a belief that (by the agent's lights) *is* sensitive. Consider, for example, White's case of the jury members who, on gaining a great deal of evidence of the defendant's guilt, believe both that the defendant is guilty and that it's not the case that "the defendant was framed and all the evidence was planted" (2010, p. 581). Insofar as the former belief is in good standing—which it is—the latter one is in good standing as well, since the jury members can be sure that any world where the former one is true is thereby a world where the latter one is true.[26] This despite the jury members' awareness that the latter belief is insensitive, since, had the defendant been framed, they'd have exactly the same evidence they now have. But sensitivity, recall, is intended to be a modal proxy for responsiveness. So the temptation, in the face of cases like this, is to abandon our responsiveness-based critique of platonism—i.e., to allow that a belief can remain in good standing even in a case in which clauses (i) and (ii) of our sufficient condition are both satisfied.

But to abandon our critique in this way would be to take entirely the wrong lesson from cases like White's. To see why this is, note that White's jury members do have one belief that, according to their own theory of the world, *is* responsive to what the relevant part of the world is like: the belief that the defendant is guilty.

thereby a view on which, on *any* reasonable way of specifying our similarity metric, the nearest worlds where the realm of abstracta is different in some respect aren't going to be worlds where the causal realm is also different.

[26] I assume, with White, that good standing is closed under this sort of obvious entailment, though this has sometimes been denied—see, e.g., Nozick, 1981, chap. 3.

What the truth of this belief requires is just that a particular part of the world be a particular way—i.e., that it be such that the defendant committed the crime—and the jury members' theory is one on which, had this part of the world not been that way, they wouldn't have gained the evidence they now have and so wouldn't believe that the defendant is guilty. But note also that, insofar as this part of the world is indeed such that the defendant committed the crime, it's thereby such that the defendant wasn't framed, so that the truth of the jury members' belief that the defendant is guilty is sufficient for the truth of their belief that it's not the case that the defendant was framed and all the evidence was planted. Furthermore, the jury members *accept* that the truth of the former belief is sufficient for the truth of the latter one. Indeed, this is part of what explains why they have the latter belief in the first place: by their own lights, the grounds on which they believe that the defendant is guilty are certain also to be sufficient to ground the belief that the defendant wasn't framed. So the latter belief, despite its obvious insensitivity, does turn out to be responsive (albeit in an inherited way) to what a particular part of the world is like such that this part of the world being the way it is suffices for the truth of that belief.[27]

In short, this just isn't a case in which the agents should conclude that the belief in question fails to be responsive to the part of the world relevant to its truth. So clause (ii) of our sufficient condition isn't satisfied. And if that's right, we can, consistently with that condition, insist that the agents aren't committed to taking the belief to be accidentally true if true at all, which means we don't have a route, via the Lucky Coincidence Thesis, to the claim that the good standing of the belief is undermined. What cases with this structure indicate, then, is not that there's something wrong with our responsiveness-based critique of platonism.[28] It's simply that sensitivity turns out not to be necessary for responsiveness.

[27] For a more thorough discussion of this point, see my "Saving Sensitivity" (2022).

[28] It's sometimes suggested (see, e.g., Korman, 2014) that there are also cases of a different sort in which beliefs remain in good standing despite failing to be sensitive: cases of beliefs formed by enumerative induction (e.g., my belief that the sun will rise tomorrow). Now: it's plausible that, in cases in which such beliefs are true, this isn't an accident. And insofar as we should think that such beliefs genuinely fail to be responsive to the parts of the world relevant to their truth—as it's standardly assumed that we should (though see, e.g., Cross, 2010; Wallbridge, 2018)—these are cases in which clause (ii) is satisfied. So we might think these cases are counterexamples to our sufficient condition. But notice that in these cases, there's a belief in the background that pretty clearly *is* responsive to the part of the world relevant to its truth: the belief that (roughly) there's some lawlike explanation of why the agent might have observed the pattern in question in the first place. (If there's no such belief in the background, there's no reason to expect the next instance to fit the pattern. And the belief is responsive to the relevant part of the world simply because, if there hadn't been anything lawlike going on, the agent wouldn't have observed any pattern.) And though there's no inherited responsiveness here, there's inheritance of another kind: if we narrow the class of possibilities to those where there the background belief is true—i.e., where there is in fact a lawlike explanation of the relevant kind—then the *vast majority* of those possibilities turn out to be ones where the next instance does indeed fit the pattern. (The vast majority, but not all: the fact that there's a lawlike explanation of why the sun has risen every day up to this point doesn't guarantee that the sun will rise tomorrow.) So it seems that, just as in the case discussed in fn. 16 above, clause (i) isn't satisfied, which means we don't have a counterexample to our sufficient condition.

We can, if we like, generate a better modal proxy for responsiveness by modifying our sensitivity condition so as to handle cases with this structure. (This is precisely what I do in my "Saving Sensitivity" (2022): I devise a recursive condition for responsiveness such that a belief satisfies the base clause just in case it's sensitive and such that the inductive clause captures exactly the sort of inherited responsiveness exhibited by the jury members' belief that the defendant wasn't framed, and I argue that this condition (with one other minor modification in place) turns out to be entirely counterexample-free.[29]) But it's not clear that we need to do so. We already have enough of a grip on the sort of responsiveness that's relevant here that we can conclude, with confidence, that agents in cases with this structure should indeed take their beliefs to be responsive to the parts of the world relevant to their truth, which means our sufficient condition isn't met. And this is all that's required in order for it to be clear that these cases aren't counterexamples to our sufficient condition and so have no tendency whatsoever to suggest that our critique of platonism fails to be cogent.[30]

---

[29] The point of the second modification is to ensure that a belief meets the condition only if it's responsive to *all* the features of the world that are required for its truth. This is needed in order to handle what Kripke (2011) calls the "absorption phenomenon": the fact that, if an agent has a conjunctive belief that (say) her left shoelace is untied and tomorrow she'll inherit a fortune from some as-yet-unknown relative, then this belief, if true, is sensitive, just because the nearest worlds where her shoelace is tied are nearer than the nearest worlds where she doesn't inherit the fortune.

[30] Clarke-Doane suggests another route to the conclusion that some beliefs (even in cases in which clause (i) is satisfied) aren't undermined despite clearly failing to be responsive. He begins by claiming that (insofar as there can be nonvacuous counterfactuals whose antecedents are false in all metaphysically possible worlds) our metaphysical beliefs about the composition conditions of macroscopic objects don't seem to be sensitive: "had particles arranged 'paper-wise' failed to compose a piece of paper, it seems that we still would have believed that they did" (2020a, p. 145). Nor would modifying our sensitivity condition be helpful here; the beliefs in question genuinely don't seem to be responsive to what the relevant bridge laws in fact are. At first sight, this seems unobjectionable—the epistemology of the metaphysics of composition seems to be mysterious in broadly the same way that the epistemology of abstracta is (see, e.g., Korman, 2014). But Clarke-Doane argues that our beliefs about composition conditions *must* be in good standing because, if they weren't, an unacceptable skepticism about everyday perceptual claims would result: "If I believe that I am looking at a piece of paper, but my belief in the metaphysically necessary bridge law that particles arranged 'paper-wise' compose a piece of paper is undermined, then it is hard to see how my belief that I am looking at a piece of paper could fail to be" (2020a, p. 145; largely the same argument also appears in his 2016a, b, 2020b, 2022). It seems to me, though, that this line of reasoning is just confused. When an agent, in her everyday life, takes herself to be looking at a piece of paper, she either is or isn't thereby taking a stand on some substantive metaphysical thesis about composition. If she isn't—if her belief, in order to be correct, requires nothing more than that she be looking at some particles arranged paperwise—then there's no failure of responsiveness here, in which case there's no danger of skepticism. (My own opinion, fairly widely shared among metaphysicians of various stripes (see, e.g., van Inwagen, 1990; Dorr, 2005; Thomasson, 2007; Chalmers, 2009; Cameron, 2010; Sider, 2013), is that it's overwhelmingly plausible that this is in fact the situation.) But if she is—if her belief, in order to be correct, requires, for example, that the substantive metaphysical thesis of mereological nihilism be false—then the resulting skepticism is not unacceptable: the epistemology of the metaphysics of composition is indeed mysterious (unless, *contra* Clarke-Doane, our beliefs here *are* responsive to the relevant metaphysical facts, as Korman (2015, chap. 7) insists they are), and so, insofar as our everyday perceptual beliefs in fact

9.5 Concluding Remarks

In the end, then, things *are* as straightforward as I've suggested they are. By our sufficient condition—a well-motivated condition that stands up to scrutiny—platonists are indeed committed to taking their beliefs about abstract objects to be accidentally true if true at all: their commitment to a view on which those beliefs are merely defeasibly in good standing ensures that clause (i) of the condition is satisfied, and their commitment to a view on which they're not in cognitive contact with abstract objects ensures that clause (ii) is satisfied. That is, the first premise of the simple argument is indeed true. So, given the second premise—which is, again, just a consequence of the Lucky Coincidence Thesis—platonists, if they want to remain in good standing in holding on to their first-order beliefs about abstract objects, must give up their platonism. The simple argument is sound after all.

What this means is that there's no hope that an easy knowledge theory can save platonism from skepticism; no amount of epistemological maneuvering will change the fact that the simple argument is sound. If we wish to avoid skepticism, we must adopt a nonplatonistic *metaphysical* picture, one on which cognitive contact with abstract objects is possible.[31] And this is just to say that we must adopt a picture on which such objects aren't both mind-independent and causally inert. Nothing less radical than this will suffice.

Acknowledgements This paper was completed with the generous financial support of the Austrian Science Fund (FWF), Grant No. P33708. I'm also grateful to Jonathan Courtney, Louis Gularte, Julien Murzi, Micaela Quintana, Mary Renaud, and an audience at the Abstract Objects: Deflationary Approaches workshop at the University of Santiago de Compostela for helpful discussion of the ideas presented here.

References

Alston, W. P. (1988). An internalist externalism. *Synthese, 74*, 265–283.
Balaguer, M. (1998). *Platonism and anti-platonism in mathematics*. Oxford University Press.

commit us to substantive metaphysical theses about composition, their epistemology is mysterious as well. In short, Clarke-Doane is presupposing *both* that our questions about the composition conditions of macroscopic objects are substantive metaphysical questions *and* that it's a trivial matter to answer those questions by appeal to our everyday perceptual beliefs. But that's absurd. I can't refute the metaphysical thesis of mereological nihilism simply by insisting that I'm looking at a piece of paper, though it would certainly be convenient if I could.

[31] Notice that merely adopting some plenitudinous form of platonism (see, e.g., Balaguer, 1998) isn't going to be of any help here. We can grant, at least for the sake of argument, that on this picture, our belief that (say) the continuum hypothesis is true is responsive to its corresponding fact, since the nearest worlds where that hypothesis isn't true are worlds where we've chosen to talk about a mathematical universe in which it isn't true. But there remains a problem: adopting plenitudinous platonism also involves having beliefs about the multiverse itself—e.g., the belief that it contains a universe corresponding to any consistent mathematical theory—and these beliefs aren't in any sense responsive to the relevant facts.

Baras, D. (2017). Our reliability is in principle explainable. *Episteme, 14*, 197–211.
Beddor, B. (2021). Reasons for reliabilism. In J. Brown & M. Simion (Eds.), *Reasons, justification, and defeat* (pp. 146–176). Oxford University Press.
Bergmann, M. (2006). *Justification without awareness: A defense of epistemic externalism*. Oxford University Press.
Berry, S. E. (2020). Coincidence avoidance and formulating the access problem. *Canadian Journal of Philosophy, 50*, 687–701.
Bhogal, H. (2023). What's the coincidence in debunking? *Philosophy and Phenomenological Research, 107*, 147–167.
Boghossian, P. A. (1996). Analyticity reconsidered. *Noûs, 30*, 360–391.
Boghossian, P. A. (1997). Analyticity. In B. Hale & C. Wright (Eds.), *A companion to the philosophy of language* (pp. 331–368). Blackwell.
Boghossian, P. A. (2000). Knowledge of logic. In P. Boghossian & C. Peacocke (Eds.), *New essays on the a priori* (pp. 229–254). Oxford University Press.
Boghossian, P. A. (2003a). Blind reasoning. *Proceedings of the Aristotelian Society, Supplementary Volumes, 77*, 225–248.
Boghossian, P. A. (2003b). Epistemic analyticity: A defense. *Grazer Philosophische Studien, 66*, 15–35.
Burgess, J. P., & Rosen, G. (1997). *A subject with no object: Strategies for nominalistic interpretation of mathematics*. Oxford University Press.
Cameron, R. P. (2010). How to have a radically minimal ontology. *Philosophical Studies, 151*, 249–264.
Chalmers, D. J. (2009). Ontological anti-realism. In D. Chalmers, D. Manley, & R. Wasserman (Eds.), *Metametaphysics: New essays on the foundations of ontology* (pp. 77–129). Oxford University Press.
Chalmers, D. J. (2011). The nature of epistemic space. In A. Egan & B. Weatherson (Eds.), *Epistemic modality* (pp. 60–107). Oxford University Press.
Chalmers, D. J. (2012). *Constructing the world*. Oxford University Press.
Clarke-Doane, J. (2016a). Debunking and dispensability. In U. D. Leibowitz & N. Sinclair (Eds.), *Explanation in ethics and mathematics: Debunking and dispensability* (pp. 23–36). Oxford University Press.
Clarke-Doane, J. (2016b). What is the Benacerraf Problem? In F. Pataut (Ed.), *Truth, objects, infinity: New perspectives on the philosophy of Paul Benacerraf* (pp. 17–43). Springer.
Clarke-Doane, J. (2020a). *Morality and mathematics*. Oxford University Press.
Clarke-Doane, J. (2020b). Set-theoretic pluralism and the Benacerraf problem. *Philosophical Studies, 177*, 2013–2030.
Clarke-Doane, J. (2022). *Mathematics and metaphilosophy*. Cambridge University Press.
Cohen, S. (2002). Basic knowledge and the problem of easy knowledge. *Philosophy and Phenomenological Research, 65*, 309–329.
Collin, J. H. (2018). Towards an account of epistemic luck for necessary truths. *Acta Analytica, 33*, 483–504.
Constantin, J. (2020). Replacement and reasoning: A reliabilist account of epistemic defeat. *Synthese, 197*, 3437–3457.
Cross, T. (2010). Skeptical success. In T. S. Gendler & J. Hawthorne (Eds.), *Oxford studies in epistemology* (Vol. 3, pp. 35–62). Oxford University Press.
Donaldson, T. M. E. (2014). If there were no numbers, what would you think? *Thought: A Journal of Philosophy, 3*, 283–287.
Dorr, C. (2005). What we disagree about when we disagree about ontology. In M. E. Kalderon (Ed.), *Fictionalism in metaphysics* (pp. 234–286). Oxford University Press.
Ebbs, G. (2022). Putnam on trans-theoretical terms and contextual apriority. In J. Conant & S. Chakraborty (Eds.), *Engaging Putnam* (pp. 131–156). De Gruyter.
Elga, A. (2008). *Lucky to be rational*. Paper presented at Bellingham Summer Philosophy Conference.

Enoch, D. (2010). The epistemological challenge to metanormative realism: How best to understand it, and how to cope with it. *Philosophical Studies, 148*, 413–438.
Enoch, D., & Schechter, J. (2008). How are basic belief-forming methods justified? *Philosophy and Phenomenological Research, 76*, 547–579.
Field, H. H. (1989). *Realism, mathematics and modality*. Basil Blackwell.
Field, H. H. (2000). Apriority as an evaluative notion. In P. Boghossian & C. Peacocke (Eds.), *New essays on the a priori* (pp. 117–149). Oxford University Press.
Goldman, A. I. (1979). What is justified belief? In G. S. Pappas (Ed.), *Justification and knowledge: New studies in epistemology* (pp. 1–23). D. Reidel.
Goldman, A. I. (1986). *Epistemology and cognition*. Harvard University Press.
Greco, J. (1999). Agent reliabilism. *Philosophical Perspectives, 13*, 273–296.
Grundmann, T. (2009). Reliabilism and the problem of defeaters. *Grazer Philosophische Studien, 79*, 65–76.
Hale, B. (1994). Is platonism epistemologically bankrupt? *Grazer Philosophische Studien, 66*, 15–35.
Harman, G. (2003). Skepticism and foundations. In S. Luper (Ed.), *The skeptics: Contemporary essays* (pp. 1–11). Ashgate Publishing.
Horwich, P. (2006). A world without isms: Life after realism, fictionalism, non-cognitivism, relativism, reductionism, revisionism, and so on. In P. Greenough & M. P. Lynch (Eds.), *Truth and realism* (pp. 188–202). Oxford University Press.
Korman, D. Z. (2014). Debunking perceptual beliefs about ordinary objects. *Philosophers' Imprint, 14*, 1–21.
Korman, D. Z. (2015). *Objects: Nothing out of the ordinary*. Oxford University Press.
Kornblith, H. (2004). Does reliabilism make knowledge merely conditional? *Philosophical Issues, 14*, 185–200.
Kripke, S. A. (2011). Nozick on knowledge. In *Collected papers: Vol. 1. Philosophical troubles* (pp. 162–224). Oxford University Press.
Lackey, J. (2006). Review of Duncan Pritchard, *Epistemic Luck. The Philosophical Quarterly, 56*, 284–289.
Leite, A. (2006). Review of Duncan Pritchard, *Epistemic Luck. Notre Dame Philosophical Reviews, 2006.04.20*.
Lewis, D. (1986). *On the plurality of worlds*. Basil Blackwell.
Liggins, D. (2010). Epistemological objections to platonism. *Philosophy Compass, 5*, 67–77.
Linnebo, Ø. (2006). Epistemological challenges to mathematical platonism. *Philosophical Studies, 129*, 545–574.
Luper, S. (2012). False negatives. In K. Becker & T. Black (Eds.), *The sensitivity principle in epistemology* (pp. 207–226). Cambridge University Press.
Miščević, N. (2007). Armchair luck: Apriority, intellection and epistemic luck. *Acta Analytica, 38*, 48–73.
Nozick, R. (1981). *Philosophical explanations*. Harvard University Press.
Nyseth, F. (2021). Linguistic conventionalism and the Truth-Contrast Thesis. *The Philosophical Quarterly, 71*, 264–285.
Pritchard, D. (2005). *Epistemic luck*. Oxford University Press.
Pritchard, D. (2007). Anti-luck epistemology. *Synthese, 158*, 277–297.
Pritchard, D. (2009). Safety-based epistemology: Whither now? *Journal of Philosophical Research, 34*, 33–45.
Pust, J. (2004). On explaining knowledge of necessity. *Dialectica, 58*, 71–87.
Putnam, H. (1983). "Two dogmas" revisited. In Realism and reason: *Philosophical papers, volume 3* (pp. 87–97). Cambridge University Press. (Reprinted from *Contemporary aspects of philosophy*, pp. 202–213, by G. Ryle, Ed., 1976, Oriel Press).
Roush, S. (2005). *Tracking truth: Knowledge, evidence, and science*. Oxford University Press.
Schechter, J. (2010). The reliability challenge and the epistemology of logic. *Philosophical Perspectives, 24*, 437–464.

Schechter, J. (2013). Could evolution explain our reliability about logic? In T. S. Gendler & J. Hawthorne (Eds.), *Oxford studies in epistemology* (Vol. 4, pp. 214–239). Oxford University Press.

Schechter, J. (2018). Is there a reliability challenge for logic? *Philosophical Issues, 28*, 325–347.

Schechter, J., & Enoch, D. (2006). Meaning and justification: The case of Modus Ponens. *Noûs, 40*, 687–715.

Schoenfield, M. (2014). Permission to believe: Why permissivism is true and what it tells us about irrelevant influences on belief. *Noûs, 48*, 193–218.

Setiya, K. (2012). *Knowing right from wrong*. Oxford University Press.

Sider, T. (2013). Against parthood. In K. Bennett & D. W. Zimmerman (Eds.), *Oxford studies in metaphysics* (Vol. 8, pp. 236–293). Oxford University Press.

Skipper, M. (2023). Can arbitrary beliefs be rational? *Episteme, 20*, 377–392.

Sosa, E. (1991). *Knowledge in perspective: Selected essays in epistemology*. Cambridge University Press.

Sosa, E. (1999). How must knowledge be modally related to what is known? *Philosophical Topics, 26*, 373–384.

Thomasson, A. L. (2007). *Ordinary objects*. Oxford University Press.

Thomasson, A. L. (2015). *Ontology made easy*. Oxford University Press.

Topey, B. (2021). Realism, reliability, and epistemic possibility: On modally interpreting the Benacerraf–Field challenge. *Synthese, 199*, 4415–4436.

Topey, B. (2022). Saving sensitivity. *The Philosophical Quarterly, 72*, 177–196.

van Inwagen, P. (1990). *Material beings*. Cornell University Press.

Wallbridge, K. (2018). Induction, sensitivity, and miracles. *Australasian Journal of Philosophy, 96*, 118–126.

White, R. (2010). You just believe that because.... *Philosophical Perspectives, 24*, 573–615.

Williamson, T. (2000). *Knowledge and its limits*. Oxford University Press.

Williamson, T. (2009). Replies to critics. In P. Greenough & D. Pritchard (Eds.), *Williamson on knowledge* (pp. 279–384). Oxford University Press.

Wright, C. (2004a). Intuition, entitlement and the epistemology of logical laws. *Dialectica, 58*, 155–175.

Wright, C. (2004b). Warrant for nothing (and foundations for free)? *Proceedings of the Aristotelian Society, Supplementary Volumes, 78*, 167–212.

Wright, C. (2014). On epistemic entitlement (II): Welfare state epistemology. In D. Dodd & E. Zardini (Eds.), *Skepticism and perceptual justification* (pp. 213–247). Oxford University Press.

Wright, C. (2016). Abstraction and epistemic entitlement: On the epistemological status of Hume's Principle. In P. A. Ebert & M. Rossberg (Eds.), *Abstractionism: Essays in philosophy of mathematics* (pp. 161–185). Oxford University Press.

Part IV
Agnosticism/Quietism and Fictionalism

Chapter 10
Objects, Thin Objects, Thin Access to Objects

Otávio Bueno

Abstract Deflationist approaches to abstract objects typically have a distinctive advantage over non-deflationist counterparts on the epistemological front, given the avoidance of commitment to the existence of these objects (Azzouni, Deflating existential consequence: a case for nominalism. Oxford University Press, 2004; Bueno, Stud Hist Phil Mod Phys, 36:465–490, 2005) or given the light-weight character that is assigned to them, as thin objects (Linnebo, Thin objects: an abstractionist account. Oxford University Press, 2018). But these approaches need to accommodate whatever use abstract objects are put to in mathematics, metaphysics, and in the sciences. Two deflationist approaches should be distinguished: (a) *quantificational* approaches, which explore the role of quantification in deflating ontological commitment (Azzouni, Deflating existential consequence: a case for nominalism. Oxford University Press, 2004; Bueno, Stud Hist Phil Mod Phys, 36:465–490, 2005), and (b) *definitional* approaches, which emphasize the role of suitable abstraction principles in the characterization of the relevant objects (Linnebo, Thin objects: an abstractionist account. Oxford University Press, 2018). In this paper, these views are contrasted and assessed. In the end, there is something to be said in favor of quantificational approaches and the thin access to objects they recommend.

Keywords Objects · Thin objects · Abstract objects · Deflationism · Ontological commitment

10.1 Introduction

According to platonist views about mathematics, abstract objects, such as numbers, sets, and functions, exist. These views have the advantage of being able to take mathematical discourse literally: such discourse is about mathematical objects,

---

O. Bueno (✉)
Department of Philosophy, University of Miami, Coral Gables, FL, USA

X. de Donato-Rodríguez et al. (eds.), *Deflationist Conceptions of Abstract Objects*, Synthese Library 516, https://doi.org/10.1007/978-3-031-98717-5_10

which are taken to exist independently of what may be thought or said about them—independently, in fact, even of the existence of anyone who may have ever thought or stated anything about mathematics. This is a significant benefit of platonism, since at least in principle it offers the resources to make sense of salient features of mathematical practice.

By positing mathematical objects, nevertheless, platonists need to offer a suitable account of how one can acquire knowledge of or, at least, form justified beliefs about the existence of those posited entities. Platonists have developed various strategies to address this difficulty, depending on the particular formulation of the view. Standard platonists advanced access to mathematical entities via intuition (Gödel, 1964); neo-Fregean platonists articulated suitable abstraction principles to secure reference to and, in this way, knowledge of the relevant mathematical entities (Hale & Wright, 2001), whereas structuralist platonists have advocated suitable forms of structure and pattern identification (Shapiro, 1997; Resnik, 1997). A critical examination of epistemological challenges faced by these forms of platonism is provided in Bueno (2011).

In contrast, deflationary approaches to abstract objects avoid difficulties faced by their platonist counterparts since they do not posit the existence of abstract entities. The issues of securing reference to and knowledge of these entities that has plagued platonist views for so long simply do not emerge. This does not mean that deflationary views face no difficulties. They still need to account for the role that abstract entities play in scientific, mathematical, and metaphysical theorizing. They are often used in expressive, explanatory, and predictive roles—to formulate certain theories, to explain some phenomena, or to predict that certain states of affairs will obtain. Is it possible to account for these roles without commitment to the existence of the relevant entities? Alternatively, if an existential commitment is forthcoming, can these roles be fulfilled while still explaining the possibility of reference to and knowledge of the objects under consideration?

To address these questions, it is crucial first to be clear about the underlying conception of objects that is articulated by these different approaches. After doing that I will focus on two deflationary conceptions. According to *quantificational* approaches, the crucial component to avoid ontological commitment is found in quantification and ontologically neutral quantifiers, which do not presuppose the existence of what is quantified over (Azzouni, 2004; Bueno, 2005). In contrast, *definitional* approaches emphasize the role of abstraction principles in the characterization of the relevant objects (Linnebo, 2018) so that the postulation of abstract objects is not as costly as it may initially seem to be. In what follows, I contrast and assess these two views, and conclude that quantificational approaches and the thin access to objects they recommend should be favored in the end.

10.2 Objects: Four Clusters of Conceptions

To set the stage, four clusters of conceptions of objects will be considered. Two old; two new. The old—or, more respectfully put, classical—conceptions come in two varieties (Lowe, 1998, pp. 33–39): one semantic and another metaphysical. They are substantive in that they characterize objects in accordance with substantive criteria. The new clusters of conceptions also include semantic and metaphysical components, but they are structured differently and articulate more deflationary views. I will consider each of them in turn, starting with the classical cluster of conceptions.

10.2.1 Classical Conceptions of Objects: Substantive Views

(A) According to the *semantic conception*, objects are anything that can be referred to by a singular term (Frege, 1885/1953) or quantified over, thus being the value of a variable (Quine, 1953/1980, 1960/2013). This conception is essentially linguistic in that it characterizes objects via suitable linguistic devices, whether referential or quantificational. The conception comes in two versions:

(i) On the *Fregean version*, reference to singular terms is guaranteed by the introduction of suitable definitions and abstraction principles. For instance, Hume's Principle specifies that the number of *F*s is identical to the number of *G*s if, and only if, there is a one-to-one correspondence between them, that is, provided that *F* and *G* are equinumerous (Hale and Wright, 2001). In symbols: $\#F = \#G$ if, and only if, $F \backsim G$. On this view, if such principles are true, reference succeeds, and as a result the objects referred to exist. In other words, the existence of abstract objects is guaranteed from true reference invoking singular terms. Successful reference, on this view, presupposes existence of what is referred to.

(ii) On the *Quinean version*, if quantification ranges over indispensable objects that cannot be eliminated by suitable paraphrases in a regimented language—for Quine, it is a first-order language with an underlying classical logic—the objects in question exist (Quine, 1953/1980; 1960/2013). After all, on this view, reference to indispensable entities cannot be eliminated without seriously jeopardizing the content of the theories that invoke them. In the absence of the relevant entities, that is, without reference to them, it is impossible to formulate and express certain components of the theories and to advance predictive and explanatory strategies. In this way, indispensability and unavailability of effective paraphrases go hand in hand—that is, as long as paraphrases that preserve and convey, without loss, the content of the theories under consideration are in place, the entities that are, thus, dispensed with do not carry ontological weight. The result is a conception of objects that play a significant role in shaping scientific reasoning.

Whether in its Fregean or in its Quinean form, the underlying conception of object crucially relies on linguistic traits to specify what objects are—referentially or quantificationally, respectively. Hence, the semantic characterization of the view.

(B) According to the *metaphysical conception*, objects are anything that has determinate identity-conditions, and, thus, "any item falling under some sortal concept supplying a criterion of identity for its instances" (Lowe, 1998, p. 34). Criteria of identity are also a Fregean requirement for objecthood, but the metaphysical conception identified by Lowe differs from the Fregean view since instead of demanding the strict logicist identity conditions, something more substantive, metaphysical is proposed. Fregean requirements are logical rather than metaphysical, in the sense that they are not substantive conditions on objects but involve conditions of a more formal sort.

In fact, for Frege (1885/1953), natural numbers are objects whose identity conditions are formulated in terms of logic and definitions alone. As a result, since numbers are purely logical in nature, and not mathematical, their epistemology is substantially simplified. 0 is then identified with the number of things that fall under the concept *non-self-identical*. In symbols, $\#[x: x \neq x] = 0$ (see Boolos, 1998; Bueno, 2001). Note that Frege's formulation does *not* presuppose sets, but only identity, negation, and predication—that is, only logical items are invoked. A commitment to sets would undermine the strict logicist requirements imposed by Frege, given the resulting introduction of mathematical content. Since nothing falls under the concept *non-self-identical* and given that this concept only includes logical notions, 0 is thereby characterized.

Once 0 is available, 1 can be characterized as the number of things that fall under the concept *being identical to 0*. In symbols, $\#[x: x = 0] = 1$. Since only one object falls under that concept, 1 is thereby characterized. In order to establish that 0 and 1 are different, Frege introduces a principle of identity for numbers: Hume's Principle (Boolos, 1998). Two concepts are the same if, and only if, there is a one-to-one correspondence between them. In particular, in the case of numbers, $\#F = \#G$ if, and only if, $F \sim G$, that is, the number of *F*s is the number of *G*s if, and only if, there is a one-to-one correspondence between them. Since no object falls under the concept 0 and only one falls under the concept 1, there is no one-to-one correspondence between them. They are, thus, different.

With 0 and 1 in place, 2 can be characterized as the number of objects that fall under the concept *being identical to 0 or being identical to 1*. In symbols, $\#[x: x = 0 \text{ or } x = 1] = 2$. Since only two objects fall under that concept—namely, 0 and 1—and given that the characterization involves only logical items, 2 has been properly formulated in accordance with logicist constraints. An application of Hume's Principle then establishes the difference between 1 and 2, given the lack of one-to-one correspondence between the concepts in question as different objects fall under the corresponding concepts. The remaining numbers can be characterized accordingly.

The metaphysical conception does not restrict the formulation of the identity conditions for objects to their logical features as the logicist characterization of natural numbers demands. Instead of formal constraints on criteria of identity for objects, substantive metaphysical requirements are added to the identity conditions, depending on the nature of the objects in question. For instance, on this view, what it is to be human may require a suitable account of the essence of a human being, something that involves both the general essence shared by every human being—e.g., having a DNA of a certain kind—and the individual essence that is unique to a particular person—such as the thisness that individuates the particular human being in question (Lowe, 2002, Chap. 6). Fine grained metaphysical criteria of identity of this sort provide individuation conditions for the relevant objects. The added criteria are significantly different from and metaphysically more substantial than the logical, formal requirements discussed before.

10.2.2 New Conceptions of Objects: Deflationism

The new conceptions of objects also come in two varieties. Both are metaphysical in nature, in the sense that they provide more than merely formal, logical constraints to what it is to be an object. Nonetheless, they are significantly more deflationary than the classical views discussed so far, especially the metaphysical conception, which adds substantive identity and individuation conditions. Both conceptions also advance semantic moves, crucially relying on quantificational devices.

(A) According to the *thin conception*, thin objects are those whose existence "does not make a substantial demand on the world" (Linnebo, 2018, p. xi). This conception combines the metaphysical conception's criteria of identity with the Fregean semantic, abstractionist approach, but it restricts abstraction principles to those that are predicative and dynamic. Abstraction principles have the form:

$$\#\alpha = \#\beta \text{ if, and only if, } \alpha \equiv \beta$$

in which α and β are variables, # is a singular-term forming operator that is applied to variables, and ≡ is an equivalence relation among the items over which the variables range. Hume's Principle is a well-known instance of an abstraction principle in which the number of *F*s is the number of *G*s as long as there is a one-to-one correspondence between them. Since abstraction principles are extremely general, the concern emerges that inconsistencies may be inadvertently introduced by them. This was the case with Frege's Basic Law V, which associates concepts with extensions of objects that fall under those concepts (Frege, 1885/1953). Despite its apparent analytic character, the principle turns out to be inconsistent (for an insightful discussion, see Boolos, 1998).

Predicative abstraction principles are introduced to safeguard from this potential concern. Such principles impose a constraint on the characterization of new abstract entities so that questions about new entities that are introduced by abstraction principles can be reduced to old ones from which they are abstracted (Linnebo, 2018, p. xiii).

The predicative constraint may avoid the introduction of unwanted inconsistencies. Nonetheless, it may restrict the domain too drastically, having difficulty to generate more substantial content that is, in fact, needed. The situation is analogous to the formulation of set theory. An unrestricted comprehension schema for sets—according to which to every property P, there is a set of objects that have that property, that is, $\forall x\, \exists y\, y = \{x: Px\}$—is inconsistent, given Russell's paradox. After all, the assumption to the effect that to the property of *being a non-self-membered set* corresponds a set leads one straight into a contradiction. The Russell set $R = \{x: x \notin x\}$ is such that R is both a member of R and is not a member of R. A restricted comprehension principle for sets is then introduced in Zermelo-Fraenkel set theory, but it restricts too drastically the sets that can be obtained. Additional axioms are needed to yield a set theory with substantial content (see Jech, 2006).

Dynamic abstraction is meant to address a similar concern. As Linnebo notes:

> Since abstraction often results in a larger domain, we can use this extended domain to provide criteria of identity for yet further objects, which can thus be obtained by further steps of abstraction. [...] The successive "formation" of sets described by the influential iterative conception of sets is just one instance of the more general phenomenon of dynamic abstraction. Legitimate abstraction steps are iterated indefinitely to build up ever larger domains of abstract objects. Dynamic abstraction can be seen as a development and extension of the famous iterative conception of sets (Linnebo, 2018, pp. xiii–xiv).

The result is a conception of objects that, although thin, is still substantial in its outcomes.

(B) According to the *ontologically neutral conception*, quantification over objects does not require their existence (Azzouni, 2004; Bueno, 2005). This conception combines the quantificational approach with revised criteria of identity—in the minimal sense that quantification requires the identity of the objects that are quantified over—but without the Fregean constraint on identity criteria and leaving behind the Quinean indispensability requirement. In other words, the conception advances no requirement to the effect that only objects with well-defined identity criteria exist (as opposed to Fregean constraint in arithmetic), nor does the conception take the indispensability of an item as a condition for its existence (in contrast to Quine's demand).

The ontologically neutral conception comes in two forms: one nominalist, the other agnostic. The nominalist approach denies the existence of abstract objects, since these objects are not ontologically independent from our linguistic practices and theoretical processes (Azzouni, 2004). The agnostic stance is characterized by the suspension of judgment regarding the commitment to abstract entities, given the difficulties of properly establishing their existence or their nonexistence (Bueno,

2011, 2012, 2020). Both approaches are, therefore, deflationary as they ultimately avoid commitment to abstract objects.

(i) On the *nominalist version*, ontological independence from linguistic practices and psychological processes is taken to be a criterion of existence (Azzouni, 2004). As a result, since mathematical objects are made up and have, thus, been introduced by those who conceived of them, these objects do not exist. Regarding their ontological status, mathematical and fictional objects are not different: both kinds of entities are, on this view, entirely fabricated and equally nonexistent.

(ii) In contrast, on the *agnostic version*, the existence of mathematical objects (or lack thereof) is not settled by mathematical practice. Perhaps these objects exist, perhaps they do not: nothing in the practice settles the matter. In fact, nothing in the practice requires commitment to the existence of mathematical objects (Bueno, 2020). This aspect of mathematical practice, nevertheless, does not entail that mathematical entities do not exist. For all we know, perhaps platonists about mathematics are right and these entities, being ontologically independent from our psychological processes and linguistic practices, do exist, after all. The fact that the same criterion—ontological independence—can be applied by both nominalists and platonists to support conflicting conclusions regarding the proper ontological status of the objects in question provides an understanding of the challenge to resolve the issue. If a very reasonable criterion for existence—the ontological independence of the objects under consideration—can be used in such dramatically inconsistent ways, it is unclear how the issue could be properly resolved.

It may be argued that the conflicting conclusions generated by the criterion of ontological independence ultimately highlight the criterion's inadequacy: no proper criterion could yield such blatantly inconsistent conclusions. The criterion is just unfit for the task at hand, given that clearly numbers cannot both exist and do not exist. In response, perhaps it could be pointed out that the issue has less to do with the criterion *per se* and more with the way the criterion is being applied. Abstract objects are taken by the nominalist as the kind of things that can be made up by the intentional acts of capable agents, such as mathematicians, just as fictional characters are similarly intentionally made-up things in which creation acts play a crucial role. One can invent a number just as one can create a fictional detective.

However, this comparison between numbers and fictional characters is, according to the platonist, a category mistake, as numbers cannot, even in principle, be created: they are necessarily existing, atemporal objects. Clearly, the nominalist will insist, to assume that numbers are necessary existents begs the question against the nominalist, who claims that no such abstract objects exist. Nonetheless, the platonist will retort, to conceive of numbers as the kind of things that can be made up is to think of them in a way that flies in the face of mathematical practice, according to which numbers are not temporal entities, created at certain moments in time. The nominalist, in response, will argue that even though numbers are created in a given moment, this situation does not change the fact that the particular circumstances of creation, such as its time or location, do not affect the properties of the corresponding numbers. After all, temporal and spatial properties are not encoded in the created objects. The platonist undoubtedly agrees with the premise but contests the conclusion: spatial and temporal properties are not encoded in mathematical objects *because* these objects are *not* the kind of things that can, even in principle, bear properties of this sort in the first place.

These considerations make clear the difficulty of settling the matter, given the lack of a common, suitable, non-question-begging interpretation of the criterion of existence required to resolve the issue. It is doubtful that, given the complexity of the subject, the matter can be resolved. The evidence available in support of conflicting conclusions regarding the existence or the non-existence of mathematical entities is inconclusive given the conflicting ways of interpreting the relevant existence criterion and the question-begging attempts to overcome the differences. In this way, ontological neutrality regarding the existence of mathematical objects, and abstract entities more generally, is achieved: whether these objects exist or not has not been settled. In the end, it is better to leave the matter open (Bueno, 2005, 2012, 2016, 2020).

In what follows, I focus on and critically engage with deflationist views. After raising concerns regarding the thin conception and the nominalist version of the ontologically neutral approach, I recommend the agnostic version of neutralism instead.

10.3 The Thin Conception of Objects: Some Difficulties

Crucial to the thin conception of objects, it should be clear by now, is the availability and adequacy of abstraction principles since thin objects are characterized in terms of them. Three difficulties emerge in this context.

1. The truth of an abstraction principle does not guarantee the existence of mathematical objects. Abstraction principles are used as part of implicit definitions of certain entities. Consider, once again, Hume's Principle: The $\#F = \#G$ if, and only if, $F \backsim G$ (that is, the number of Fs is the number of Gs if, an only if, there is a one-to-one correspondence between F and G). The singular terms '$\#F$' and '$\#G$' are taken to refer to numbers and the one-to-one correspondence between the numbers is an equivalence relation on them. Suppose that $\#F = \#G$. This states the identity of the numbers in question (namely, of Fs and Gs). Thought of as objects, the numbers are given in the Fregean construction discussed earlier:

$$\#[x : x \neq x] = 0$$
$$\#[x : x = 0] = 1, \text{and so on.}$$

Nevertheless, Hume's Principle itself does not presuppose any such numbers. On the contrary, the principle is used to show that a key relation holds among the numbers: their difference (Boolos, 1998). In fact, the logicist proof that $0 \neq 1$ (and related numerical differences) requires Hume's Principle. Since nothing falls under the concept $\#[x: x \neq x]$ (which is identical to 0) and exactly one object falls under the concept #[x: $x = 0$] (which is identical to 1), there is no one-to-one correspondence between these concepts. By Hume's Principle, they are different. Hence, $0 \neq 1$. The truth of Hume's Principle does not entail the existence of mathematical objects, since even if mathematical objects did not exist, #F and #G could be satisfied by physical objects—provided that there are enough of them, an unending supply of knives and forks, say, or perhaps, modally, possible concrete objects, such as possible spacetime regions—and a mapping between them, which pairs, one-to-one, knives and forks (or possible regions of spacetime). The same worry applies to predicative and dynamic abstraction principles.

It may be argued that the fact that abstract principles, such as Hume's Principle, do not require the existence of abstract objects in fact supports the principles in question. After all, this fact is an expression of their ontological lightness. In response, there is much to be said in favor of ontological lightness. The concern is that such lightness is not compatible with realism about the objects in question, given that, as formulated, ontological lightness is compatible with the *non*existence of thin objects—provided that the world cooperates in relevant ways (e.g., by having enough concrete entities). In the end, the considerations regarding Hume's Principle above indicate that the principle is compatible with abstract objects not existing at all.

2. Abstraction principles seem to be artificial and do not play a significant role in mathematical practice. They are unable to *illuminate* the practice and fail to help us to understand it better. In fact, it is not even clear how common mathematical objects can be characterized in terms of abstraction principles. In contrast to the characterization of numbers *via* abstraction, consider the way in which a metric or a group is characterized.

   A metric is a function from the Cartesian product of a set X of points onto the real numbers, such that:
   (a) $d(p, q) > 0$ if $p \neq q$; $d(p, p) = 0$;
   (b) $d(p, q) = d(q, p)$;
   (c) $d(p, q) \leq d(p, r) + d(r, q)$, for any $r \in X$ (Rudin, 1976, p. 30).

It is unclear that a metric has the proper form to be cast in terms of an abstraction principle: (i) It is not an identity between objects, but a two-place function. (ii) It does not provide identity criteria for objects in terms of an equivalence relation between such objects.

Consider the characterization of a group: A group is a set together with a binary operation on it (denoted by '$\circ$') so that if a and b are elements of the group, so is ($a \circ b$); moreover, the following axioms are satisfied:

(a) *Associativity*: $(a \circ b) \circ \mathrm{c} = a \circ (b \circ \mathrm{c})$.
(b) *Identity element*: There is an identity element, that is, an element e of the group such that for every element a of the group, $(e \circ a) = a$ and $(a \circ e) = a$.
(c) *Inverse element*: For each element a in the group, there is an element b in the group so that $(a \circ b) = e$ and $(b \circ a) = e$, where e is the group's identity element.

A group also does not have the proper form to be cast in terms of an abstraction principle: (i) It is not an identity between objects, but it is a set and an operation. (ii) It does not offer identity criteria for objects *via* an equivalence relation between such objects.

It may be objected that this complaint does not go through. If set theory can be formulated in terms of suitable abstraction principles (Linnebo 2018, Chap. 12, offers a modal set theory), we can recapture the characterizations of groups and metric spaces in it. Even if the formulation of set theory *via* abstraction principles succeeded (a big *if*!), this would be, however, beside the point. The difficulty raises a concern regarding the extent to which the use of abstraction principles *illuminates* mathematical practice. To reconstruct mathematical theories within an abstractionist mold is to create a parallel discourse to mathematical practice and even if the project were successful, it would fail to illuminate the practice *in its own terms*.

3. The modal status of thin objects is also troublesome. As Linnebo notes:

> *The less that the existence of an object demands of the world, the more modally robust the object is.* Since my desk makes a substantial demand on the world, for example, there are nearby circumstances in which it fails to exist: the carpenter who made the desk might instead have used the wood to make some chairs. So my desk is modally fragile. A pure abstract object, by contrast, makes no demand on the world and therefore exists by necessity. The cardinal number 0 would have existed however the circumstances had been, as there would still have been empty collections from which this number could be obtained by abstraction (Linnebo, 2018, p. 46).

Since thin objects, thought of as pure abstract objects, make no demand on the world, they exist "by necessity" and would have existed "however the circumstances had been".

However, are (thin) abstract objects necessary existents? (a) It is unclear that they exist in the first place. The fact that thin abstract objects make no demand on the world only establishes that nothing about the world matters for their existence or not, or for what properties they have or lack. It does not follow, however, that thin objects exist. Predicative abstract principles can still be true even if no abstract objects exist at all, given that, with enough physical objects in the world set in an appropriate relation, such principles can be satisfied.

(b) Even if thin objects did exist, it is unclear that they necessarily exist. In the modal set theory that Linnebo advances (2018, Chap. 12), he recaptures ZF, but *not* ZFC, as the axiom of choice is not included. The axiom of choice states that any collection of non-empty sets has a choice function, which maps each set of that collection to a member of that set. Is it the case that every set can be well-ordered or that every vector space has a basis? If the axiom of choice is true, there is a well-ordering for each set and there is a basis for each vector space; otherwise, none of these results go through. So, whether there are such well-orderings or such bases depends on whether the axiom of choice is the case. Despite thin objects not making any demands on the world, whether they exist or not depends on the existence of other thin objects. This means that thin objects are contingent rather than necessary, since they are contingent on the existence of other thin objects: whether there are well-orderings for every set and bases for every vector space depends on whether choice functions for collections of non-empty sets exist.

(c) Abstract principles presuppose the existence of certain abstract objects that, presumably, have not been characterized *via* abstract principles. Equivalence relations, on their usual characterization, are themselves abstract objects, and given that they are required for the formulation of abstract principles, they need to be in place in order for abstract principles to get off the ground. But this requires the existence of abstract objects that arguably have not been formulated in terms of abstract principles. As a result, it seems that an abstractionist approach to mathematics cannot provide a complete, independent view of the subject, as it requires an additional view of abstract objects to support it (or, at least, a nonstandard account of equivalence relations).

10.4 An Alternative Neutral Approach

In order to articulate an alternative deflationary approach, it is important to disentangle quantification and existence (Azzouni, 2004; Bueno, 2005). We often quantify over objects we do not believe to exist. Consider, for instance:

(*) Some sets are too big to exist.

If "some" is understood in an ontologically committing way, we immediately obtain a contradiction: "There exist sets that do not exist". Clearly, this fails to express what was intended with (*), since no contradiction is, in fact, involved.

This tension is avoided if one adopts quantifiers that have no commitment to existence: ontologically neutral quantifiers (Azzouni, 2004; Bueno, 2005). Existence is then expressed *via* an existence predicate, whose conditions are not a matter of quantification, but depend on the relevant metaphysical assumptions. In fact, (*) can be easily formulated in terms of ontologically neutral quantifiers as follows: $\exists x\ (Sx \wedge \neg Ex)$, where '$S$' stands for sets and '$E$' is the existence predicate.

In this way, ontologically neutral quantifiers offer a strategy to avoid commitment to mathematical objects, given that one can quantify over them while leaving the issue of their existence undecided. In fact, even those objects that are indispensable to our best theories of the world need not exist, as indispensability often involves linguistic issues regarding what can be expressed in a theory rather than ontological questions about what exists in the world (Bueno & Shalkowski, 2020). Once it becomes clear that quantification does not require the existence of the objects that are quantified over, the universal quantifier just ranges over the entire domain of quantification, whereas the existential quantifier ranges over a portion of that domain—with no assumption being made about the existence of any of the relevant objects.

A significant feature of the proposal is that mathematical discourse can be taken literally. There is no need to rewrite mathematical theories so that they fit one's preferred ontology. Theories can be taken exactly as they are formulated in mathematical practice, leaving entirely open the issue of the existence of mathematical objects, properties, or structures.

As noted above, the nominalist version of the neutral view (Azzouni, 2004) insists that mathematical objects do not exist given that they do not satisfy the criterion of existence that is favored by that conception, namely, ontological independence from one's psychological processes and linguistic practices. However, as a criterion for what there is, ontological independence is at best radically indeterminate and at worst incoherent as a standard to determine what exists. On the one hand, the criterion justifies nominalism since, it is argued, mathematical objects are entirely made up by the relevant practitioners and, thus, are ontologically dependent on the latter. On the other hand, ontological independence also justifies platonism and the corresponding existence of mathematical objects. After all, according to the platonist, mathematical objects do *not* depend on what anyone thinks or says about them, since they are indeed ontologically independent. Thus, ontological independence is unable to determine which of these incompatible views—nominalism or platonism—is true: the criterion is radically indeterminate.

Alternatively, if one insists on applying the criterion to both rival views, it follows that mathematical objects exist and that they do not. What results, clearly, is incoherence. In either case, the criterion of existence that is central to deflationary nominalism is highly problematic and is better left behind.

This is not a difficulty faced by the agnostic form of neutralism, the one I favor, since the stance does not advance necessary conditions for the existence of objects (Bueno, 2020). In order to avoid begging the question against rival views, the agnostic neutralist suggests only sufficient conditions for existence, such as being causally active or located in space-time—objects that satisfy conditions of this kind arguably exist. No incoherence or indeterminacy results.

The three challenges that were raised against thin objects can also be accommodated by the neutral approach. (1) It was argued that the truth of an abstraction principle does not guarantee the existence of mathematical objects. The neutral approach does not rely on abstraction principles to formulate mathematics: it follows the informal, but rigorous, way in which mathematics is, in fact, done. Metric functions and groups are characterized informally, but rigorously, in natural language (plus some special symbols), and quantification ranges, unperturbed, over the relevant domains, including sets, functions, numbers, and spaces. One has, in this way, a thin access to these objects, independently of the objects' existence.

Crucial to mathematical practice is what follows from the relevant mathematical principles, which typically are *not* abstraction principles. If every metric space is a topological space, that is, if given a metric, *there is* a corresponding topology, this is the case because one can construct, from the metric, open balls that induce a topology. If, however, not every topological space is a metric space, this is because *there is* a particular topology, such as the indiscrete one, for which no metric can be constructed. Existence is not understood in any metaphysically robust way: nothing in the practice settles the matter as to whether the objects are abstract, universals, tropes, properties, individuals in a possible world, nodes in a structure, modal-structural constructs, or Fregean concepts. These are metaphysical additions to the practice. Ultimately, they are not needed to resolve issues about the practice, and it is unclear how to settle these metaphysical issues in any case (see Bueno, 2020).

(2) It was also complained that abstraction principles seem to be artificial and do not play a significant role in mathematical practice. They are unable to *illuminate* the practice and do not help us to understand it better.

Given that there is no place for abstraction principles in the neutral approach, the objection does not apply either. Mathematical practice is understood *in its own terms*. The informal, but rigorous, language of mathematics suffices to characterize the objects of investigation. The neutral approach urges one to resist the temptation of adding a layer of metaphysical interpretation to the practice, rewriting mathematical theories so that they will fit one's preferred ontology. It is worrisome that abstraction principles are not even formally adequate to characterize basic mathematical objects (such as metrics and groups), but perhaps this should not be surprising, since the goal of the thin ontological approach was not to make sense of the practice, at least not of large portions of the practice that could not be cast in the abstractionist mold. Yet, this is still a cost of the view.

By taking mathematical practice literally, the neutral approach can make sense of it: Why is the practice informal but rigorous? Because this is a simple and effective way of proceeding. There is typically no need to cast mathematics in a more formal setting. Why is mathematical practice neutral on ontological issues? Because these issues are typically irrelevant to establish mathematical results, which is a key goal of the practice.

(3) Finally, it was complained that the modal status of thin objects is troublesome, given that (a) it is unclear that thin objects exist in the first place, and (b) even if they did, it is unclear that they necessarily exist. The neutral approach allows for quantification over mathematical objects without commitment to their existence (as ontologically independent entities). This is just the way in which mathematicians operate. Being neutral on ontological issues is an advantage since it allows one to accommodate the corresponding neutrality of mathematical practice.

Furthermore, the neutral approach welcomes the contingency of mathematical objects (Bueno, 2020). In the end, the presumed necessary existence of mathematical entities is an artifact of the rigid presentation of mathematical theories as though classical logic and classical mathematics were the only game in town. An acknowledgement of the pluralism typical within mathematics should give us reason to pause and reconsider the alleged necessary existence of mathematical objects (for further discussion, see Bueno, 2020).

10.5 Conclusion

After mentioning two older (Fregean and Quinean) views about objects, I contrasted two more recent ones: an *abstractionist* conception that favors the commitment to thin objects *via* abstraction principles, and a *neutral* approach that, in closer proximity to mathematical practice, takes objects as those items that are quantified over, but without the corresponding ontological commitment. Three concerns were raised to the abstractionist conception, and these concerns, I argued, can be accommodated by the neutral approach.

Far more, of course, needs to be said about the issues examined here, but I hope enough was said to suggest that the neutral approach is, at least, worth considering carefully.

References

Azzouni, J. (2004). *Deflating existential consequence: A case for nominalism.* Oxford University Press.

Boolos, G. (1998). *Logic, logic, and logic*. Harvard University Press.

Bueno, O. (2001). Logicism revisited. *Principia, 5*, 99–124.

Bueno, O. (2005). Dirac and the dispensability of mathematics. *Studies in History and Philosophy of Modern Physics, 36*, 465–490.

Bueno, O. (2011). Logical and mathematical knowledge. In S. Bernecker & D. Pritchard (Eds.), *Routledge companion to epistemology* (pp. 358–368). Routledge.

Bueno, O. (2012). An easy road to nominalism. *Mind, 121*, 967–982.

Bueno, O. (2016). An anti-realist account of the application of mathematics. *Philosophical Studies, 173*, 2591–2604.

Bueno, O. (2020). Contingent abstract objects. In J. L. Falguera & C. Martínez-Vidal (Eds.), *Abstract objects* (pp. 91–109). Springer.

Bueno, O., & Shalkowski, S. (2020). Troubles with theoretical virtues: Resisting theoretical utility arguments in metaphysics. *Philosophy and Phenomenological Research, 101*, 456–469.

Frege, G. (1885/1953). *The foundations of arithmetic: A Logico-mathematical enquiry into the concept of number* (2nd revised ed.; J. L. Austin, Trans.). Harper and Brothers.

Gödel, K. (1964). What is cantor's continuum problem? In P. Benacerraf & H. Putnam (Eds.), *Philosophy of mathematics: Selected readings* (Second ed.). Cambridge University Press.

Hale, B., & Wright, C. (2001). *The reason's proper study: Essays towards a neo-Fregean philosophy of mathematics*. Oxford University Press.

Jech, T. (2006). *Set theory* (3rd revised and expanded ed.). Springer.

Linnebo, Ø. (2018). *Thin objects: An abstractionist account*. Oxford University Press.

Lowe, J. (1998). *The possibility of metaphysics: Substance, identity, and time*. Clarendon Press.

Lowe, J. (2002). *A survey of metaphysics*. Oxford University Press.

Quine, W. V. (1953/1980). *From a logical point of view: Nine logico-philosophical essays* (2nd revised ed.). Harvard University Press.

Quine, W. V. (1960/2013) *Word and object*. (New ed.). MIT Press.

Resnik, M. (1997). *Mathematics as a science of patterns*. Clarendon Press.

Rudin, W. (1976). Principles of mathematical analysis (Third edition.) New York: McGraw-Hill.

Shapiro, S. (1997). *Philosophy of mathematics: Structure and ontology*. Oxford University Press.

Chapter 11
Deflating Fictionalism

Suki Finn

Abstract This paper describes how fictionalism can be coupled with the quietism of Carnap, to create the position 'Neo-Carnapian Fictionalism'. Carnap argues that it is not meaningful to question reality in an external sense to assess what really exists, rather it is only meaningful to talk in an internal sense within a framework about what exists according to the framework rules. I use the concept of fictions in place of frameworks to argue that we ought to treat much seemingly ontologically committing language as consisting in nothing more than a useful heuristic and as being fictional. This reframes ontological debates as being based around the practical advantages of utilizing a way of speaking in a fictionalist manner. The aim of this paper is thus to resurrect Carnap's metaontology in the form of a deflated global fictionalism, that is divorced from the antirealism usually associated with fictionalism, where the fictionalism is based on quietism instead of antirealism.

Keywords Fictionalism · Quietism · Metaontology · Carnap · Yablo · Thomasson

11.1 Introduction

In this paper I put forward a novel account in metaontology that combines fictionalism and quietism, which I call 'Neo-Carnapian Fictionalism'. I argue that Carnap's quietism can be helpfully reinterpreted as a type of fictionalism, and also that fictionalism is best developed as a form of quietism, such that they meet in the middle in a mutually beneficial place. The benefits of the combination are as follows: the fictionalist version of quietism that results avoids Quine's charge of dependence on a problematic analytic/synthetic distinction, and; the quietist version of fictionalism that results avoids dependence on antirealism and the Quinean assumptions that lead fictionalism into an unnecessary error theory. This Neo-Carnapian position,

S. Finn (✉)
Royal Holloway University of London, London, UK

X. de Donato-Rodríguez et al. (eds.), *Deflationist Conceptions of Abstract Objects*, Synthese Library 516, https://doi.org/10.1007/978-3-031-98717-5_11

despite being deflated into a quietist position, is still continuous enough with traditional fictionalism in maintaining the key elements and motivations of such fictionalist positions to be deserving of the title of 'Fictionalism': hence, the name 'Neo-Carnapian Fictionalism'.

In Sect. 11.2 I describe Carnap's metaontology and the effect that this has on traditional fictionalism in order to motivate a quietist version of fictionalism. In Sect. 11.3 I follow Yablo in resurrecting Carnapian metaontology in the form of fictionalism in order to avoid Quine's charge of dependence on an analytic/synthetic distinction, and I put forward my own fictionalist account of quietism inspired by Yablo. Section 11.4 outlines common features of traditional fictionalisms to show how the Neo-Carnapian position satisfies them, despite being quietist (rather than antirealist). Finally, Sect. 11.5 compares the resulting Neo-Carnapian Fictionalist position with Thomasson's deflationary Simple Realism, in order to demonstrate its unique position in the debate.

11.2 Quietist Account of Fictionalism

The type of quietism I am interested in developing was made famous by Carnap, in his seminal 'Empiricism, Semantics, and Ontology' paper from 1950. In that paper, Carnap distinguishes between internal (IQ) and external (EQ) questions with regard to linguistic frameworks.[1] These frameworks provide the meaning and rules for usage of our language for terms denoting particular kinds of entities and are adopted on a pragmatic basis where "efficiency, fruitfulness, and simplicity of the use of the thing language may be among the decisive factors."[2] To use language internally to the framework is to speak about how things are "based on the rules for the new expressions"[3], whereas externally is independent of the framework, "prior to the acceptance of the new framework".[4] Carnap claims that "the concept of reality occurring in these internal questions is an empirical scientific non-metaphysical concept"[5] yet in external questioning we ask whether entities "have a certain metaphysical characteristic called reality".[6]

Take the following example of an internal and external reading regarding whether there is a prime number between 2 and 4:

Internal reading: Is there a prime number between 2 and 4?

---

[1] Carnap (1950, p. 21). Carnap does not use the term 'quietism', but it is often ascribed to him. E.g. See Carus (2018, p. 456), Leng (2010, pp. 28–31), Allen (2016, p. 156), and Price (2011, pp. 13 & 313). See Macarthur (2008) for an account of metaphysical quietism.

[2] Ibid., p. 23.

[3] Ibid., p. 24.

[4] Ibid., p. 24.

[5] Ibid., p. 22.

[6] Ibid., p. 24.

External reading: Are there really numbers, like 3?.[7]

Here, the IQ will be answered internally according to the rules of the number framework to show whether 3 satisfies the definition of being a prime or not. Alternatively, the EQ questions the existence of numbers like 3 externally and independently of the number framework. Note that the ontological debate in metaphysics is traditionally concerned with the external questioning of what really exists, rather than what exists internally according to the framework rules. But Carnap rejects the ontological debate rendered in this way (and the EQs it is based on) as being meaningless, since it attempts to utilize language externally from where it gets its meaning. As such, Carnap "feel[s] compelled to regard the external question as a pseudo-question", and therefore reframes and salvages the EQ as "not a theoretical question but rather the practical question whether or not to accept those linguistic forms"..[8]

With the metaphysical EQ turned pragmatic, the ontological debate would thus be transformed from what really exists into an evaluation of the practical virtues of accepting a framework based on whether it is "fruitful [and] conducive to the aim for which the language is intended."[9] The EQ 'are there Xs?' is now the question 'is it useful to adopt the X framework?'. It is not the existence of X, but the usefulness of an X framework, that is at the heart of Carnap's 'ontological' debate. Ontology thus becomes the study of which frameworks are useful to adopt, and these frameworks can then be utilized to provide internal existence answers. Any further metaphysical question about whether something really exists is rendered meaningless, hence the resulting 'quietism' about metaphysics. Metaphysical debates concerning external reality are rejected as meaningless, and the only ontology remaining is internal to frameworks which are chosen pragmatically.

The realist about X, in a metaphysical sense, holds that Xs exist—that there really are Xs. For Carnap, this is meaningless. A charitable, meaningful, Carnapian interpretation takes it that the realist answers the EQ positively (to adopt the framework for X) and answers the IQ positively (so as to state that Xs exist according to the framework). An antirealist about X in a metaphysical sense holds that Xs don't exist—that there really are no Xs. Again, meaningless for Carnap. There are two options for how to charitably and meaningfully revise antirealism for the Carnapian: (1) the antirealist answers the EQ negatively and thus does not adopt the X framework, or; (2) the antirealist answers the EQ positively to adopt the X framework and then answers the IQ negatively to deny the existence of X. But both ways are paradoxical.[10]

---

[7] This example is similar to one from Carnap's 'system of members' in his (1950, p. 24).

[8] Ibid., p. 32.

[9] Ibid., p. 29.

[10] This is similar to the problem of true negative existentials, also known as the paradox of non-being or Plato's beard. In that problem, one cannot meaningfully deny the existence of X because in order to do so one has to successfully refer to X (which on a Quinean understanding requires X to exist). This successful reference in order to deny existence results in a contradiction. A similar issue arises for the antirealist on a Carnapian model, in that either they have to reject the X frame-

Carnap states that only "once we have accepted the thing language with its framework for things, we can raise and answer internal questions."[11] As such, someone who does not adopt the framework for X cannot take any stance on X at all, for they cannot meaningfully speak of X without its framework, not even to deny the existence of X. Any attempt to deny the existence of X via an external statement about reality is going to be meaningless, as any "alleged [external] statement is a pseudo-statement without cognitive content".[12] And accepting the framework just to deny X also seems problematic, as it would imply both that X-discourse is useful and that it is useful to deny Xs. This is not viable to the Carnapian.

Traditionally, fictionalism is motivated by antirealism:[13] Eklund states that "the typical motivation for fictionalism is ontological."[14] But this ignores what Eklund calls a "crucial distinction"[15] between linguistic and ontological fictionalism. He describes the linguistic view "according to which utterances of sentences of the discourse are best seen not as efforts to say what is literally true, but as useful fictions of some sort."[16] On the other hand, he describes the ontological view as holding "that the entities characteristic of the discourse do not exist, or have the ontological status of fictional entities."[17] He goes on to claim that the linguistic and ontological views often get conflated due to the linguistic being motivated by the ontological. Given that this conflation often happens, when I speak of 'traditional' fictionalism, I will mean a combination of the linguistic and ontological fictionalism, whereby the reason for the linguistic part of the fictionalism is the ontological part.

Now, the Neo-Carnapian Fictionalist position that I will outline in this paper divorces fictionalism from antirealism and thus may well turn out to be something like the linguistic fictionalism described by Eklund. But there is a difference that serves to show why the Neo-Carnapian Fictionalist position is unique: the reason the discourse is not 'literally true' for the Neo-Carnapian Fictionalist is not because it is literally false (as the traditional fictionalist would have it, especially if it is ontologically motivated) but because it is meaningless (due to quietism).

---

work to deny Xs existence (thereby robbing themselves of the language required to meaningfully deny the existence of X), or they accept the X framework only to deny that X exists.

[11] Carnap (1950, p. 22).

[12] Ibid., p. 29.

[13] Fictionalists who are antirealists include Leng (2010) on mathematics, Joyce (2001) on morality, van Fraassen (1980) on scientific theories, Rosen (1990) on possible worlds, Brock (2002) on fictional entities, Kroon (2001) on temporal parts, and Yablo (2000) on abstract entities. But there are also agnostic fictionalists (Bueno, 2009), realist fictionalists (Jay, 2011), and indifferent fictionalists (Eklund, 2005), and these 'alternative' styles of fictionalism will not fall under what I call 'traditional'.

[14] Eklund (2005, p. 558).

[15] Eklund (2015).

[16] Eklund (2015).

[17] Eklund (2015).

Ontological fictionalism cannot be understood if we are to take Carnapian quietism seriously, given that the antirealism that underlies that position is meaningless or a somewhat paradoxical position to hold. And linguistic fictionalism cannot be understood as being motivated by the ontological view for the same reasons.[18] Fictionalism must then be removed from any antirealist foundation.

Traditional fictionalists hold that it is acceptable to say (i) 'there is a prime between 2 and 4' as an answer to the IQ, even though they strictly regard this answer as literally false given that they too say (ii) 'numbers do not exist' as an answer to the EQ. However, since the Carnapian rejects this metaphysical understanding of the EQ, this type of fictionalist position cannot be meaningfully formed. One could not answer the EQ because to do so is meaningless unless construed as the acceptance of a framework, and so in answering the EQ negatively the traditional fictionalist makes a meaningless claim and does not accept the framework within which they would then go on to answer the IQ. In denying the EQ they deny themselves the framework within which they would answer the IQ, and thus say meaningless things. What distinguishes fictionalism from other antirealist positions is that the fictionalist continues to utilize a false discourse because of its usefulness (as opposed to eliminating the discourse, as an eliminativist would do for a false discourse). So, in some sense, fictionalists do adopt the framework that provides meaning for X (by talking about X) even though they answer the EQ negatively by being antirealist about X.

This now does not make sense on the Carnapian model, since answering the EQ negatively is simply to not adopt the framework.[19] The traditional mathematical fictionalist, for instance, is antirealist since they deny that there are objects in their ontology that are numbers, but is distinguished by further saying that we should still talk as if there are objects in the ontology that are numbers since it is useful to do so. So, they strictly answer the EQ negatively (that there are no such things existing to which their talk refers to, and so are antirealist about those things) but continue talking about such things for its use and so in some sense 'pretend' to answer the EQ positively in order to utilise the framework for X. In effect, they talk like a realist without being a realist—they want to use the language of a framework without adopting the framework (to have their cake and eat it too).

My quietist version of fictionalism, then, reduces fictionalism down to a position that looks very similar to quietism itself. It states that one is quiet about the metaphysical ontology, and utilizes language for its usefulness. Indeed, Carnap's

[18] As I shall show later in Sect. 11.4.2, the linguistic fictionalist view also cannot be understood once we remove the Quinean assumption that to talk truthfully of X requires successful reference to X, where X is an existent thing. Without that assumption, ontological fictionalism would not entail linguistic fictionalism, as the nonexistence of X would not lead to talk of X being not literally true. Linguistic fictionalism then would require different motivation for its error theory (in thinking that talk of X is not literally true) that doesn't appeal to the nonexistence of X.

[19] Thanks to Amie Thomasson who argues that it's not clear that the fictionalist answers the EQ negatively given that they utilize the framework, and it is hard to fit the fictionalist into Carnap's model, since they seem to adopt the talk yet deny the entities, like some sort of external antirealist whilst endorsing some internal realism.

motivation for his quietism was to allow people to speak as they feel is useful without ontological commitments, in order to be tolerant of 'free speech' (not in a political sense) without ontological scruples getting in the way. Therefore this description of Carnap makes him sound like a fictionalist—the value of language as being useful independent of existence, and to have the freedom of speech (talking like a realist) without being ontologically committed to the things that you talk about (not being a realist). These typical characteristics of fictionalism will be discussed in Sect. 11.4. Next, I shall demonstrate how to formulate a fictionalist account of quietism.

11.3 Fictionalist Account of Quietism

Famously, Quine attacks Carnap's internal/external distinction by guilt of association with the analytic/synthetic distinction.[20] Quine believes that Carnap's notion of a linguistic framework needs an analytic/synthetic distinction due to the framework rules being analytic, and so without the distinction there can be no notion of a framework, which is central to Carnap's account. Quine argues that no definition of analyticity can be given,[21] and this is part of his argument against there being an analytic/synthetic distinction. Quine further argues that Carnap's derivation of the meaninglessness of the EQ (other than being a pragmatic matter of framework choice) cannot be reached without admitting such an analytic/synthetic distinction.

Yablo has since responded to this critique and argues that one does not need to hold that the rules making up linguistic frameworks are analytic in order to be able to understand the idea that some ways of talking can be adopted for merely practical reasons.[22] Yablo defends Carnap by calling for a change in how frameworks are understood, and in this paper I follow Yablo's suggestion and redefine frameworks as fictions without analytic rules, inspired by his replacement of the internal/external distinction with a metaphorical/literal distinction which can be made independently of an analytic/synthetic distinction.[23] Yablo says we could think of this distinction as between assertions in make-believe fictional games and those assertions outside of the game:

The usual charge against Carnap's internal/external distinction is one of 'guilt by association with the analytic/synthetic'. But it can be freed of this association, to become the distinction between statements made within make-believe games and

[20] Quine (1951a).

[21] Quine (1951b).

[22] Yablo (1998).

[23] Note that this is not a position that Yablo currently holds or followed through with in the way that I have—he more recently calls himself a 'quizzicalist'. For example, see his (2014) and (2017) which I have argued against in: Finn (2017b).

those made outside them—or rather, a special case of it with some claim to be called the metaphorical/literal distinction.[24]

We can free the internal/external from the analytic/synthetic distinction once we remind ourselves of the purpose and role of Carnap's IQ/EQ distinction in the first place. The IQ was meant to provide a way in which we can assert a sentence S (which will become true relative to the framework it is internal to), without presupposing or requiring an answer to the EQ as to whether S is really true (external to the framework). The internal/external distinction can be understood as a way of saving our internal assertions from being bound by external truth-values, so that we may freely assert for purely pragmatic purposes instead. Yablo describes the enterprise of utilizing language for its pragmatic features rather than its truth-value as such:

> Now, what is our usual word for an enterprise where sentences are put at the service of something other than their usual truth- conditions, by people who may or may not believe them, in a disciplined but defeasible way? It seems to me that our usual word is 'make-believe game'.[25]

These make-believe games are what I will take as fictions. The same analogy can be made for fictions, since in fictional contexts we make assertions internally to fiction without a care for whether the assertions are true independent of the fiction (and sometimes we even know them to be false). This is much like how Carnap wishes ordinary language to be utilized internal to linguistic frameworks—carefree of independent external truth-values. The parallel between Yablo's distinction and the IQ/EQ is clear with regard to the meaninglessness of questioning truth external to frameworks or make-believe games:

> Numerical calculation does not answer to external facts about numbers for the same reason that players of tag don't see themselves as answerable to game-independent facts about who is really 'IT'; just as apart from the game there is no such thing as being 'IT', apart from the framework there is no such thing as being 'the sum of seven and five'.[26]

When we speak of numbers and make numerical calculations we are not committed to the ontological reality of numbers (at least for the mathematical fictionalist, like Leng[27]), just like when we are involved in a game or in a fiction we are not committed to the reality of what we are making believe to be the case. The analogy is between the game (or fiction) and Carnap's frameworks, with regard to what is external to them—being external to the fiction or framework is to be external to where the game or language gets its meaning. As such, in Yablo's words, there is 'no such thing' as the game- (or fiction-) or framework- independent realm. So the

---

[24] Yablo (1998, p. 229).

[25] Yablo (1998, p. 243).

[26] Yablo (1998, p. 240). A parallel here to fictional works is that it would be meaningless to question something from within the fiction in an external manner, i.e. 'Does Sherlock smoke a pipe externally to the Holmes stories?'. So, questioning external to frameworks is just as meaningless as questioning external to fictions.

[27] Leng (2010).

realm independent of the game (or fiction) becomes as inaccessible and meaningless as the external is for Carnap. It is just as meaningless to ask the EQ of whether the sum of seven and five is really twelve independent of the number framework as it is to ask if I am really 'IT' (a.k.a 'tag') independent of the game of IT.

These make-believe games, or fictions as I will take them, are chosen in much the same way as Carnap proposes to choose frameworks: by being the most 'apt' or having the most 'cognitive promise',[28] where Carnap's choice of framework is pragmatic by being the most 'useful'. Fictions can thus be used in place of frameworks to make Carnap's quietist point without reference to analyticity. The analytic rules that Quine found problematic are replaced by presuppositions of fiction. The fictions are generated by these presuppositions, whose status is such that we choose not to question them. The principles of generation of fictions are treated as being not up-for-grabs, just like presuppositions are, and this is how we should treat frameworks. If we presuppose the framework, then things will follow from that framework's rules, much like when we presuppose a fiction with a story that follows. The Carnapian now has no need for analyticity in their internal/external distinction or in their formulation of a framework as there is no special status of rules—the frameworks just have things following from presuppositions, as the rules themselves are the presuppositions.[29]

My idea is therefore to interpret Carnapian linguistic frameworks as fictions, thus creating a fictionalist account of Carnapian quietism, inspired by Yablo's construal of the internal/external distinction being made with regard to make-believe games in order to show that it is free of the analytic/synthetic distinction. Where there were frameworks for entities for Carnap (e.g. the 'number' framework), now there will be fictions for entities.[30] When we utilize number-discourse, we employ a number fiction, and our numerical calculations will be as ontologically uncommitting as the assertions made within fictional contexts are. Our employing of fictions is metaphysically insignificant, and is only evidence of us finding that fiction as being practical to employ in order to speak of the entities internal to that fiction. Just as we presuppose a fiction without taking it on in any metaphysically serious way, we also presuppose linguistic frameworks in metaphysically neutral/quiet way. The Neo-Carnapian Fictionalist therefore construes frameworks as fictions, and is quiet about the area outside of them (the metaphysical external reality).

---

[28] Yablo (2005, p. 101).

[29] Thanks to Mary Leng for pointing out that the Quinean may object to such presuppositions and principles of generation being not up for grabs. For Quine, on his alternative confirmational holistic picture, even those claims that strike us as most plausibly analytic are at least in principle revisable in the light of experience (and so are not analytic after all). If that's the problem with the analytic/synthetic distinction it looks like it may also be a problem for fictions as generated by presuppositions and principles of generation that aren't revisable. In response, the 'revision' would be just to adopt another fictional framework instead with different principles of generation.

[30] Darren Bradley (private correspondence) notes a Quinean objection that we may not be able to individuate fictions (nor frameworks) without a notion of analyticity. Yet Quine's denial of analyticity needn't be considered decisive, as even Quine states analyticity "undeniably has a place at the commonsense level" (Quine, 1991, p. 270).

But note that the Neo-Carnapian is not a fictionalist about this external reality, and so does not treat talk of external reality as fictional. Rather the fictionalism manifests with the idea that all our internal talk is internal to fictions, and is therefore presented as a way of construing the frameworks as fictions. Given that Carnapian quietism is global, so will the fictionalism be global, as all frameworks will be considered as fictions. Therefore, this is a fictionalist way of dismissing all EQ's concerning what really exists, and treating all IQ's as relative to fictions. All talk is fictional and the contrast class is lost.

All frameworks are treated as fictions, and sentences can be true or false internally to them.[31] Internal talk becomes truth-evaluable in a quietist way. So the IQ 'is there a prime between 2 and 4?' is a fictional question, and when one answers that 3 is a prime number they are answering according to the mathematics fiction. Talk of 3 will not be about the number 3 externally, as this is meaningless. If our talk about numbers is not about numbers in the external world, then we may ask what it is about, and the Neo-Carnapian Fictionalist responds that it is about the mathematics fiction, referring to things in that fiction. Sentences will be made true in an internal sense in virtue of fiction presuppositions, just as Carnap's frameworks had rules that made sentences true according to it. Truth becomes relative to fiction. This will be discussed in Sect. 11.4.3, distinguishing 'true' for internal truth from 'True' for external truth.

I will now briefly describe an important difference between (1) fictionalism about frameworks and (2) frameworks as fictions. I am endorsing (2), only. Fictionalism about frameworks is the idea that we adopt the fictional story that there is an X framework. Whereas, to treat frameworks as fictions is the idea that the X framework is a fictional story about X. When it comes to (1), making an ontological claim about the status of the frameworks is something that a quietist Carnapian would not (and could not) do, since it would require a framework-of- frameworks to talk internally to in order to question the existence of frameworks and thus assess whether we should be fictionalist about the frameworks themselves.[32] As such, the way to combine fictionalism and quietism, (to meet in the middle, so to speak), is to take a fictionalist account of quietism by treating frameworks as fictions, and to take a quietist account of fictionalism to divorce it from antirealism. I will now describe how the quietist fictionalism of Neo-Carnapian Fictionalism compares to traditional fictionalism.

---

[31] This means we get truth-in-a-fiction as opposed to truth-in-a-framework. Thanks to Darren Bradley for coining such a slogan and suggesting that I put it in these terms.

[32] This creates a problem for the Carnapian in general as to how they motivate their quietism and what they take the frameworks to be. I have discussed this as a dilemma in Finn (2017a) which asks whether the Carnapian position itself should be considered as internal or external to frameworks. Treating frameworks as fictions avoids this dilemma by insisting that we start with the metaontology to inform what we mean when we say something ontological about fictions, rather than starting with our ontology of fictions and applying this at the metaontological level to inform what we mean by fictionalism or quietism. This is also discussed further in Sect. 11.4.

11.4 Traditional Features of Fictionalisms

A fictionalist about entity X sometimes describes her fictionalism as treating X as a fiction. It then remains to be said what it is like to treat X as a fiction, and what being a fiction is like. These fictionalists may look to the metaphysics of fiction in order to clarify their own position, as in understanding their fictionalism in terms of fiction they must first understand their metaphysics of fiction. I argue that the strategy of looking to the metaphysics of fiction first (in the way that, say, Sainsbury does)[33] is the wrong starting point. A better direction is to do the metaontology first and this will then tell you what you can say about the metaphysics of fiction (so, to look to fictionalism before fictions). Just as getting one's metametaphysics clear can clarify one's metaphysics, likewise getting one's metaontology clear can clarify one's ontology. With an understanding of our metaontological position of fictionalism, we can then understand our ontological position on fiction. So we clarify meta-P before clarifying P: the 'meta' has priority.

A Carnapian quietist denies the meaningfulness of the debate between realists and antirealists, for all things X, including fictions. This is because these quietists reject all metaphysical ontology as meaningless. Included in the quietist's list of rejected debates, then, is the debate over whether fictional entities really exist or not. The entire debate surrounding the metaphysics of fiction (whether to be realist or antirealist about fictional entities) is faulty according to the Neo-Carnapian (unless understood internally). Therefore, they cannot apply the results of that debate to inform their fictionalism, but rather, they apply their quietism to that debate in rejecting it. The metaphysics (of fiction) does not influence the metametaphysics (of fictionalism), rather it goes the other way round. The sort of metaontology you buy into will prescribe the sort of ontology you will be able to do, and since the Neo-Carnapian buys into quietism, this entails that the Neo-Carnapian cannot take a metaphysical ontological stand on any X (including fictions). Therefore, the Neo-Carnapian cannot look to the metaphysics of fiction to inform their fictionalism, since they are quietist and thus will deny the meaningfulness of the metaphysics of fiction.

Traditional fictionalism, as described in Sect. 11.4.1, tends to be based on antirealist foundations. The antirealist aspect of these traditional fictionalist positions plays an important role in motivating the fictionalist ideology of using a discourse for its usefulness. It is influenced by the idea that one cannot speak truthfully of an entity X unless X exists, and so if X is non-existent then any discourse purporting to refer to X will be false. The traditional fictionalist move is to then continue to utilize this discourse even though it is false, because it is deemed as being useful. The fictionalist motivation is thus derived in four steps, from antirealism to pragmatism, as such:

Antirealism: Xs do not exist.

---

[33] Sainsbury (2009).

Quinean Assumption: If Xs do not exist then X-discourse is not True.[34]
Error Theory: If X-discourse is not True, then the X-discourse (user) is in error.
Pragmatism: Despite being in error, X-discourse is useful.

Traditional fictionalisms are thus united in the following common features: (i) Antirealism; (ii) the Quinean Assumption; (iii) Error Theory; and (iv) Pragmatism. Traditional fictionalisms also take on the following features: (v) involving something less than absolute truth; and (vi) involving something less than belief. The truth involved in (v) can be relative truth, or true under a presupposition. The attitude involved in (vi) can be acceptance, where we accept what is presupposed and believe what it derives. I will go through each of these common features in turn to show how the motivation for traditional fictionalism is somewhat undermined by the quietist, yet the Neo-Carnapian Fictionalist can still maintain the ideology of traditional fictionalism in many ways. Most importantly, the positive component of traditional fictionalism (the Pragmatism where discourse is valued as useful) is retained by the Neo-Carnapian despite lacking the negative component (the Antirealism).

11.4.1 Antirealism

The typical motivation for traditional fictionalisms is ontological, namely to be an antirealist but talk like a realist, so that one may reap the benefits of the discourse without "paying an ontological bill".[35] A traditional fictionalist about a type of entity X claims that X is like a fictional entity in being unreal, and so we are ontologically uncommitted to the fictional claims about X. Fictionalism is thus traditionally antirealist because statements about X are considered fictional and X is considered unreal rather than factual and real. Despite being antirealist in theory, this fictionalist appears as a realist in practice, because in practice they use the language as they would if they were realist for its practical advantages. However, the fictionalist need not be realist in practice all of the time as they may assert their antirealist beliefs in critical contexts. As Berkeley said (though in a different context), they 'think with the learned, and speak with the vulgar'.[36] The defining aspect of the fictionalist is that they adopt a fiction some of the time in order to continue using a realist-like discourse despite their antirealist beliefs, so as to talk without ontological commitment to something that they consider unreal. The discourse is seen as useful where dispensing of it due to Antirealism is not preferable. The

[34] The reason for the 'True' being capitalized in the Quinean Assumption and Error Theory will be made clear later on in Sect. 3.3. Basically, it is to emphasize that the truth appealed to when capitalized is external (what is really true), rather than internal (which will be written with a small 't'), to linguistic frameworks or fictions.

[35] Woodward (2008, p. 274).

[36] Berkeley (1710, Part I Section 51).

fictionalist engages in this realist talk but in the form of a fiction to speak without commitment.

Despite taking antirealism to be meaningless, the Neo-Carnapian may still fulfill a good proportion of this fictionalist ideology by arguing for the value of utilizing a discourse without ontological commitment, by being ontologically quiet (rather than by being antirealist). The Carnapian has no ontological commitments since they are quiet about ontology, and so like an antirealist fictionalist they can talk about all sorts of things without being ontologically committed to the things talked about. All commitments to a metaphysical reality are merely apparent, and so all the discourses have the same status as fictional discourses. The fictional discourse is not antirealist but rather is non-committal ontologically, which is what quietism provides. The role antirealism plays in traditional fictionalism (to refrain from ontological commitments) can thus be successfully performed by quietism, demonstrating that the Neo-Carnapian can be construed as a fictionalist despite rejecting antirealism and being quietist instead. They can be said to retain the fictionalist rationale of adopting a discourse for its practical utility without paying ontological bills.

11.4.2 Quinean Assumption

The 'Quinean Assumption' says 'If Xs do not exist then X-discourse is not True'. The reason I have named this assumption as being Quinean is due to Quine's slogan 'to be is to be the value of a bound variable',[37] which leads to the view that without an object in the domain of quantification for your language to refer to, such language will be deemed false. Traditional fictionalism may be wrong-headed in having this assumption centrally built in to it, since it is an assumption that is heavily challenged in the history of philosophy. For example, there are Meinongians[38] and free logicians[39] who would deny the Quinean Assumption and accept that certain

[37] Quine (1948, p. 33).

[38] Meinongians believe that there are nonexistent things, and that we can truly speak of such things and predicate properties of them when referring to them. Meinongians can say, for example, that some numbers are prime even though numbers do not exist, and thus can truly say '3 is a prime number' whilst referring to the nonexistent object '3'. The ability to successfully refer to nonexistent things is explained by domains of quantification containing both existent and nonexistent objects where quantifiers are treated as ontologically neutral in order to quantify over both. Successfully referring to an object in the domain will not entail the existence of that object, and truthfully describing a thing to be a certain way will not entail the existence of that object. For examples of Meinongianism, see Parsons (1982) and Routley (1982). For examples of neutral quantification, see Berto (2012), Priest (2008), and Hofweber (2007).

[39] One feature of some systems of free logic is that they allow for successful reference without referents. The positive free logician could make a true statement without referring to anything whatsoever (not even to a nonexistent thing, as in the Meinongian case). (The negative free logician could make a false statement, as opposed to a meaningless statement, without a referent.) They can talk truthfully of X even when there is no such referent for 'X'. This is because such free logicians allow for quantification over empty domains, and so it is not always required that there

truth-apt statements about X need not require that X exists. Take the following two statements:

φ: Some numbers are prime.
ψ: Numbers exist.

Yablo describes φ as being hostage to ψ, in that φ entails ψ.[40] This is what he describes as a 'hostage crisis', because we may not want our ordinary assertions like φ to entail something ontological like ψ. The traditional fictionalist reaction to this crisis is to say either that both ψ and φ are literally false but true in the fiction, or that at least φ is true in the fiction (but otherwise false) and that it is helpful to continue to talk in ways like φ. But, as Finn points out, to infer ψ from φ, one must take quantification to be ontologically committing.[41] Hence the hostage crisis only occurs for those who endorse a certain view of the relation between truth, quantification, and existence (namely, something like the Quinean Assumption) and are otherwise completely avoided. For example, the Meinongian would state that when one speaks of a thing as in φ, they may refer to an existent or a nonexistent thing, and as such φ will not entail ψ. And for some free logicians, statements like φ can be true whilst a statement like ψ is false, because φ can be truthful without any referent at all, let alone an existent referent as ψ demands. Traditional fictionalism thus loses a central motivator without the Quinean Assumption, since there would be nothing entailing a certain truth evaluation of X-discourse just because X does not exist.

The Neo-Carnapian would understand φ as an internal claim, and ψ as an external claim. The entailment is thus only that internal claims require the acceptance of the framework (which is how external claims should be interpreted), rather than the acceptance of an ontology. The Neo-Carnapian does not make any claims about the metaphysical existence of X, however this does not prevent the Neo-Carnapian from making true (albeit a thin notion of truth, as described in Sect. 11.4.3) and meaningful claims about X. The Neo-Carnapian therefore undercuts the motivation for traditional fictionalism from antirealism, since not only is antirealism not a viable position but also this antirealism would not lead to the conclusion that our discourse is false. The connection between Antirealism and Error Theory, via this Quinean Assumption, is broken, since not existing does not entail not being able to be truthfully or meaningfully spoken of. Regardless of whether Antirealism is true, we

---

be a member of a domain to serve as a referent. Hence, again (like in the Meinongian case) quantification will be ontologically neutral since to quantify over a domain will not necessitate the existence of members, as there may be no members at all to quantify over. For a discussion of reference without a referent, see Sainsbury (2005).

[40] Yablo (2017). A key place where I deviate from Yablo (and also Thomasson) is that I take the inference from φ to ψ to be invalid (unless when merely quantificational), whereas Yablo (2014) and Thomasson (2014) agree that the inference is valid. The difference between Yablo and Thomasson is that Yablo sees the EQ as legitimate but unanswerable, whereas Thomasson sees the EQ as illegitimate but answerable when revised as an IQ. I am more aligned with Thomasson here (see Sect. 11.5 to compare).

[41] Finn (2017b)

cannot arrive at Error Theory since the Quinean Assumption that connects them can be rejected.

The very point of Carnap's quietism was to retain our ability to speak about things regardless of their ontological status, as we are to be quiet about this status, yet it need not follow that we then stop talking about things altogether. The point was to be able to continue to talk in a useful way without ontological constraint, but it was not the case that without the ontology the talk is automatically false. This is very much in tension with the Quinean Assumption that without entity X existing the X-discourse will be faulty in some way, and in need of rescuing on the grounds of it being useful. However in denying that existence has any effect on a truth evaluation we need not conclude that a discourse is faulty just because its terms do not refer to existent things. So, having denied Antirealism and the Quinean Assumption, we then lack the error that motivates traditional fictionalism. Without this Quinean Assumption, there is no entailment from Antirealism to Error Theory, and without reason to be in error there is no reason to be a fictionalist in the traditional sense. In order for the Neo-Carnapian to be a fictionalist then they will need to arrive at Error Theory independently of the Quinean Assumption and the Antirealism that they deny.

11.4.3 Error Theory

The traditional fictionalist, motivated by an antirealism that leads the discourse about X to be untrue (according to the Quinean Assumption), will state that the discourse user is in error, and as such traditional fictionalism is based on an Error Theory. If all of the X-discourse is untrue, then when we use it to make positive claims about X we say something false. The error arises in different ways: Our discourse is in error (by being false), and the discourse user is in error (by believing the discourse to be true or at least by not believing that it is fictional). It is these two elements of error that then motivates the traditional fictionalist to find something positive about the discourse in order to allow us to still be justified in using it, despite the error. Importantly, then, it is the Error Theory that motivates the Pragmatism.

As shown above, the Neo-Carnapian does not arrive at Error Theory through Antirealism, since they reject Antirealism and the Quinean Assumption that links it to Error Theory. However the Neo-Carnapian still locates error in the discourse user. The Neo-Carnapian thus fulfills this criterion for being a fictionalist by locating the error in language users either by them thinking that they are speaking about the external world (but actually they are being metaphysically quiet), or by not treating linguistic frameworks as fictions. Their error is in taking realism to be true and thinking they are speaking in a realist way accordingly, or by not acknowledging that linguistic frameworks are fictions and their talk is fictional. The Neo-Carnapian can thus retain the Error Theory of fictionalism (and qualify as a fictionalist) by treating ordinary discourse users as being in error in some way.

An important distinction is required here to make the point: between internal truth (true according to the rules of the framework) and external Truth (with a capital T, Truth in the metaphysical realm). The traditional fictionalist takes a discourse to be in error by not being True (although it will be true). Such fictionalists talk about 'true in the fiction' for what the Neo-Carnapian calls 'internal truth', so a claim such as $2 + 2 = 4$ is not True (and in fact will be literally False), but is considered to be true in the fiction. The Neo-Carnapian proposal is to call 'true in the fiction' simply 'true', and to reject the meaningfulness of Truth talk (as it is external to the framework).[42] To return to Yablo's hostage crisis example, φ (some numbers are prime) will be true for both the Neo-Carnapian and the traditional fictionalist yet False for the traditional fictionalist, whereas ψ (numbers exist) will be False for the traditional fictionalist and meaningless for the Neo-Carnapian.

The Neo-Carnapian can meet the traditional fictionalist criterion of Error Theory if being 'in error' is just being not True, since the frameworks will not be True. The Neo-Carnapian treats the frameworks themselves as non-cognitive because they are adopted for purely pragmatic reasons and not based on Truth. Given the rejection of Truth, the frameworks have nothing to be true according to, and so they are not True (or False). The discourse itself can also be seen to be 'not True' on the Neo-Carnapian model, just as it is with traditional fictionalism. The difference between the traditional fictionalist and the Neo-Carnapian is the reason for the discourse not being True: for the traditional fictionalist it is due to being False, whereas for the Neo-Carnapian it is due to not being Truth-apt (as Truth is meaningless). So the Neo-Carnapian follows the traditional fictionalist in saying that claims within a framework are true in the sense of being 'true in the fiction', but rejects the assumption behind such fictionalism that there is a value of Truth that goes beyond that. So, like the traditional non-global fictionalist, the Neo-Carnapian rejects the Truth of these claims, but unlike the non-global fictionalist, they do not think that any discourse meets the external Truth standard.[43] So all that is left is the fictionalist assessment of the pragmatic value (see next Sect. 11.4.4) of speaking in a particular way that does not appeal to Truth—this value is usefulness. Neo-Carnapian Fictionalism therefore does not have the feature of taking the discourse to be False (as the traditional fictionalist would have it). Rather, they refuse to answer the question as to whether the discourse, when taken literally, is True or False, and reject such a question as meaningless. But this doesn't prevent the Neo-Carnapian Fictionalist from having sufficient qualities of fictionalism, since the Error Theory that results from

[42] For this reason, it is hard to see how a Carnapian can make any sense of an external Truth standard, as a framework (which is a set of rules of use for talking) cannot be thought of as True or False. It is only internally (given the rules for the language) that sentences can be truth-apt. Thanks to Amie Thomasson for this point.

[43] Thanks to Mary Leng for pushing the question 'what's the point of all these fictions if none of it is True?'. The non-global fictionalist can say that some fictions are useful because they help discover what follows from True empirical theories. But if none of our theories are True, then something else needs to be said about what they're useful for, perhaps in terms of organising experience and formulating predictions. The Quinean may push back that such usefulness is evidence for Truth, but it needn't be.

the traditional fictionalist taking their claims to be False is met by the Neo-Carnapian (where the error is located not in the discourse being False but in the attitude in thinking the language is anything other than pragmatic).

11.4.4 Pragmatism

The key, central positive aspect of being a traditional fictionalist about an entity X is that X-discourse does not have to be True in order to be valuable. For the traditional fictionalist, X-discourse will be False because X doesn't exist, yet they retain X-discourse as it is still useful to speak as if X exists. The ideology here is that Truth is not the only value to discourse, and usefulness is a value that does not depend on (or imply) such Truth. Fictional discourse is considered useful, and so the fictionalist about X can say that discourse about X is fictional, and is useful, regardless of whether X exists (which entails whether X-talk is True according to the 'Quinean Assumption'). A discourse about X can be classed as valuable simply for being useful. It is this positive Pragmatism of traditional fictionalism that is the most important aspect to retain to be a fictionalist, which the Neo-Carnapian can meet.

The Neo-Carnapian meets this traditional fictionalist criterion by holding that the frameworks we speak within are useful to employ even though they are not Truth-evaluable. Therefore, the Neo-Carnapian is fictionalist about a discourse that is not Truth-apt, whereas the traditional fictionalist takes the discourse to be False. The frameworks are considered useful in order to utilize their language and have value independent of Truth. The discourse internal to the framework is deemed useful despite not being based on an existing external reality to make it True, and so the discourse is valued as useful independent of whether it refers to existent things. So the Neo-Carnapian, like the traditional fictionalist, agrees that the discourse in question does not refer to existing things in the metaphysical realm. For the Neo-Carnapian this is because such a realm is meaningless, and for the traditional fictionalist this is because that realm does not contain the entities in question. Thus despite lacking the traditional route to Pragmatism, the Neo-Carnapian still maintains the fictionalist ideology by valuing a discourse for its usefulness.[44]

The Neo-Carnapian therefore holds the Pragmatism of traditional fictionalism by saying the only virtue that matters for evaluating a discourse is the pragmatic one. There is thus no need to find fault with the discourses, but rather, the Neo-Carnapian can provide a fictionalist account of what is valuable about them—namely their usefulness. The discourses in question have a value that is independent of their Truth, since for the Neo-Carnapian this Truth is external and thus meaningless. This is enough to maintain the fictionalist pragmatic ideology to show how the Neo-Carnapian can be seen as a fictionalist.

[44] Thanks to Mary Leng for pointing out that for Quine, though, this pragmatic virtue gives us reason to believe a fiction or framework is true and thus that its objects exist.

The Neo-Carnapian Fictionalist has demonstrated their claim to the title of 'fictionalism' by (i) maintaining the appeal to lack of ontological commitment (through quietism rather than antirealism); (ii) their attribution of error to the discourse and the discourse user (through it being quiet and not realist); and (iii) valuing a discourse for being useful (despite it not being True). The apparent tension between the fictionalist encouraging talk of X and the quietist discouraging talk of X is resolved in the following way: quietism about Xs means we should not talk about whether Xs really exist, but the fictionalism means we should continue to utilize X-discourse regardless of whether Xs really exist. This is a consistent and unique position, distinct from Carnap's quietism (by being based on a metaphorical/literal distinction rather than an analytic/synthetic distinction, where frameworks are fictions), and distinct from traditional fictionalism (by being quiet rather than antirealist, and by rejecting a controversial Quinean Assumption). It is also unique in being global, in that fiction loses its contrast class (of the real). This is similar to Thomasson's deflationary position whereby there is no sense to the external side of the IQ/EQ distinction, leaving the internal with no meaningful contrast—hence her 'internal' realism. I will now compare and contrast my view with that of Thomasson's.

11.5 Comparison to Thomasson's Simple Realism

Thomasson's position is called 'Simple Realism', and is made with regard to her 'easy ontology' found in her book Ontology Made Easy.[45] Simple realism is motivated by Carnap's deflationism (or quietism as I have called it—though the reason for usage of the word 'deflationism' instead of 'quietism' will soon be made clear), and as such could be seen as a fellow Carnapian position. However my position differs from Thomasson's Simple Realism in a few important respects. And interestingly, Thomasson cites fictionalism as a competitor with her Simple Realism, showing that the comparison between Neo-Carnapian Fictionalism and Simple Realism is one that is worthy of making.

For Thomasson, ontological questions are easy to answer, so ontology is deflated. The ontological questions that are meaningful are akin to Carnap's IQ's, despite Carnap stating that it is misleading to call such internal use 'ontological': "I should prefer not to use the word 'ontology' for the recognition of entities by the admission of variables. This use seems to me to be at least misleading."[46] This is the main reason why my Neo-Carnapian Fictionalism differs from Simple Realism—for me, endorsing a Carnapian position, answers to IQ's are ontologically quiet, whereas for Thomasson, answers to IQ's are ontological answers (as such she calls herself a

[45] Thomasson (2015).

[46] Carnap (1956, p. 43). See also Carnap (1950) note 5: "With respect to the basic attitude to take in choosing a language form (an 'ontology' in Quine's terminology, which seems to me misleading), there appears now to be agreement between us".

'realist' and not a 'quietist').[47] For Thomasson, we can get our ontological answers from trivial inferences in our language usage (where certain 'application conditions'[48] are fulfilled). We can derive an ontology including numbers from the sentence 'I have two hands' by making an analytic inference to another sentence where numbers are quantified over, such as 'the number of hands I have is two'. Note that this is much like the inference from (φ) 'Some numbers are prime' to (ψ) 'Numbers exist' that we saw in Yablo's hostage crisis.

Thomasson is clear that the ontology she derives is one not to be fretted over—it is the not the heavy-duty metaphysical ontology that a Carnapian avoids commitment to. But as I (2017b) argue elsewhere the commitments derived from such inferences, when valid, are merely quantificational and not ontological, so any use of 'ontology' or 'realism' is misleading as a Carnapian. It may be because of this that she avoids using the term 'quietism' to describe her Carnapian position, as she is not actually quiet when she calls herself a realist. However she regards this realism as being sufficiently deflated to justify her using the term 'deflationism' to describe her position as being Carnapian. Not only do I believe it to be misleading to call her position a simple 'realism' doing easy 'ontology', but also I believe it to be misleading for her to regard her position as deflationary. Thomasson tells us that commitments in her Simple Realism are not heavy-duty, yet she also rejects them being 'thin'[49]—she argues instead that the existence derived is existence *full stop*, thus denying a distinction between the heavy and the thin commitments. If the commitments from Simple Realism are not thin, then ontology is not deflated but rather merely redefined. In that sense, Thomasson is more Quinean than Carnapian, due to doing 'ontology'.

Thomasson regards her easy approach to ontology as in competition with traditional fictionalism. Yet she regards her position as trumping such fictionalism, and being ontologically on a par with fictionalism. However since Simple Realism derives commitments about existence full stop, then clearly they are not on a par since fictionalism is antirealist (and uncommitted) about the entity in question. But Thomasson also recognizes the similarities between fictionalism and her Simple

---

[47] As Thomasson notes in private correspondence, a lot hinges on what one means by 'ontological': to the extent that ontological questions are supposed to be about what exists, Thomasson says the IQs are ontological questions in the only sense that has sense, but notably not in some deep(er) metaphysical sense (which is senseless). To my ears, this makes her more like Quine, who similarly holds an internal realism.

[48] For Thomasson (2015, p. 90), a term refers if its application conditions (the semantic rules of use for terms we master) are fulfilled, but the conditions under which things exist are various, and so she denies that there are any shared criteria of existence.

[49] Thomasson (2015, p. 146) revealingly states: "we should not suggest that the entities to which we become committed via trivial inferences are in general 'thin and inconsequential', 'ontologically shallow' or that their existence is somehow to be understood in a deflationary manner. Instead we should simply say that such entities exist—full stop." If the existence is not deflated, though, then she is not a deflationist. However, she argues that it is the debates, and not the objects, that are deflated, such that her position is metaontologically, rather than ontologically, deflationary. This to me makes her position somewhat inconsistent, as Schaffer (2009, p. 142) argues.

Realism, which could motivate my marriage of Carnap's deflationary quietist view with fictionalism. As Thomasson states:

> In many ways, the fictionalist approach and the easy approach are similar. Both are equally opposed to both traditional Platonism and to traditional nominalism or eliminativism about disputed entities. Both bring to ontological debates a 'no worries' attitude that suggests that we can preserve the discourse in question without saddling ourselves with a heavy-duty ontology... Both reject the assumption that the function of, say, mathematical discourse is to track objects... And they tend to appeal to the same sort of philosopher: someone who suspects the heavy-duty realist of taking the discourse in question too seriously, and suspects the eliminativist of overreacting by rejecting a perfectly functional range of discourse.[50]

So it is clear to see how traditional fictionalism could be seen as similar to a 'deflationary' view like Thomasson's easy realism or Carnapian quietism, and also clear to see how fictionalism could itself be seen as deflationary. Yet Thomasson notes where the positions crucially differ:

The views are clearly in conflict, for taking the easy approach leads us to the simple realist view that there are entities of the disputed kind, while the fictionalist does not assert that there are such entities.[51]

This takes the difference to be that fictionalism is antirealist, and Thomasson's position is realist. But Neo-Carnapian Fictionalism is neither antirealist nor realist. Traditional fictionalists thought that realists took ontology too seriously as the language usage is actually just fictional, whereas Thomasson says that fictionalists take it too seriously as there is no need to avoid the commitments derived from language usage. Since Carnap did want to avoid such commitments, and avoided realist and antirealist branding, I believe my Neo-Carnapian position to be closer to Carnap's quietism than Thomasson's Simple Realism is. As such, my position has important differences to Thomasson's Simple Realism, yet since my position is not traditionally fictionalist (as it is quietist) the differences do not include those Thomasson notes between traditional fictionalism and deflationary approaches. I take my Neo-Carnapian Fictionalism to be closer than Thomasson's Simple Realism to Carnapian motivations, and I also take it to be an improvement on both traditional fictionalist and Carnapian positions due to avoiding Quine's objections based on the analytic/synthetic distinction and also avoiding Quinean assumptions that existence is required for truth.

---

[50] Thomasson (2015, p. 178). See also her (2013) where she argues her view trumps the fictionalist by better making sense of the discourse. Yet many of her arguments rest on the fictionalist regarding the discourse as False and to be contrasted with the 'real' content, which thus does not apply to my Neo-Carnapian view as it is globally quiet.

[51] Thomasson (2015, p. 180). However, this is in tension with Thomasson's claim that "where ontological issues are concerned the two views are on a par" Thomasson (2015, p. 182). They're not on a par if one denies existence of X and the other affirms it.

11.6 Conclusion

In this paper I have described the Neo-Carnapian Fictionalist position. After demonstrating how a Neo-Carnapian study of ontology would be pursued, I showed the effects of quietism on fictionalism to show that it required being divorced from antirealist foundations. Then I provided a fictionalist picture of the Neo-Carnapian position, married to quietism instead. The Neo-Carnapian is quiet with regard to metaphysics, and fictional with regard to frameworks. To construe the frameworks as fictions is to say they have value by being useful to adopt, and that they are to be treated similar to how we treat fictions—presupposing them in order to engage in them and assess claims relative to them, and accepting what we presuppose in order to believe the claims that are true relative to them. The account is typically fictionalist in its escape from ontological commitment whilst maintaining the discourse that may seem to sound realist; in its attribution of error to the discourse users in thinking it is realist; and of course its citing value in the discourse by being useful to adopt independent of its Truth. This concludes my description of Neo-Carnapian Fictionalism to show that fictionalism needn't be antirealist, which has important similarities yet crucial differences to Thomasson's Simple Realism.

References

Allen, K. (2016). *A naïve realist theory of colour*. Oxford University Press.

Berkeley, G. (1710). A treatise concerning the principles of human knowledge. In A. A. Luce & T. E. Jessop (Eds.), (1948–57) *The works of George Berkeley*. Thomas Nelson.

Berto, F. (2012). *Existence as a real property*. Springer.

Brock, S. (2002). Fictionalism about fictional characters. *Noûs, 36*(1), 1–21.

Bueno, O. (2009). Mathematical fictionalism. In O. Bueno & Ø. Linnebo (Eds.), *New waves in the philosophy of mathematics* (pp. 59–79). Palgrave.

Carnap, R. (1950). Empiricism, semantics, and ontology. *Revue Internationale de Philosophie, 4*, 20–40.

Carnap, R. (1956). *Meaning and necessity: A study in semantics and modal logic* (2nd ed.). University of Chicago Press.

Carus, A. W. (2018). Going global: Carnap's voluntarism and price's expressivism. *The Monist, 101*, 441–467.

Eklund, M. (2005). Fiction, indifference, and ontology. *Philosophy and Phenomenological Research, 71*(3), 557–579.

Eklund, M. (2015). Fictionalism. In E. N. Zalta (Ed.), *The Stanford encyclopedia of philosophy*. https://plato.stanford.edu/archives/win2017/entries/fictionalism/

Finn, S. (2017a). Metametametaphysics and dialetheism. *Australasian Journal of Logic, 14*(1), 128–146.

Finn, S. (2017b). Dissolving Yablo's hostage crisis: In defence of defiance. *Australasian Philosophical Review, 1*(2), 184–188.

Hofweber, T. (2007). Innocent statements and their metaphysically loaded counterparts', philosophers. *Imprint, 7*(1), 1–33.

Jay, C. (2011) *Realistic fictionalism*. PhD Thesis found at discovery.ucl.ac.uk/1356103/1/Realistic%20Fictionalism.pdf. 05/2015.

Joyce, R. (2001). *The myth of morality*. Cambridge University Press.

Kroon, F. (2001). The semantics of 'things in themselves': A deflationary account. *The Philosophical Quarterly, 51*, 203, 165–181.

Leng, M. (2010). *Mathematics and reality*. Oxford University Press.

Macarthur, D. (2008). Pragmatism, metaphysical quietism, and the problem of normativity. *Philosophical Topics, 36*(1), 193–209.

Parsons, T. (1982). Are there nonexistent objects? *American Philosophical Quarterly, 19*(4), 365–371.

Price, H. (2011). *Naturalism without mirrors*. Oxford University Press.

Priest, G. (2008). Closing of the mind: How the particular quantifier became existentially loaded behind our backs. *The Review of Symbolic Logic, 1*(1), 42–54.

Quine, W. V. O. (1948). On what there is. *The Review of Metaphysics, 2*(5), 21–38.

Quine, W. V. O. (1951a). On Carnap's views on ontology. *Philosophical Studies, 2*(5), 65–73, reprinted in Quine, W. V. O. (1976). *The ways of paradox & other essays* (Revised ed., pp. 203–211). Columbia University Press.

Quine, W. V. O. (1951b). Two dogmas of empiricism. *The Philosophical Review, 60*(1), 20–43.

Quine, W. V. O. (1991). Two dogmas in retrospect. *Canadian Journal of Philosophy, 21*(3), 265–274.

Rosen, G. (1990). Modal fictionalism. *Mind, 99*(395), 327–354.

Routley, R. (1982). On what there is not. *Philosophy and Phenomenological Research, 43*(2), 151–177.

Sainsbury, R. M. (2005). *Reference without referents*. Oxford University Press.

Sainsbury, R. M. (2009). *Fiction and fictionalism*. Routledge.

Schaffer, J. (2009). The deflationary metaontology of Thomasson's ordinary objects. *Philosophical Books, 50*(3), 142–157.

Thomasson, A. (2013). Fictionalism versus deflationism. *Mind, 122*(488), 1023–1051.

Thomasson, A. (2014). Quizzical ontology and easy ontology. *The Journal of Philosophy, 111*(9–10), 502–528.

Thomasson, A. (2015). *Ontology made easy*. Oxford University Press.

van Fraassen, B. (1980). *The scientific image*. Oxford University Press.

Woodward, R. (2008). Why modal fictionalism is not self-defeating. *Philosophical Studies, 139*(2), 273–288.

Yablo, S. (1998). Does ontology rest on a mistake? *Aristotelian Society Supplementary, 72*(1), 229–262.

Yablo, S. (2000). Apriority and existence. In P. Boghossian & C. Peacocke (Eds.), *New essays on the a priori*. Oxford University Press.

Yablo, S. (2005). The myth of the seven. In M. Kalderon (Ed.), *Fictionalism in metaphysics* (pp. 88–115). Clarendon Press.

Yablo, S. (2014). Carnap's paradox and easy ontology. *The Journal of Philosophy, 111*(9–10), 470–501.

Yablo, S. (2017). If-thenism. *Australasian Philosophical Review, 1*(2), 115–132.

Chapter 12
To Be (Thin) or Not to Be? Are Thin Objects Fictional Objects, or Are Fictional Objects Thin?

Mary Leng

Abstract Øystein Linnebo (2018) offers an account of mathematical objects that promises to bypass some of the features that can make standard Platonism seem unpalatable. Although Linnebo's mathematical objects are abstract and indeed exist of necessity, their existence is lightweight: they make little by way of demands on the world. As such the costs of realism about mathematical objects are, in Linnebo's account, much lower than is standardly assumed. Worries about how we can know about mathematical objects, or indeed refer to them, given what Linnebo calls their 'shallow nature', fall away. Why then persist, as the mathematical fictionalist does, with nominalist scruples about accepting the existence of abstract mathematical objects, if the existence of objects turns out to be so metaphysically undemanding?

Keywords Platonism · Existence · Realism · Fictionalism · Metaphysically undemanding

12.1 Introduction

This chapter has two main aims. First, by considering some key features of Linnebo's (2018) account of mathematical objects as 'thin', it raises the question, 'Are thin objects really just fictional objects?'. For all Linnebo's insistence on the necessary existence of mathematical objects, it will be argued that there are features of the 'objects' he introduces that suggest that they fall short of really existing and should instead be thought of as merely fictional. Second, it considers the rejoinder that, rather than thin objects being merely fictional (and so not existing at all), the bar to existence has been lowered so that fictional objects are shown to exist and to be thin (so that fictionalism turns out not to be a nominalist position at all, with fictional

M. Leng (✉)
Department of Philosophy, University of York, York, UK
e-mail: mary.leng@york.ac.uk

X. de Donato-Rodríguez et al. (eds.), *Deflationist Conceptions of Abstract Objects*, Synthese Library 516, https://doi.org/10.1007/978-3-031-98717-5_12

objects turning out to be thin objects). Conceding that the difference between Linnebo's thin realist perspective and the fictionalist perspective might be more a matter of taste (concerning how to use the word 'exists') than of substance, it will be noted that Linnebo's account is therefore likely to be vulnerable to similar challenges to those faced by the fictionalist. Specifically, Linnebo will have to defend an account of modality that resists reduction to facts about set theoretic models, and will have to meet the challenge raised by explanatory indispensability arguments that charge that mathematical objects play a more substantial explanatory role in empirical science than could be played by objects of a shallow nature.

12.2 Are Thin objects Fictional Objects?

It is important to distinguish our first question (are thin objects fictional objects?) from our second (are fictional objects thin?). Linnebo himself has entertained the possibility that the metaphysics of fiction could conclude that fictional entities exist as thin objects ('Thin Objects: A Discussion with the Author' https://www.youthttps://www.youtube.com/watch?v=TnU8DiUVPes&t=3814s (at 01:03:34)). If so, then to hold as mathematical fictionalists do that mathematical entities have no more reality than fictional characters would not amount to a denial of abstracta: both fictional characters and mathematical objects would exist as thin objects. I will turn to this question (of whether fictional objects should be thought of as enjoying a thin existence) in Part II, but for now let us focus on what I mean by my rather different question, 'Are thin objects fictional objects?'. From the perspective of mathematical fictionalism, to say that an object is merely fictional is to say that it doesn't exist at all. So the question of whether thin objects are fictional objects, as posed here, is the question 'Are thin objects not really objects at all?' To motivate a positive answer to this question I will point to some features of Linnebo's account of thin objects that suggest that when we speak of thin objects we are best thought of as merely speaking 'as if' there are such things, rather than truly speaking of objects that really exist (albeit thinly).

12.2.1 Thin Objects: An Overview

We will start by running through some key features of Linnebo's account of mathematical objects as thin objects, before pointing to elements of his account that might suggest that we are best understood as merely speaking 'as if' there are thin objects, rather than being committed to their actual existence. For Linnebo, thin objects are objects whose "existence doesn't make a substantial demand on the world". Linnebo recognises that the view that the existence of mathematical objects is an insubstantial matter can be developed in different ways, one historical example being the position suggested by Hilbert in his correspondence with Frege, according to which the existence of mathematical objects requires no more than the consistency of the theories used to characterise such objects (with axioms read as implicit definitions of their primitive terms). While allowing that an account of mathematical objects as thin objects could be developed along these lines, Linnebo's own account is somewhat different, focussing on *abstraction principles* rather than axioms as implicit definitions as a way of introducing thin objects. Whereas for the Hilbertian picture, consistency is enough for the thin existence of mathematical objects, Linnebo's account is more restrictive, supporting only those mathematical objects that can be introduced through a careful step by step process of domain expansion.

Starting from an existing domain of objects as given, suppose we accept the position (defended by Boolos (1984) and others (Florio & Linnebo, 2021)) that plural quantification over this domain by itself commits us to no more things than the individual objects, taken plurally. The general form of a plural abstraction principle takes the form.

$$\S xx = \S yy \text{ iff } xx \sim yy$$

where § is a term-forming operator, ~ an equivalene relation, and xx, yy plural variables (so the principle reads, roughly: the § of xs = the § of ys if and only if the xs are equivalent (via ~) to the ys. If we start from a position that plural quantification and the second order notions required to characterise the equivalence relation '~' are no more committal than first-order quantification, then the claim on the right hand side that the xs and ys are equivalent (with respect to '~') is committed to no more objects than the xs. But the left hand side, introducing as it does an identity between objects, brings with it a commitment to an object corresponding to the xs and the ys. What kind of object depends on the particular equivalence relation we use. Thus, if the equivalence relation is 'x is one of the xs if and only if x is one of the ys' then §xx can be thought of as introducing the *set* of the xs (as identity conditions for sets are given by extensionality). On the other hand, if the equivalence relation is 'the xs are equinumerous with the ys', then §xx can be thought of as introducing the *number* of xs.

It is tempting, depending on one's initial prejudices, to read the biconditionals involved in abstraction principles in one of two ways. Those with Platonist leanings will be tempted to say that, despite their initial appearance of innocence, the equivalences on the right hand side of the biconditional already require the existence of abstract objects for their truth: the claim that the xs are equinumerous with the ys may look to quantify only over xs and ys, but its truth requires the existence of numbers too. On the other hand, for those of us with a taste for desert landscapes, if we accept the abstraction principles at all, we will do so only because we think that the equivalence of their apparently ontologically committing left hand sides with their ontologically innocent right hand sides suggests that the left hand sides do not really introduce a commitment to objects at all: we can speak *as if* there is an object called 'the number of xs', but really claims about numbers introduced this way via abstraction principles should be thought of as simply shorthand for more complicated claims expressed in terms of equivalences. Either way, though, the temptation for the Platonist and the nominalist alike is to see each side of the abstraction principle as no more—or no less—ontologically committed than the other. The Platonist will say, since the left hand side of HP is committed to numbers and since we accept that the left hand side is equivalent to the right, then the right hand side of HP is likewise committed (this is the position taken by Hale and Wright (2001)). On the other hand, the nominalist will say, since the right hand side of HP is *not* committed to numbers, we can only accept HP as a true equivalence if (despite appearances) the left hand side is likewise not so committed. (Typically, nominalists who wish to preserve a standard semantics will therefore not accept HP as true, preferring to retreat to a hypothetical form of HP: if there were numbers (or: according to the fiction in which there are numbers), the number of xx = the number of yy iff the xx are equinumerous with the yy. (See e.g., Field 1984a.)

Linnebo's approach is rather different. He takes the left hand side of abstraction principles to be committed to *new* objects that are not already (implicitly) committed to in the equivalences on their right hand side, so that abstraction principles are asymmetric in terms of their ontological commitments. This raises the interesting question of what happens via the process of abstraction: are new objects *created* out of old when we introduce the abstraction §xx out of an equivalence relation on the xs? By introducing a new term-forming operator '§', do we somehow bring into existence objects to which the new term can refer?

While it is tempting to think that, through abstracting we somehow magic something new into existence, and while Linnebo's discussion of the results of abstractions as 'new' objects is suggestive in this regard, this cannot be Linnebo's considered view. Linnebo holds that, given that their existence doesn't require any specific concrete objects, numbers and pure sets are *necessary existents*, existing at every possible world. Perhaps a better way of thinking of what happens when we abstract is that we introduce into our language a way of referring to objects that we could not previously talk about. On this picture, numbers and sets were there all along waiting to be referred to, so that all that is created via the introduction of an abstraction principle is a way to refer to things about which we couldn't previously (when we only had the resources of the right hand side of the principle) speak. There remains, however, something odd about this picture of how abstraction can be used to extend our domains. Quantification typically allows us to speak of objects even without having found a way of referring/pinpointing those objects directly. If numbers and sets already exist prior to our finding a means to refer to them (by abstraction principles) what stops them from being in the domains of our quantifiers, amongst the xs on the right hand side of our abstraction principles when we talk about the xs? After all, Hamlet tells Horatio "there are more things in heaven and earth…than are dreamt of in your philosophy", and most of us are happy to concede that we can quantify over things that we haven't even dreamt of ways of referring to directly. If thin objects are real objects, why do they have to await our introduction of ways of referring to them via abstraction principles before they can appear in the domains of our quantifiers?

A further issue, albeit somewhat related, arises for the position that holds that what is secured in abstraction is a new way of referring to objects to which we couldn't previously refer. The issue arises because it turns out that, on Linnebo's view, we don't *actually* have to have carried out the required abstraction steps in order to be able to talk about numbers and sets. Note that Linnebo's abstraction principles are *predicative*: the right hand side of our abstraction principles do not quantify over the objects introduced by the abstraction step. But if we take our equivalence relation to be one of equinumerosity, predicative abstraction on a finite initial domain of n objects will only introduce n + 1 new objects (the numbers 0,...,n, corresponding to all the equivalence classes of this relation), and if we take it to be extensional equivalence, for n objects it will introduce 2^n subsets as new objects. To get *all* the numbers and *all* the sets we need to iterate and indeed to take limits. No one has actually done more than finitely many abstractions, so if we think (as Linnebo does) that it is safe to quantify over *all* the numbers, then it cannot be that it is having *actually* introduced additional objects to our domain via an abstraction step that secures our ability to talk about/quantify over such objects.

Linnebo's solution is to modalize, introducing into his logical system the modal operators '$\Box$' and '$\Diamond$' "to describe how the interpretation of language can be shifted—and the domain expanded—as a result of abstraction. Thus, the intended meaning of '$\Box\varphi$' is 'no matter how we abstract and thereby shift the meaning of the language, φ." Here it is tempting to draw a parallel with an older approach to demystifying our talk of mathematical objects, Philip Kitcher's (1983) development of a Millian approach to arithmetic. For Kitcher, what ultimately grounds the truth of claims of arithmetic is facts about our collecting and segregating activities ('2 + 3 = 5' is grounded in the fact that if we collect together two things, and three further things, we've collected five things). But even very large *finite* arithmetic truths cannot be grounded in facts about our actual collecting activities. Rather, for Kitcher, these truths are grounded in facts about what collecting we *could* do, had we but world enough and time (and, to get as far as completed infinities, had we the ability to complete supertasks). As such, the truths of arithmetic are not to be thought of as grounded in our actual manipulations of physical objects, but in the possibilities for manipulation afforded by the world:

> To coin a Millian phrase, arithmetic is about 'permanent possibilities of manipulation.' More straightforwardly, arithmetic describes those structural features of the world in virtue of which we are able to segregate and recombine objects: the operations of segregation and recombination bring about the manifestation of underlying dispositional traits. (Kitcher 1983, p. 108)

By analogy, we might say that for Linnebo, arithmetic and set theory are about 'permanent possibilities of abstraction'.

It is notable then that one critique (Hoffman, 2004) that has been made of Kitcher's account is that, in moving from actual manipulations to ground arithmetic to *possible* manipulations (which he describes in terms of the ideal collecting and segregating activities of an ideal agent), the account is perhaps best thought of as grounded in a *fiction*: a story about what an ideal agent *could* do with world enough and time. As such, Kitcher's account of the *truth* of mathematical claims is perhaps better thought of as an account of their fictional *assertibility*, so that Kitcher's view of mathematics turns out to be a form of fictionalism. Similarly, we might think, for the idealizations introduced by Linnebo when he moves from actual abstraction steps to abstraction steps we could take to expand our domains, had we world enough and time. Is Linnebo's account of mathematical truth grounded in a story about the abstraction steps taken by an ideal abstractor? And as such is Linnebo's account likewise best thought of as another form of fictionalism?

How about Linnebo's iteration process itself? The idea that we can iterate the abstraction process and each time introduce new objects into our domain to be swept up in our next round of abstraction depends on our being willing to accept that the new 'objects' introduced by abstraction are available to us alongside the objects we started with. And this is a non-trivial assumption. To draw an analogy, if we were to take a pack of 52 playing cards, a familiar Fregean thought is that there is no one way of answering the question, 'how many?'. There are 52 cards, but 13 different card values, 4 suits, 1 pack, and, for that matter, who knows how many molecules. Setting aside the molecules, if asked 'how many?' we could equally well answer '52', '13', '4', and '1', depending on whether we are asking about cards, card values, suits, or packs. But could we also equally well answer '$52 + 13 + 4 + 1 = 70$', noting that we have in front of us 52 cards *and* 13 card values *and* 4 suits *and* 1 pack? A natural response to such an answer is that this is double counting, that while it is fine to think of the objects in front of us *either* as 52 cards *or* as 4 suits, once we have decided to count them as cards, we cannot also think of them as suits and think of ourselves as having *extra* objects on top of all of the cards. Yet iterated abstraction requires just that: we start with a domain of n objects and choose to think of pluralities of those objects instead as 'equinumerosities', so that from n objects we have $n + 1$ different sizes of collections of those objects. This may seem innocuous enough. But if we then say that we now have a domain consisting of $2n + 1$ objects—those being the original n that we started with and the $n + 1$ new objects obtained by abstraction, mightn't we not baulk at this as an egregious form of double counting? For Linnebo's iterations to occur we have to think of our domain as genuinely extending as we add in the new objects formed via abstraction. But why on earth should we think *that*? Why not think that abstraction simply provides us with new ways of talking about old things, not with new things that can be talked about alongside the old?

This last question raises the issue of whether to adopt a reductionist vs non-reductionist semantics for the terms introduced by abstraction. A reductionist semantics would say that, although our abstraction principles introduce new term forming operators, we need not afford these operators a standard, referential, semantics: 'the number of *x*s' looks like it should refer to an object, especially given that expressions of this form appear on either side of identity claims. But, for the reductionist, this appearance is simply an illusion: the apparent reference to new objects occurring on the left hand side of our abstraction principles is reduced to reference to the old objects involved on the right hand side. The reductionist and non-reductionist agree that the sentences on the left hand side of abstraction principles are true whenever the sentences on their right hand side are true, and indeed they agree that the sentences on the left hand side of abstraction principles are true *in virtue of* the truth of the sentences on the right. Where they disagree is that only the non-reductionist thinks that the singular terms on the left hand side of abstraction principles are genuine singular terms, to be understood using a standard referential semantics. The reductionist, by contrast, thinks that the left hand side identities should be thought of as perhaps convenient ways of expressing the equivalence facts on the right hand side.

Why adopt a non-reductionist semantics rather than the reductionist alternative, given that reductionist and non-reductionist alike agree about the fundamental facts in virtue of which claims apparently about numbers or sets are true? In support of non-reductionism, Linnebo cites the importance of adhering to the principle of compositionality, that is, the principle that "the semantic value of a complex expression is functionally determined by the semantic values of its components and their syntactic mode of combination". Give up on the idea that the semantic values of the identities on the left hand side of abstraction principles are a function of the referents of the terms and the meaning of the '=' sign, as the reductionist does, and we abandon compositionality. Better, Linnebo thinks, to stick to a standard semantics and accept, given that we accept the truth of the statements on the left hand side of abstraction principles that speak *as if* there are objects to which they refer, that they do indeed refer to objects.

However, there is something rather odd here in Linnebo's case for the neat solution of adopting a standard semantics. Typically our semantics connects meaning to truth conditions, but note that the reductionist and the non-reductionist alike agree that the ultimate truth-makers for claims about what appear to be 'new' objects on the left hand side of our abstraction principles are facts about the old objects on their right hand side. The reductionist disagrees, of course that there *are* really any new objects introduced by this way of talking, as talk in terms of the 'new' objects introduced by taking equivalence classes is reducible to talk about the equivalence relations between the old objects referred to on the right hand side; nevertheless, both sides agree on what ultimately grounds the truth of the left hand side claims. It's true that the non-reductionist's 'neat' semantics can say that a sentence of the form *Fa* is true in virtue of the referent of '*a*' satisfying the predicate *F*. But bear in mind that Linnebo holds that, if *a* is a thin mathematical object, reference to *a* is also grounded in facts that don't involve *a* itself. So simply to say that '*Fa* is true in virtue of the referent of '*a*' satisfying the predicate *F*' isn't to give the full story as concerns truth. When '*a*' refers to a thin mathematical object, the fact that the referent of '*a*' satisfies the predicate *F* will be grounded in further facts that don't involve the object *a* at all.

Thus, Linnebo tells us, a feature of thin mathematical objects is their shallow nature. That is,

> To explain why mathematical objects have their properties, *there is no need to invoke these objects*; rather, an explanation can be based on any specifications of the objects and certain properties of these specifications. (Linnebo, p. 196, my emphasis)

A neat semantics for Linnebo is secured by holding that sentences of the form '*Fa*' are true in virtue of the objects they refer to having the properties ascribed to them. But since, when '*a*' refers to a thin mathematical object of shallow nature, *a* has property *F* in virtue of the features of some entirely unrelated objects, it seems that underneath the non-reductionist's neat, standard, semantics, is what looks like a messy, non-standard account of truthmaking.

Given their shallow nature, then, thin mathematical objects, while invoked in the course of offering a standard semantics for the truth of mathematical claims, play no role in the most fundamental explanations of the truth values of sentences in which they occur. But if this is the case then, despite the simplicity of a standard semantics, it is tempting to see this as a case of our speaking 'as if' there are objects, for practical purposes, so as to hide a more complex relation between sentences and the features of the world that make them true? If we do wish to adopt a standard semantics for the sentences in question, rather than a more complicated but perhaps more honest reductionist semantics that links those sentences directly to the facts in virtue of which they are true, wouldn't it be more honest to say that these sentences, as standardly construed (as referring to objects) *aren't true*, but are nevertheless *assertible* in virtue of the facts about basic objects that the reductionist sees as grounding their truth? If so, then again we are back to fictionalism about thin objects: there are no such things, but we merely speak 'as if' there are so as not to have to complicate our semantics.

12.3 Are Fictional Objects Thin?

We have seen a few reasons to question whether Linnebo's 'thin' mathematical objects are worthy of being considered genuine objects at all. Unlike ordinary objects, they are not available for us to quantify over without having been introduced at some stage of abstraction; when we consider what grounds their availability for quantification it is not that we have *actually* carried out an abstraction but only that we consider that an ideal abstractor *could have* done so, so that their 'existence' seems dependent on what's true in a story about the possible activities of an ideal abstractor; and finally when we consider their contribution to providing the fundamental grounds for the truth of sentences in which they occur, none is to be found. All of these features, to me as one already inclined to fictionalism, speak in favour of thinking that there are no such things as thin objects: we merely speak *as if* there are such things because it is advantageous (for example in terms of allowing for a standard semantics for mathematical theories) for us to do so.

However, these considerations have rested on looking for features of thin mathematical objects which make them look analogous to fictional entities, and as ever there are inflationary and deflationary ways of taking the analogy. Whereas on my reading I start from the assumption that there are no such things as fictional entities and draw the conclusion that the similarities of alleged thin objects to fictional ones means that there are no such things as thin objects too, a defender of thin objects could easily draw the analogy in the opposite direction, and argue that since fictional objects share many features with thin objects, these objects do exist—and exist thinly. Indeed, the very features noted above that led me to question the reality of Linnebo's thin mathematical objects are embraced by Linnebo as advantages of his Platonist picture. We can easily avoid the traditional problems with Platonism if the mathematical objects to which we are committed make such insubstantial demands on the world. Why not simply embrace the quite bearable lightness of being offered by the thin objects picture and stop worrying about commitment to abstract mathematical entities?

If the defender of thin objects wished to push this response I'm not sure that there's anything that can be said to take us beyond a clash of intuitions concerning what it should take for something to count as really existing. The thin realist has the advantage that they can say that our ordinary mathematical assertions are not only to be given a face value semantics, but can be taken to be true; the fictionalist has the advantage that they only admit as existing those objects that play a genuine role in explaining the truth of the sentences in which they occur, thus offering a more traditional understanding of the link between semantics and truth. But ultimately we might concede that the difference here is not a substantial metaphysical one, but rather a difference in opinion on how to use words such as 'exists', 'refers', 'is true'. The thin realist and the fictionalist alike agree that mathematical objects play no substantial role in explaining the truth of mathematical sentences. Perhaps Linnebo's account and that of the fictionalist are, to borrow a phrase from Penelope Maddy, simply "alternative ways of expressing the very same account of the objective facts that underlie mathematical practice" (Maddy, 2011, p. 112).

Rather than argue against the idea that fictional objects exist and are thin, I would like to consider the consequences of bringing thin realism and fictionalism so closely together as alternative ways of expressing the same underlying facts. Particularly, I would like to note that, if Linnebo's thin realism and fictionalism are so close together, then thin realism inherits some of the major challenges faced by fictionalism. I have argued elsewhere (Leng, 2009) that the central challenges faced by fictionalism and related views are two. First, the problem of modality, and second, the problem of mixed claims. If Linnebo's thin realism and fictionalism are as close together as I have suggested, then it is worth considering whether Linnebo's account can fare any better than fictionalism does in responding to these claims.

12.3.1 The Problem of Modality

Fictionalists, in replacing commitment to the face-value truth of our mathematical theories to commitment to the truth-in-the-story of their claims, make a conscious trade of ontology for ideology, where specifically the 'ideology' that is relied on is a commitment to modal facts. What is 'true-in-the-story' of standard arithmetic, for example, will typically be understood as a matter of what *follows from* the second-order Peano axioms.[1] Given that the fictionalist does not believe in mathematical objects, a standard reductive account of logical modality to model theory is not available to them. Instead the fictionalist has to swallow a commitment to primitive modal facts concerning logical consequence. This is a problem for the fictionalist as it may seem that nominalist qualms about our ability to know facts about sets will simply transfer to equivalent qualms about our ability to know primitive modal truths. Fictionalists respond by noting that there are some routes to modal knowledge that make our ability to obtain at least some logical knowledge less mysterious than knowledge of abstract mathematical objects, and that at any rate Platonist claims to have successfully reduced modality to set theoretic ontology are in doubt (Field, 1984b, 1991; Leng, 2007), so that it is arguable that the platonist's account requires *both* primitive modality and a Platonist ontology, rather than a simple trade of one for the other.

As noted above, Linnebo too requires modality to ground his account of thin objects as what I have called 'permanent possibilities of abstraction', and a reduction to model theory is no more available to Linnebo to explain the modality underlying his permissible abstractions/extensions of a language than it is to the fictionalist. So a question arises about Linnebo's modal commitments. Is he committed to a primitive notion of logical possibility equivalent to that assumed by the fictionalist? And if he is going to include this commitment into his ideology why stick with the predicative, abstractionist, approach to introducing 'thin objects', rather than simply grounding the 'truth' of claims about mathematical objects in modal truths about what follows from our mathematical axioms (as per the 'Hilbertian' approach to thin objects that he mentions but sets aside).

12.3.2 The Problem of Mixed Claims

The problem of mixed claims arises for the fictionalist once we try to integrate a mathematical 'fiction' (where the 'truths' in the fiction are a matter of what follows from its axioms) into empirical science. Pure mathematics is relatively easy to account for from a fictionalist perspective, as its 'stories'—at least once axioms have been developed—are easily delineated. But once mathematics makes its way into applications, the question arises as to how to understand a 'story' in which mathematical and empirical vocabulary are entangled together, and in particular

[1] There are fictionalists who hope to manage with less than full second order logic, while still going beyond the limits of first-order axiomatic theories so as to avoid incompleteness. For example in his updated preface to Field (1980), Field considers supplementing first-order logic with cardinality quantifiers or a 'finitely many' quantifier. And Mark Balaguer (1998) considers using non-axiomatic methods to pin down the intended interpretation of the first-order Peano axioms (via a broader 'full conception of natural numbers (FCNN)). I will set aside these alternatives here as I follow Boolos (1984) (and indeed Linnebo) in holding that the commitments of second order quantification take us no further than those of first-order quantification.

how to justify our use of such a story to draw out empirical conclusions which we believe to be true.

Two options arise for the fictionalist. One is the 'hard road' attempted by Hartry Field (1980), of showing that the mathematical and empirical components of our empirical scientific theories can be disentangled, so that the content of those theories can be expressed in purely nominalistic terms. In this case, the usefulness of mathematics can be explained via its conservativeness, as a tool to draw out the nominalistic consequences of nominalistically stated theories. The alternative so-called 'easy road',[2] is pursued by Balaguer (1998), Leng (2010), and others, and involves arguing that it is reasonable to expect that, by engaging in the fiction that there are mathematical objects such as numbers and sets, and using this to help to formulate hypotheses about what the physical world *would* be like, *were* there such objects (that didn't interfere in any way with how things actually are) we can produce theories that, while not literally true, are correct in their nominalistic consequences. One of the strongest challenges pressed to those wishing to travel this 'easy road' points to the explanatory role apparently played by mathematical objects, including cases where the apparent explanatory work done by mathematics is not easily accounted for as residing in the nominalistic content of our mathematical theories.[3]

Typically it is assumed that realist/Platonist philosophies of mathematics avoid this challenge as, believing that they do that mathematical objects exist and that our theoretical claims about them are typically true, they have no problem accepting the entanglement of mathematical vocabulary into empirical theories, and no problem explaining why theories of that sort (which after all—at least if they are scientific realists—they believe to be true) should be expected to have true predictions and to deliver good explanations. However, the 'thinness' of Linnebo's thin objects raises a question for whether *he* can be as sanguine as the typical Platonist is about the role of mathematical objects in scientific theories. Recall, in particular, Linnebo's characterisation of the 'shallow nature' of thin mathematical objects. To quote again:

> To explain why mathematical objects have their properties, *there is no need to invoke these objects*; rather, an explanation can be based on any specifications of the objects and certain properties of these specifications. (Linnebo, p. 196, my emphasis)

[2] The labels 'easy road' and 'hard road' are due to Colyvan (2010).

[3] Much has been written on this lively debate, starting with Alan Baker's influential (2005) challenge to the nominalist which has received a wide range of nominalist responses. My own considered response (Leng, 2021) is to agree that the explanatory work of mathematics in science does not typically reside in the nominalistic content of our mathematically-stated scientific theories, but argue that neither does it reside in the mathematical objects apparently referred to in formulating mathematical explanations of empirical phenomena. Rather, mathematical explanations are typically structural explanations, that explain by showing the explanandum to hold as a consequence of the mathematical structure instantiated in a physical system. I argue that such an account of mathematical explanation as involving the (approximate) instantiation of axiomatically-characterised mathematical structures is compatible with a fictionalist understanding of mathematical theories.

The shallow nature of mathematical objects mean that they do not themselves feature in explanations of truths about them: truths about the numbers, for example, are explained in terms of the truths about the equinumerosities that ground our introduction of numbers by abstraction.

What does this mean when it comes to the alleged explanatory role of mathematical objects in empirical science? In presenting the shallow nature of mathematical objects Linnebo doesn't claim that they can never play *any* explanatory role, but only that they need not feature in explanation of mathematical facts. So it is entirely open for this account to claim that numbers never feature in explanations of ordinary facts about numbers and their properties, but nevertheless do sometimes feature in explanations of some empirical facts. However, while this option is open to the thin mathematical realist, on the face of it it looks unattractive. When Baker (2005) argues that the life cycles of cicadas are explained by facts about the numbers (particularly, the fact that 13 and 17 are prime numbers), it is natural to ask, are Linnebo's thin objects substantial enough to do the kind of substantial explanatory work Baker envisages? The worry from shallow nature arises because the shallow nature of mathematical objects means that, for Linnebo, the numbers 13 and 17 are not even part of our best explanation of the fact that 13 and 17 are prime numbers ('shallow nature' means that the explanation of this fact must be grounded in facts about the equinumerosities used to specify 13 and 17). If mathematical objects are so explanatorily anaemic that they fail to feature even in explanations of facts about their own properties, it is hard to see how 13 and 17 could, on Linnebo's picture, play a genuine explanatory role in accounting for the facts about cicada life-cycles that Baker thinks require us to accept Platonism.[4]

This is not to say that there aren't options for Linnebo's account of 'thin objects' in considering how such objects play the role they do in our empirical scientific theories. After all, the nominalist literature shows many routes to resisting the Platonist claim that mathematical objects play a genuine explanatory role. Rather, my point is that (in part because of the closeness of thin objects to objects that the typical fictionalist holds not to exist at all) Linnebo's account inherits this challenge to fictionalism. It is not sufficient to rest on the fact that he is comfortable with the *existence* of mathematical objects and so simply accept Baker's claim that mathematical objects play an explanatory role in science. Linnebo must either show that the merely *thin* form of existence had by his mathematical objects is sufficient to account for the explanatory role they play in our scientific theories, or alternatively follow the nominalist in resisting the claim that the objects of our mathematical theories play a genuine explanatory role.

[4] In line with this assessment, Concepción Martínez Vidal's contribution to this volume suggests that Linnebo's mathematical objects, as thin relative to physical objects, can play only a lightweight explanatory role. As, Martínez Vidal points out, defenders of the explanatory indispensability argument (EIA) will insist on mathematical entities playing a *heavyweight* role in our explanations, then despite his Platonism Linnebo must follow the fictionalist in resisting this aspect of the EIA.

12.4 Conclusion

Are thin objects merely fictional objects, or are fictional objects thin? Maybe there isn't a determinate answer to this question: maybe it rests simply on, as Carnap (1950, p. 29) put it, 'a matter of decision rather than assertion', in this case the decision being how we extend our use of the concept 'exists'. What I hope, however, to have shown is that Linnebo's thin objects are close enough to the merely fictional objects that fictionalists claims to not really exist for Linnebo's account to inherit some of the standard challenges to mathematical fictionalism. My guess is that Linnebo will have to make use of the very same resources that the fictionalist uses in responding to these challenge, including adopting substantial modalist commitments and showing how the explanatory work done by mathematical theories in empirical science does not equate to a substantial explanatory role for mathematical objects. And if so, one wonders what the advantage is of Linnebo's abstractionist route over simply accepting the objects of any logically possible mathematical theory?

Acknowledgements I presented this material to audiences at the University of York, the University of Konstanz, and the University of Oxford, and I am grateful to audiences at those presentations, and especially to Øystein Linnebo, for very helpful discussions. A am grateful to Bahram Assadian and Concha Martínez Vidal for their generous and helpful comments on an earlier draft.

References

Baker, A. (2005). Are there genuine mathematical explanations of physical phenomena? *Mind, 114*, 223–238.

Balaguer, M. (1998). *Platonism and anti-platonism in mathematics*. OUP.

Boolos, G. (1984). To be is to be the value of a variable (or to be some values of some variables). *The Journal of Philosophy, 81*, 430–449.

Carnap, R. (1950). Empiricism, semantics, and ontology. *Revue Internationale de Philosophie, 4*(11), 20–40.

Colyvan, M. (2010). There is no easy road to nominalism. *Mind, 119*(474), 285–306.

Field, H. (1980/2016). *Science without numbers: A defence of nominalism* (2nd ed.). OUP.

Field, H. (1984a). Critical notice of Crispin Wright: *Frege's conception of numbers as objects. Canadian Journal of Philosophy, 14*, 637–662. Reprinted as *'Platonism for cheap? Crispin Wright on Frege's Context Principle' in Field* (1989), pages 147–170.

Field, H. (1984b). Is mathematical knowledge just logical knowledge? *Philosophical Review, 93*(4), 509–552. reprinted in Field (1989).

Field, H. (1991). Metalogic and modality. *Philosophical Studies, 62*(1), 1–22.

Florio, S., & Linnebo, Ø. (2021). *The many and the one: A philosophical study of plural logic*. OUP.

Hale, B., & Wright, C. (2001). *The reason's proper study*. OUP.

Hoffman, S. (2004). Kitcher, ideal agents, and Fictionalism. *Philosophia Mathematica, 12*(1), 3–17.

Kitcher, P. (1983). *The nature of mathematical knowledge*. OUP.

Leng, M. (2007). What's there to know? A fictionalist account of mathematical knowledge. In Leng, Paseau, & Potter (Eds.), *Mathematical knowledge* (Vol. 2007). Oxford University Press.

Leng, M. (2009). Algebraic approaches to mathematics. In Bueno & Linnebo (Eds.), *New waves in philosophy of mathematics* (pp. 117–136). Palgrave Macmillan.
Leng, M. (2010). *Mathematics and reality*. OUP.
Leng, M. (2021). Models, structures, and the explanatory role of mathematics in empirical science. *Synthese, 199*, 10415–10440.
Leng, M., Paseau, A., & Potter, M. (Eds.). (2007). *Mathematical knowledge*. OUP.
Linnebo, Ø. (2018). *Thin objects: An abstractionist account*. OUP.
Maddy, P. (2011). *Defending the Axioms: On the Philosophical Foundations of Set Theory*. OUP.

Chapter 13
Fictional Characters in Fictional Discourse

Manuel García-Carpintero

Abstract Fictional discourse is, primarily, discourse used to produce literary fictions; there is also the "metafictional" discourse used to talk about fictions, to report their contents or other features. On a traditional view classically articulated by Searle, primary fictional discourse doesn't have a specific semantics; sentences there just have the semantics they would in their standard uses. Fiction-makers convey their fictions by pretending to use them in their standard ways without doing so, and without what they do giving them a specific, dedicated representational point with a semantics of its own. This view is less popular nowadays than it used to be. A question the now more popular alternative view raises is, what is the contribution of intuitively empty names to such a dedicated semantics? Many supporters of the traditional *mere pretense* view, including Searle, argued that, in the metafictional uses to which they granted a semantics, apparently empty names are not in fact empty, but refer to some more or less exotic entity. Some of those who grant a dedicated semantics to primary fictional discourse, like Salmon and others, extend this *realist* view to it. The paper aims to uphold considerations already given by irrealists against these proposals, by highlighting its counterintuitive features and explaining why they are bad in this theoretical setting.

Keywords Fiction · Fictionality · Reference in fiction · Fictional entities · Fictional content

---

Financial support for my work was provided by the DGI, Spanish Government, research project PID2020-119588GB-I00, and the award *ICREA Academia* for excellence in research, 2018, funded by the Generalitat de Catalunya. The paper was presented at talks and workshops at XXX, where it received helpful comments. Thanks to Greg Currie, Andreas Stokke, and Ed Zalta for useful comments and suggestions, and to Michael Maudsley for the grammatical revision.

M. García-Carpintero (✉)
LOGOS/BIAP-Departament de Filosofia, Universitat de Barcelona, Barcelona, Spain
e-mail: m.garciacarpintero@ub.edu

X. de Donato-Rodríguez et al. (eds.), *Deflationist Conceptions of Abstract Objects*, Synthese Library 516, https://doi.org/10.1007/978-3-031-98717-5_13

13.1 Introduction: The Phenomenon of Coordination

In recent work, I have surveyed proposals in contemporary philosophy about the semantics of fictional names and the ontology of fictional characters (García-Carpintero, 2019a, 2022a). I didn't hide my allegiance to irrealist views on the latter topic and the view on the former that, in fictional discourse proper (more on this momentarily), fictional names are empty—which I had already defended in previous work (García-Carpintero, 2010a, b, 2021a). But the aim these works had as reviews of the literature kept the considerations for my views in the background. Here I want to straightforwardly motivate them, and to highlight what I take to be a main consideration in its favor.

To set the stage, I will make a distinction within what I called above "fictional discourse", implicit in works from the 1970s that shaped current debates, by Lewis (1978), van Inwagen (1977, 1983), Kripke (2013), and Walton (1990), made explicit later by others (cf. Currie (1990, pp. 158, 171), Thomasson (2003, p. 207)).[1] I rely on Bonomi's (2008) apt terminology, which I regularly use in discussions of these issues. By default, declarative sentences convey assertions, which we evaluate as correct or otherwise depending on whether they are true (García-Carpintero, 2021b). Consider three uses of declaratives in connection with fiction:

1. Emma Woodhouse had lived nearly twenty-one years in the world.
2. (According to/In *Emma*) Emma Woodhouse had lived nearly twenty-one years.
3. Emma Woodhouse is a fictional character.

Take first Austen's utterance of (1), as part of her longer utterance which for now we can think of as the act of putting forward *Emma* for readers to enjoy. Following Bonomi, I will call uses like this *textual*. Even when, if taken literally as assertions, they conflict with what we believe, we do not typically find any tension in accepting them in this context and we would not find it plausible to criticize Austen on this regard.[2] The other types lack this feature. There is, first, the *paratextual* use of sentences such as (2)—with or without a prefix—to report on what goes on in a fiction concerning the character of its *fictional world*, its *plot* or *story*. Readers of *Emma* would consider (2) in such a use true, while they would reject as false the result of substituting "fifty-one" for "twenty-one" in it. Finally, I will call uses of sentences

[1] Related classifications are sometimes made by resorting to Lamarque's (1996) distinction between *internal* and *external* perspectives (Semeijn & Zalta, 2021, §2.2). I find them less taxonomically unhelpful because it is difficult to locate paratextual utterances in them, but I'll help myself to the terminology below.

[2] The "typically" hedge acknowledges cases of *imaginative resistance*, cf. Stock (2017, ch. 4). To illustrate, I learnt that fact-checkers at the *New Yorker* asked Antonya Nelson to correct the "inaccurate" color of the Wichita police cars for her 1992 fiction piece *Naked Women*. They must have thought that Wichita readers would find the corrected proposal to imagine jarring. Most of us feel this incongruity experience on similar issues, as I did when (for another illustration) a well-known art nouveau flat in Barcelona is implicitly located in a different neighborhood than it really is in Almodóvar's 1999 *All About My Mother*.

such as (3) *metatextual*; they are also intuitively truth-evaluable relative to the actual world but not content-reporting, in that they are not (obviously) equivalent to explicit content ascriptions like the prefixed version of (2).[3]

Most writers who discuss our issues (the semantics of fictional names and the ontology of fictional characters) focus on paratextual and metatextual uses.[4] At its core, semantics aims to assign truth-conditions to sentences in a systematic way. Being intuitively truth-evaluable, paratextual and metatextual uses appear to have a semantics in this sense. Thus Lewis's (1978) aims to provide truth-conditions for paratextual uses like (2). He acknowledges metatextual uses like (3), but chooses to put them aside (1978, p. 263). The main explanatory goal he sets up for himself concerns the behavior of *prima facie* empty names like "Emma Woodhouse" in utterances like (2): how could they be true or even merely truth-evaluable, if we reject the "Meinongian" view that the names (against first appearances) do refer after all, to some exotic thing? Lewis uses 'Meinongian' in a capacious sense, encompassing all realist views that take fictional names to refer to some such entity, including Kripke (2013) and van Inwagen (1977). Their arguments for positing such entities focus primarily on the metatextual uses that Lewis casts aside, but van Inwagen at least extends the account to paratextual uses.[5]

In my view, however, textual uses play a fundamental, grounding role and the theoretical focus should be on them. Kripke (2013, p. 24), Lewis (1978, pp. 266–267) and van Inwagen (1977, p. 306) casually assumed an account articulated by Searle (1975), which was then common among analytic philosophers: the view that textual uses are *mere pretense* ('MP' henceforth).[6] MP suggests that they do not have a semantics, a view that both Kripke (2013, pp. 24–25) and van Inwagen (1977, p. 306)—but not Lewis (1978)—explicitly embrace.[7]

---

[3] The hedge is here due to the appeal to "extended fictions" envisaged by Walton (1990, ch. 10) and other irrealists like Brock (2002) and Everett (2013). Negative existentials like "Emma does not exist" count as metatextual (and true) in their more intuitively accessible use; "Emma exists" contradicts it in that use, but it is intuitively taken as paratextual and true. Such baffling impressions (Reimer, 2001) motivate pragmatic, metalinguistic treatments (van Inwagen, 1977, p. 308, fn. 11), (Kripke, 2013, p. 159), (Yablo, 2020), (Felappi, 2021).

[4] Yablo (2020, p. 78) does not mention textual uses; Collins (2019, §1) has a category of *in-fiction* uses, examples of which appear to be all paratextual uses. Semeijn and Zalta (2021, pp. 173–4) note the neglect, but they argue that "the [textual] use of fictional names … distinctively differs" from others (175), on account of which they also brush off the semantics of names in that use. On the view I defend here, the differences they mention are not due to the semantics of names, which is uniform for textual and paratextual uses.

[5] Cf. fn. 20 below. Van Inwagen (1977, pp. 299–300) rejects *Meinongianism*, because he uses the term in a more restricted sense than Lewis's, applying only to views on which fictional names refer to the specific sort of exotic entity that Meinong is supposed to have envisaged.

[6] Proponents include Macdonald (1954), Ohmann (1971), and many others. Predelli (2020) offers the best defense I am familiar with of a view in the vicinity. García-Carpintero (2022b) and von Solodkoff (2022) critically engage with it.

[7] Lewis (p. 267) says: "at our world, the storyteller only pretends that 'Sherlock Holmes' has the semantic character of an ordinary proper name. We have no reason at all to suppose that the name, as used here at our world, really does have that character. As we use it, it may be very unlike an

Searle (1975) rejects the alternative view—which Currie (1990) elaborated later along Gricean lines—that, in proffering declarative sentences, authors of fiction are "not performing the illocutionary act of making an assertion but the illocutionary act of telling a story or writing a novel" (p. 323); I will refer to this representational act as *fictionalizing*. I call Searle's view *mere* pretense because the proponent of the alternative view (which I will call the *dedicated representation* view, 'DR') agrees that fictionalizers may resort to pretense to perform the fictionalizing act: they may have actors pretending or themselves play a role; Lewis (1978, p. 266) aptly describes Doyle as playing the role of Doctor Watson, "pretending to be" him. But they need not do so, which distinguishes DR from MP: they might convey the contents of their dedicated fictionalizing act directly, without any pretense. On DR, fiction need not be "parasitic on 'serious' discourse" (Walton, 1983, p. 84)..[8]

13.2 The Semantics of Textual Uses

The view that textual uses do not have a semantics is motivated by Millianism.[9] This is the view that "a proper name is, so to speak, *simply* a name. It *simply* refers to its bearer, and has no other linguistic function. In particular … a name does not describe its bearer as possessing any special identifying properties" (Kripke, 1979, pp. 239–240). This appears to entail that (1) does not have a content—does not express a proposition. If "Emma Woodhouse" has "no other linguistic function" than to "simply refer", then it appears not to have any meaning.[10] This creates a problem. Walton (1990) describes Austen's use of (1) as a "prop in a game of make-believe", inviting a fiction-constituting imagining. On this view, fictions have contents conveyed by the textual uses constituting them, which readers are to imagine. It would be difficult to deny this; Predelli (2020, pp. 37–38) and Recanati (2021, p. 26) do grant it. Such imaginings have propositional contents (Ryle, 1933, p. 29;

---

ordinary proper name". In conceding that fictional names in textual discourse are not names Lewis is granting too much to Millians, see fn. 25 below; but he assumes that they do have a use with its specific semantics. Wolterstorff (1980, p. 154) and Currie (1990, p. 131) defend a view like Lewis's.

[8] Recanati (2021) argues for the *asymmetric dependence thesis* that "fictional reference and fictional assertion are parasitic on genuine reference and genuine assertion" (p. 19); Szabó (2020, p. 62, fn. 5) also assumes it. This is what Walton correctly rejects. But Walton (1983) agrees on the rejection of DR. He (1990, p. 36, fn. 24, 391, 396, 417) also agrees with van Inwagen and Kripke that textual utterances do not express propositions, see fn. 10 below.

[9] Cf. Recanati (2021, pp. 19, 20).

[10] Kripke (2013) argues that "the propositions that occur in a work of fiction would only be pretended propositions" (*ibid.*, p. 24) on the basis of a "Pretense Principle" (roughly, Recanati's *asymmetric dependence thesis*, fn. 8). This is a non-sequitur, see below, fn. 16. Van Inwagen (1977, 1983) makes the intuitively problematic (cf. fn. 2 above) claims that Dickens is not "writing about anything" and is "asserting nothing" (1977, p. 301) to establish the even more questionable contention that he does not "express any proposition" (*ibid.*, p. 306; 1983, p. 73).

Stock, 2017, pp. 4–9). This raises the question, how does the sentence found in Austen's text help to fix what audiences are to imagine? (2021, pp. 77, 89). How does its occurrence help to determine the propositional content of the relevant imaginings? Unlike Kripke and van Inwagen, Walton at least confronts this question, but he does not answer it convincingly (Zemach, 1998; García-Carpintero, 2010a, pp. 286–287; von Solodkoff & Woodward, 2017, pp. 413–414).[11]

Putting aside Millianism, is there any reason to think that fictional names in textual discourse are only used in pretense? Predelli's (Predelli, 2020, p. 88) justification is that the fiction-maker's "aim was not that of putting forth a referential device, but rather that of making things up".[12] The adversative 'but rather' is unwarranted: "to make things up" the way fiction-makers do *involves* deploying genuine referential devices. They make things up *by*, among other things, putting forth referential devices. For a central aspect of making up stories is that readers put together the information to be imagined about fictional bearers of names by "trading on identity", in the way that this is done with real names.[13] This is a crucial role that ordinary names have: to help *de jure* co-identification, i.e., to help speakers to put together information or misinformation on their purported referents (cf. Recanati, 2016). They play this same role in fictions (García-Carpintero, 2020a).[14] Note also that launching a proper name does not need to be a momentous act that we should pay much attention to. We do it when we put a number beside a sentence to help to refer to it, as I did with (1)–(3) at the outset, or when we nickname somebody on the spur of the moment; we may even do it inadvertently when a name we use changes its reference without our realizing it.[15]

A second reason to reject the view that Watson's pretend use of "Holmes" excludes Doyle's real use comes from the use of indexicals in fictions, like Cortázar's *The Continuity of Parks* in which they are used instead of names.[16] The view that

---

[11] Predelli (2020) does a better job—but see García-Carpintero (2022b) for criticism.

[12] Russell writes that the "fundamental falsehood in the play is the proposition: the noise 'Hamlet' is a name"—quoted by Recanati (2021, p. 28). Recanati's related justification, like Kripke's (2013), is that fiction-makers pretend to use names. This is not conclusive, cf. fn. 16.

[13] Fictionally, "Holmes" is a real name, which has been launched (Predelli, 2020, p. 26). But given DR, utterances of declarative sentences in the Holmes stories have a *dual* role. They are a vehicle for Watson's fictional assertions; but also one for Doyle's act of fictionalizing. "Holmes" has a corresponding dual role. It is for this second role that I claim in response to Predelli that Doyle has launched the name. Realists about fictional characters discussed below, 3.1 who, like Salmon (1998), Predelli's (2002) previous self, or Abell (2020) extend their view to fictional discourse share my assumption.

[14] Most of Davis's (2005, pp. 233–245) compelling arguments that proper names are true words apply to textual uses; cp. Hunter's (1981, pp. 28–9) contrast between "Holmes" and a truly unlaunched name.

[15] García-Carpintero (2018, §4.1) accounts for "Madagascar"-like cases by defending a metasemantic view that underwrites this.

[16] Some theorists contend that in pretending to do A one disbelieves that one does A (Picciuto & Carruthers, 2016, p. 317). This is wrong (Saucelli, 2021). Some actors pretend to drink whisky by drinking tea, but others (knowingly) drink the stuff. The *New Yorker*'s fact-checkers assumption (fn. 2) that one might pretend to assert *p* in putting forward a fiction and really assert it is not just

fiction-makers are not using indexical expressions with their standard semantics in such cases looks hopeless.[17]

Given DR, the sentence-like expressions used in textual uses may well be sentences, their fictionalizing meaning their literal, semantic content, and the expressions occurring in them, including names, literally used referring expressions. In Abell's (2020) account, acts of fictionalizing belong in the speech-act category of *declarations*, "illocutionary acts that are governed by the rules of extra-linguistic institutions", like giving out players or adjourning meetings. Legitimate doubts can be raised about whether the meaning of 'I name this ship the Generalissimo Stalin.' as a declaration coincides with its literal, semantic meaning (Alston, 2000, pp. 89–91). But I do not find Abell's arguments for identifying acts of fictionalizing with declarations compelling. In my own, equally social-Austinian—in contrast to Currie's psychological-Gricean (1990)—speech-act version of DR, I follow Currie in identifying fictionalizings with directive speech acts, *invitations to imagine* (García-Carpintero, 2013, 2019b). If, as Ryle (1933, p. 33), Carroll (1995, pp. 98–9), Wolterstorff (1980, pp. 219–234) and many other early writers assumed, we take the propositional imaginings invited by fictions to be just acts of entertaining propositions for their specific purposes, it is very natural to think of textual uses as having fictionalizing acts as their literal meanings. The default use of declarative sentences is to make assertions, but we also use them literally to make weaker acts like conjectures, as when we hedge, either explicitly or contextually: 'It will rain, I think.' We can literally invite imaginings in the same way: 'S, let us imagine.' The hedge may remain implicit, contextually suggested—say, by the full utterance being declared a novel.

On this view, textual uses as much as paratextual and metatextual uses have a semantics, even though, given the character of the speech act such uses are meant to make, we don't in general feel apt to evaluate them as true or false. The question then is, what is the contribution of empty names to such semantics? In the final section, I contrast realist and irrealist accounts. Both assume that, as I have been arguing, intuitively "empty" proper names in textual uses behave exactly as their intuitively non-empty correlates in assertoric discourse. The difference lies in that the former—usually sympathetic to Millian views—argue that they are only apparently empty; in fact they also refer, albeit to entities exotic to some degree. The latter rejects Millianism and assumes some account of names on which empty ones (even those used in the shared knowledge that they are so) can be perfectly meaningful. I will offer an argument for the version of the second sort of view that I favor.

---

intuitively correct, it *is* correct (Friend, 2008). By the same token, one may pretend to refer with a referential expression *and* truly use it referentially, fn. 13.

[17] See García-Carpintero (2022b) and von Solodkoff (2022) for further development of these points, and further criticism of Predelli's (2020) *tour de force*.

13.3 Realism and Textual Uses

As said, the main arguments for realism about fictional characters rely on metatextual uses. Kripke (2013) argues that names like "Emma Woodhouse" in (3) refer to fictional entities. In a similar vein, van Inwagen (1977) provides an influential Quinean argument for realism about fictional entities, arguing that it allows for a straightforward explanation of the validity of arguments involving apparent reference to and quantification over them in the metatextual discourse of literary critics.[18] *Ficta* are argued to be concrete Meinongian non-existent entities (Priest, 2011), or (as both Kripke and van Inwagen recommend) abstract existing entities of various sorts, Platonic *abstracta* like Wolterstorff's (1980) or Currie's (1990) *roles*, or created artifacts, Thomasson (1999, 2003). Theorists of both sorts think of fictional characters as having the ontological status of the fictional works in which they occur (Kripke, 2013, p. 72; van Inwagen, 1977, pp. 302–3, 1983, p. 75; Thomasson, 1999, p. 143, 2003, p. 220).

We don't need to go into the ontological issues in any depth here. Van Inwagen (2003) is noncommittal on the two abstractionist lines, mentioning costs for both. Friedell (2021) presents responses to worries about artifactualism, congenial to the social-Austinian view that both Abell (2020) and I (García-Carpintero, 2013, 2019b) defend. Fictional works result on this view from the communicative acts of fiction-makers; they are social constructs, abstract created artifacts with norm-regulated functions. Fictional works have an identifiable "text" (Currie, 1991); Borges's *Pierre Menard* shows that two different works can share text. This "text" is the meaning-vehicle, constituted by words in literature, sounds, performances or images in other media. Contents are abstract; on the *possible worlds* semantic framework that I'll assume, along with most current semanticists, they are structureless properties selecting states of affairs. A crucial reason in favor of this view is that it allows contents to be shared by utterances in different media—in different languages but also in depictive, or dramatic media. This captures the narratological distinction between the *narration* or *sjužet* and the *narrative*, *plot*, *story*, or *fabula* that it conveys; and it allows that a film may be a faithful adaptation of a novel or a play. Works thus have a complex structure, grounded in that of the vehicles that express them; they are in part composed of singular representations.

On the artifactualist version of realism I thus find more compelling, and the one I'll thus assume for my critical purposes here, fictional characters are such singular representations:[19] "Characters, together with their settings and situations, are parts of a story", (MacDonald, 1954, p. 177). Like the works of which they are parts, they must be appropriately ontologically grounded, but this can hopefully be secured, establishing their existence. Terms like "Emma Woodhouse" in metatextual uses

[18] Cf. Brock and Everett's (2015) "Introduction" and Kroon and Voltolini (2018) for helpful presentation, discussion, and further references.

[19] The objection I will raise applies equally well to other forms of realism, Meinogianism or Platonism. It is helpful however to address the argument to a specific version of the view.

like (3) may thus have as semantic value a singular representation associated with that name, which is a constituent of Austen's *Emma*.

On this proposal, the commitment to fictional characters clearly arises first and foremost in metatextual discourse. The account assumes that textual uses have an independent semantics of their own. This is because fictional characters are identified with the representations associated with referring expressions, proper names in particular, occurring in those uses in fictional works. Once we see this, we realize that, left to their own, arguments for realism based on metatextual uses cut little ice. If they establish a form of realism, it cannot go beyond the "easy ontology" advocated by Thomasson. Following Yablo (2014), I myself interpret this account of metatextual discourse as ultimately compatible with straightforward irrealism (García-Carpintero, 2019a, 2021a). In a nutshell, metatextual uses of fictional names are just a form of *hypostatization*, a dead reifying metaphor that shouldn't be taken as seriously as van Inwagen's Quinean considerations need.

The decisive issues, then, lie with textual uses; if it can be established that their semantics lacks realist commitments, we have gone a long way to defend full irrealism about fictional discourse. Now, paratextual discourse is a good intermediate case to consider. On the one hand, like metatextual discourse it is truth-evaluable, and hence it should have a truth-conditional semantics. On the other, the use of fictional names in it doesn't feel "external" to the fiction as it does in metatextual uses (fn. 1), which is what renders the latter intuitively analyzable along artifactualist lines. 'Emma Woodhouse' is on the envisaged realist view taken to refer in (3) to a representation—a part of Austen's work seen from an external point of view as an artifact. In (2) however it intuitively seems to refer to a putative human being—the one we imagine while adopting an internal perspective, imaginatively engaging with Austen's work as we are invited to by it. What do realist say about such paratextual uses?

Van Inwagen (1977, p. 307, 1983, pp. 75–76) extends the realist account to the truth-conditions of paratextual uses.[20] There is an obvious obstacle to this, created by the contrast between metatextual and paratextual uses just highlighted. While the entities that realists posit are good candidates to intuitively instantiate the properties predicated of them in metatextual uses like (3), this is not so for the two other uses. Such entities do not seem to be the sort of thing capable of living in the world; for that requires, say, breathing, which the envisaged exotic entities are incapable of. Van Inwagen (1977, 1983) deals with this by distinguishing two types of predications, *having* and *holding*.[21] The subject-predicate pattern in unprefixed (2) does not

[20] Perhaps Kripke (2013, pp. 57, 74) has a similar view in mind, but his tortuous discussion manifests his misgivings (see pp. 155–160, and Salmon, 1998, pp. 297–298). Cf. Yablo (2020) for a good critical discussion.

[21] See Wolterstorff (1980, pp. 159–160). Zalta's *Object Theory* makes a similar distinction between *exemplifying* and *encoding*, cf. Semeijn and Zalta (2021, §3). Zalta takes abstract objects to be constituted by the properties ascribed to them in fictions; an object is then said to *encode* the properties constituting it. The definition that follows is better attuned to the representational nature ascribed here to fictional characters.

indicate that the semantic value assigned to the subject-term instantiates (*has*) the property expressed by the predicate, but rather that it *represents or characterizes an item to which the property is ascribed* in its encompassing fiction (*holds*). In brief: *o* holds *F* if and only if, in its encompassing work, *o* represents something that *has F*.

To the extent that we can make good sense of the definiens (as DR plus the irrealist view described below allows), I have just offered a straightforward, reductive definition of *holding* in terms of *having*—which is to be desired from a foundational viewpoint, for the latter is the basic, primitive sense of predication. However, van Inwagen claims that this cannot be done. In support, he (1977, pp. 306–307) just offers good criticisms of two different, uncompelling definitions. Von Solodkoff and Woodward (2017, pp. 410–411)), who offer a definition that I will criticize shortly, conjecture that van Inwagen was prevented from coming up with the right one because he was assuming Lewis's (1978) account of fictional content. I disagree. Lewis's view can be elaborated so that it offers a better account of these matters than the one that realists like von Solodkoff and Woodward provide, consistent with my definition above (García-Carpintero, 2022d). I suspect that the true explanation lies in that van Inwagen's view that textual discourse is contentless makes unavailable definiens like theirs or mine.[22]

Following Salmon (1998) and Thomasson (1999), recent realist work extends van Inwagen's account of paratextual uses to textual ones; cf. Abell (2020, ch. 5), Glavaničová (2021), Orlando (2021), Stokke (2021), Terrone (2021) and von Solodkoff and Woodward (2017). There are some differences among these realism-all-the-way-down views. A first choice point concerns whether to assume an implicit or explicit prefix corresponding to "according to/in *Emma*", which (2) presents as optional in paratextual uses. Lewis (1978) is usually regarded as committed to the prefix in paratextual uses at logical form, and Devitt (1981, p. 170 ff) follows suit for textual uses. Von Solodkoff and Woodward (2017) and Abell (2020, p. 138) support this view for paratextual uses. The formers' discussion of textual ones (*op. cit.*, pp. 412–4) is less clear-cut on this, but their rejection that predication is ambiguous in favor of implicit sentential operators suggests they embrace them in both uses.[23]

[22] Van Inwagen's neglect of the possibility of extending his account of paratextual uses to textual ones has I think the same roots as Kripke's (2013) tortuousness, fn. 20—von Solodkoff and Woodward (2017, pp. 411–2) appear to agree on this. Predelli (2020) is again much better on this issue. In Chap. 5 he discusses paratextual uses, offering two irrealist ways of handling them. On one, the utterer of paratextual uses goes along with the fiction-maker pretense, merely pretending to make related assertions. On the other, when embedded, the expressions are in fact quoted (Recanati (2021, Sect 13.2) suggests a similar account). This is a clear improvement on van Inwagen's and Kripke's discussion, but the reasons in §2 for ascribing a straightforward semantics to textual uses extend to the paratextual and tell against Predelli's and Recanati's views. If textual uses have a straightforward semantics, any decent semantics for attitude and speech reports should handle paratextual discourse as a matter of course; cf. Maier (2017) and Stokke (ms) for accounts close to the views advocated here.

[23] Abell argues that textual uses "cannot themselves be understood as implicitly prefixed by a fiction operator". These are her reasons: "The contents of fictive utterances determine what is fictional in the works they help to produce. Fiction operators do not feature as part of the contents of fictive utterances" (*op. cit.*, p. 138). This is right, but irrelevant. As suggested in Sect. 13.2, the

Kripke (2013) takes a noncommittal view: "One can regard this as a form of ellipsis, or not, as one pleases" (p. 58). Prefixes raise worries, cf. Bertolet (1984). They may be avoided by adopting the contextualism of Predelli's (1997, 2002) earlier realism. On this view, textual uses of unprefixed sentences are true or false in their contexts not relative to how things are at the actual world, but at "the" fictional world portrayed by the relevant story.[24]

A second choice point concerns the specific nature of the entities that these theories posit. I have talked of representations in fictions. The rough idea is clear enough and has been developed through precise formal modeling in semantics. Discourses like those constituting fictions include devices that purport (or pretend) to refer to entities of different sorts and co-refer among themselves, thereby ascribing bundles of properties to such purported referents and generating descriptions for them. These descriptive contents are modeled as *discourse referents* in dynamic traditions, naturally understood as representations (Sainsbury, 2021, p. 45). They specify *roles* (Glavaničová, 2021, Stokke, 2021, 2023) as descriptions like *the president of the USA* or *the mayor* do. Such roles can be identified as sets of the properties by means of which they pick out their occupiers, like Wolterstorff's (1980, §3.6) role-kinds, or Carnapian individual concepts, Fregean senses or "mental files" picking out referents relative to worlds. For our purposes, we don't need to take a stand on these choice-points. The criticism I'll raise in the final section would hold not matter which one is preferred.

13.4 A Main Consideration for Irrealism

Let me first briefly outline the irrealist picture for textual and paratextual discourse I support, and then move on to make clear a subtle but fundamental difference with realism, on which my objection will rely. The irrealist view that I think best abandons Millianism. Rigid designators like names and indexicals convey *singular propositions*, perhaps gappy ones (Braun, 2005). But they do so by triggering reference-fixing descriptive presuppositions, including metalinguistic ones like *being called Emma Woodhouse*. The difference between empty fictional names like "Emma Woodhouse" and non-fictional names (including "Vulcan") is just that the common ground (the background of shared propositions) that is meant to accommodate the presuppositions triggered by a textual use of "Emma Woodhouse" is one

operator may be "it is hereby to be imagined that" (Devitt, 1981, p. 172), an explicit performative that makes a declaration, as Abell wants, or an invitation to imagine, as I prefer, whose content is that of the embedded that-clause. In contrast to paratextual uses, the primary act conveyed by the performative will thus not be truth-evaluable (Devitt, *ibid.* 171).

[24]There is no unique such world, hence the scare quotes, García-Carpintero (2019b). Reimer (2005) advances a similar account. It can be interpreted as assuming that textual uses are a sort of assertion; although, as Everett (2013, p. 48) suggests, it can be also seen as a way of modeling a DR view of their specific fictionalizing force (declarations or invitations).

to-be-imagined, afforded by the fiction; while in non-fiction cases it is instead the standard common ground of what is accepted as true.[25]

Views like this share with realist accounts the notion that "Emma Woodhouse" is associated with a representation afforded by the fiction. If we put aside ontological qualms about fictions, there should not be any about parts thereof like the "Emma Woodhouse"-representation, the *role* it specifies, related *mental files* or *modes of presentation*. The name "Emma Woodhouse" can thus be used in metatextual discourse to refer to that representation, making claims like (3) true—indeed, it will be thus enlisted in the metatextual discourse of semanticists to describe fictional discourse. But there is a crucial difference with realism: the just sketched irrealist view does not take "Emma Woodhouse" in textual and paratextual discourse to refer to the abstract representation that is part of the fiction. Like Lewis (1978), the view assumes that characters, which are just representations, represent in textual discourse entities in the represented stories (in the situations whose imagining they invite) *that truly instantiate* the properties ascribed to them in fictions; and paratextual uses report on them. Of course, on Lewis's modal ontology these are real *possibilia*, inhabitants of causally isolated worlds in his pluriverse. But we do not need to buy this ontology (Goodman, 2010; Sainsbury, 2021); we just need an adequate non-Millian semantics.

In contrast, self-conscious realists like Salmon (1998), von Solodkoff and Woodward (2017) and Abell (2020) take referring expressions to signify across-the-board throughout (1)–(3) the very abstract artifacts their accounts posit. Thus, consider von Solodkoff and Woodward's (2017) definition of *holding* in terms of *having*. For them, (4) just means (5):

4. Emma Woodhouse holds *having lived nearly twenty-one years in the world*
5. According to *Emma*, Emma Woodhouse had lived nearly twenty-one years.

The definition I offered in the previous section may sound identical if the difference I am pointing out is overlooked. But in their account, in contrast to mine, "Emma Woodhouse" also has as semantic value in (5) and in the reported textual discourse a representation—a real abstract artifact, not an actually non-existent young person. For them, thus, what *Emma* invites us to imagine, and what is true according to the work, is a *category mistake*: something that is not just false, but conceptually and hence necessarily so—an impossibility.

This crucial difference between my definition and the one by serious realists is easily overlooked. Sainsbury (2021, p. 53 fn. 7) argues that to define *holding* or *encoding*, the realist *must* appeal anyway to an irrealist notion of representation, on which the represented entities do not exist but truly instantiate at the represented

[25] On this view names in fiction are entirely ordinary, in contrast to what Lewis (1978, p. 268) suggests, and Wolterstorff (1980, p. 154) and Currie (1990, p. 131) defend, fn. 7 above. Maier (2017) and Zucchi (2021) provide different formal semantics implementations, which make their standard character manifest. Salis (2021) offers a picture similar to mine, which she takes to be compatible with Millianism because, unlike me, she considers the descriptive elements "pragmatic", non-semantic; see also Sainsbury (2021).

situations the properties ascribed to them. This is indeed so on the most natural and less problematic definition, which is the one I gave in Sect. 13.3 above. But Sainsbury is overlooking the account that consistent Millian realists like Salmon (1998) advance, and von Solodkoff and Woodward (2017) develop.[26]

This view is coherent, but it raises the worry that Klauk (2014) aptly calls the *wrong kind of object* problem. Following Salmon (1998, p. 316 fn. 45) and Thomasson (2003, p. 212), von Solodkoff and Woodward (2017, pp. 414–7) and Abell (2020, p. 138) confront it by claiming that the ascriptions are *de re*; reference to fictional characters in textual invitations to imagine and in paratextual content-ascriptions has the same status as the underwritten description in R:

6. Sam believes that *the person who slanders her* is her best friend

Here a singular proposition is reported to be believed by Alex, but the description only provides the ascriber's way of picking out the *res* that the belief is about, not Alex's. Semeijn and Zalta (2021) develop in their framework a reply to Klauk along these lines.

But even though the view is thus defensible, it raises a serious worry. For (6) to be true, Alex must have some conception of the person at stake that makes sense of the ascribed singular belief. Correctly in my view, Abell (2020, p. 60) takes fictional content to be in principle epistemically accessible to competent users. This is required on normative views like hers and mine because the relevant norms must be able to guide and rationalize behavior. But it is independently plausible; fictions are intuitively "response-dependent" entities, which supports Abell's accessibility principle.[27] This means that we should allow competent authors and audiences of fictions to have adequate conceptions of their contents. Unlike Lewis's, the semantic picture briefly outlined above allows that fictions may have impossible contents (García-Carpintero, 2022d). There is nothing impossible or unintelligible in entertaining propositions involving the paradoxical, category-mistake-assuming contents that realists envisage.[28] But they should sound as funny to reflective people as what

[26] Stokke (2021) also appears to overlook the distinction. He describes his view as a form of realism on which fictional names denote abstract roles (p. 7830). But aside from some metatextual uses ("The president is an office established by the Constitution.", p. 7848) in which the referring expressions truly denote roles on the semantics that he provides, in textual and paratextual uses (pp. 7835, 7846) the expressions do not denote them, but their actually non-existing occupants at the situations at which the story is true. Stokke (ms) provides a unified account, single-handedly encompassing metatextual uses. From my perspective, it is only realist in *prima facie* ontologically committing to fictional characters as abstract entities (roles) in articulating the semantics, i.e., for metatextual uses. Unlike the accounts by Salmon and others I am questioning, it still doesn't take predicates in textual and paratextual uses to be instantiated by roles, but rather their occupants. Also, the unified account of metatextual uses that Stokke (ms) offers is fully consistent with the irrealist view I hold, see below.

[27] Of course, as always not all philosophers would agree with this; cf. García-Carpintero (2020b) for a critical discussion of a work which wouldn't.

[28] I had a very rational and competent logician colleague who told me that for his research he imagined himself jumping in a helicopter from one transfinite number to the next.

Pirandello's *Six characters in search of an author*, Flann O'Brien's *At swim-two-birds*, Keaton's 1924 *Sherlock Jr.*, or Allen's 1985 *The Purple Rose of Cairo* invite audiences to imagine, in ways that *Emma* does not (Klauk, 2014, p. 244; Collins, 2021, 2.4).

The challenge is thus this. There are good reasons to think that kinds like artworks should have natures sufficiently accessible to reflective participants in the practices constituting them. This constitutes a serious cost for realism, which the form of irrealism I outlined avoids. It leaves realists with two as far as I know unrelieved related burdens. First, to explain why the funny, jarring-impression that reflective audiences of the works I have mentioned have is absent in less postmodern, more standard cases like XIXth century realist novels and so on. Second, to explain what the undeniable difference between the cases in which it is present and those in which it is absent comes to. Otherwise, the view is taking for granted a widespread error among competent, reflective fiction consumers that there is little reason to grant here.

A related second cost is a version of the traditional *indeterminacy* objection to fictional realism that concerns Lewis (1978, pp. 261–2), cf. Bueno and Zalta (2017). As said, the relevant characters might be Meinongian concrete objects, abstract Platonic entities or created artifacts. Realists need to make a decision; but even after favoring the latter, favored here, many issues remain open. For instance, it should be established whether the relevant artifacts ultimately are concrete psychological entities, as recent versions of conceptualist nominalism envisage (Everett & Schroeder, 2015). It should be determined when the relevant entities start existing, if ever. Or which among the properties ascribed to them are constitutive of them. Thus, is Jekyll identical to Hyde?[29] What role do the intentions of authors play in their individuation? Should we think of these objects along hylomorphic lines? If so, exactly along which ones? There are several options here too (Evnine, 2016).

Assuming realism, *prima facie* different responses to questions like this give us different candidate-referents for fictional terms in textual and paratextual uses. For the epistemological reasons mentioned above, I doubt that answers to them can be adequately motivated on the basis of data accessible to competent consumers of fictions (Bueno & Cumpa, 2021). Note that the worry is not indeterminacy *per se*; in my view, most of our claims about *moderate-sized specimens of dry goods* like Kilimanjaro have indeterminate contents (García-Carpintero, 2021b). The worry is rather that, from the perspective of people competently engaging with fictions, seeking replies to them sounds like asking "silly questions" about fictions (Lewis, 1978, p. 270); but this is the appropriate perspective from which to address them. Once more, at the very least this is a cost that irrealism spares us.

I have helped myself to fictional characters to understand metatextual uses, in particular those with which I myself have peppered this essay. I have even suggested allegiance to a particular form of realism. Doesn't this apparent acceptance of

[29] Abell (2020, p. 141) and García-Carpintero (2020b, p. 185) argue that they are not: they *hold* identity (the entities they represent in the fiction are the same), but do not *have* it (they are different representations, even if of the same fictitious entity).

realism for metatextual discourse raise similar worries? But I don't think it does; at most that acceptance commits me to Thomasson (2021) "easy ontology", which as she shows is unburdened by the need to make the decisions I have mentioned.[30] As anticipated, I myself interpret metatextual discourse along the lines suggested by Yablo (2014), as ultimately compatible with straightforward irrealism (García-Carpintero, 2019a, 2021a). But for my purposes here I don't need to try to resolve this debate.

In this paper I have given some considerations for the view that the discourse used to create fictions has a dedicated semantics, which should ground that of discourses in which we report about them—in particular when it comes to the proper understanding of referential expressions like proper names throughout such uses. I have then considered a realist view, which takes fictional names to uniformly refer to exotic entities, and I have elaborated on previous arguments that it inadmissibly conflicts with the understanding of reflective consumers of fictions.

References

Abell, C. (2020). *Fiction*. Oxford University Press.
Alston, W. P. (2000). *Illocutionary acts and sentence meaning*. Cornell University Press.
Bertolet, R. (1984). On a fictional ellipsis. *Erkenntnis, 21*, 189–194.
Bonomi, A. (2008). Fictional contexts. In P. Bouquet, L. Serafini, & R. Thomason (Eds.), *Perspectives on context* (pp. 213–248). CSLI Publications.
Braun, D. (2005). Empty names, fictional names, mythical names. *Noûs, 39*, 596–631.
Brock, S. (2002). Fictionalism about fictional characters. *Noûs, 36*, 1–21.
Brock, S., & Everett, A. (2015). *Fictional objects*. Oxford University Press.
Bueno, O., & Cumpa, J. (2021). Resisting easy inferences. *Philosophy and Phenomenological Research, 102*(3), 729–735.
Bueno, O., & Zalta, E. (2017). Object theory and modal meinongianism. *Australasian Journal of Philosophy, 95*(4), 761–778.
Carroll, N. (1995). Critical study: Mimesis as make-believe. *The Philosophical Quarterly, 45*, 93–99.
Collins, J. (2021). "The Diversity of Fiction and Copredication: *An Accommodation Problem", Erkenntnis, 86*: 1197–1223
Currie, G. (1990). *The nature of fiction*. Cambridge University Press.
Currie, G. (1991). Work and text. *Mind, 100*, 325–340.
Davis, W. (2005). *Nondescriptive meaning and reference*. Clarendon Press.
Devitt, M. (1981). *Designation*. MIT Press.
Everett, A. (2013). *The nonexistent*. Oxford University Press.
Everett, A., & Schroeder, T. (2015). Ideas for stories. In S. Brock & A. Everett (Eds.), *Fictional objects* (pp. 275–293). Oxford University Press.
Evnine, S. (2016). *Making objects and events*. Oxford University Press.
Felappi, G. (2021). Empty names, presupposition failure, and metalinguistic negation. *Journal of Philosophy, cxviii*(5), 270–287.

[30] Thomasson (2003) in fact eventually gave up the appeal to the entities that her realism affords to understand textual and paratextual uses, and hence she is also unaffected by the objections I have raised to Salmon and von Solodkoff and Woodward.

Friedell, D. (2021). Creating abstract objects. *Philosophy Compass, 16*, e12783.
Friend, S. (2008). Imagining fact and fiction. In K. Stock & K. Thomson-Jones (Eds.), *New waves in aesthetics* (pp. 150–169). Palgrave.
García-Carpintero, M. (2010a). Fictional singular imaginings. In R. Jeshion (Ed.), *New essays on singular thought* (pp. 273–299). Oxford University Press.
García-Carpintero, M. (2010b). Fictional entities, theoretical models and figurative truth. In R. Frigg & M. Hunter (Eds.), *Beyond mimesis and convention – Representation in art and science* (pp. 139–168). Springer.
García-Carpintero, M. (2022a). Semantics of fiction. *Mind and Language, 38*, 604–618. https://doi.org/10.1111/mila.12412
García-Carpintero, M. (2022b). Predelli on fictional discourse. *Journal of Aesthetics and Art Criticism, 80*, 83–94.
García-Carpintero, M. (2022d). *Truth in fiction* reprised. *British Journal of Aesthetics, 62*(2), 307–324.
García-Carpintero, M. (2013). Norms of fiction-making. *British Journal of Aesthetics, 53*, 339–357. https://doi.org/10.1093/aesthj/ayt021
García-Carpintero, M. (2018). The Mill–Frege theory of proper names. *Mind, 127*(508), 1107–1168.
García-Carpintero, M. (2019a). Semantics of fictional terms. *Teorema, 38*(2), 73–100.
García-Carpintero, M. (2019b). Normative fiction-making and the world of the fiction. *Journal of Aesthetics and Art Criticism, 77*, 267–279. https://doi.org/10.1111/jaac.12660
García-Carpintero, M. (2020a). Co-identification and fictional names. *Philosophy and Phenomenological Research, 101*(1), 3–34.
García-Carpintero, M. (2020b). Referential indeterminacy in fiction. *Journal of Applied Logic, 7*(2), 177–190.
García-Carpintero, M. (2021a). Models as hypostatizations. In A. Cassini & J. Redmond (Eds.), *Models and idealizations in science: Fictional and artifactual approaches* (pp. 179–197). Springer.
García-Carpintero, M. (2021b). Metasemantics: A normative perspective (and the case of mood). In P. Stalmaszczyk (Ed.), *The Cambridge handbook of the philosophy of language* (pp. 401–418). Cambridge University Press.
García-Carpintero, M. (2022). Fictional narrators and normative fiction-making. In P. Engisch & J. Langkau (Eds.), *The philosophy of fiction: Imagination and cognition*. Routledge.
Glavaničová, D. (2021). Rethinking role realism. *British Journal of Aesthetics, 61*(1), 59–74.
Goodman, J. (2010). Fictionalia as modal artifacts. *Grazer Philosophische Studien, 80*, 21–46.
Hunter, D. (1981). Reference and meinongian objects. *Grazer Philosophische Studien, 14*, 23–36.
Klauk, T. (2014). Zalta on encoding fictional properties. *Journal of Literary Theory, 8*(2), 234–256.
Kripke, S. (1979). A Puzzle about belief. In A. Margalit (Ed.), *Meaning and use* (pp. 239–283). D. Reidel.
Kripke, S. (2013). *Reference and existence*. Oxford University Press. (based on the John Locke lectures originally delivered in 1973).
Kroon, F., & Voltolini, A. (2018). Fictional entities. In E. N. Zalta (Ed.), *The Stanford encyclopedia of philosophy* (Winter 2018 ed.). https://plato.stanford.edu/archives/win2018/entries/fictional-entities/
Lamarque, P. (1996). *Fictional points of view*. Cornell University Press.
Lewis, D. (1978). Truth in fiction. *American Philosophical Quarterly, 15*, 37–46. Reprinted with postscripts in Lewis, D. (1983). *Philosophical papers* (Vol. 1, pp. 261–280). Oxford: Oxford University Press.
Macdonald, M. (1954). The language of fiction. *Proceedings of the Aristotelian Society, Supplementary Volume, 28*, 165–184.
Maier, E. (2017). Fictional names in psychologistic semantics. *Theoretical Linguistics, 43*(1–2), 1–45.

Ohmann, R. (1971). Speech acts and the definition of literature. *Philosophy and Rhetoric, 4*(1), 1–19.
Orlando, E. (2021). Fictional names and fictional concepts: A moderate fictionalist account. *Organon F, 28*(1), 107–134.
Picciuto, E., & Carruthers, P. (2016). Imagination and pretense. In A. Kind (Ed.), *The Routledge handbook of the imagination* (pp. 314–325). Routledge.
Predelli, S. (1997). Talk about fiction. *Erkenntnis, 46*, 69–77.
Predelli, S. (2002). "Holmes" and Holmes – A Millian analysis of names from fiction. *Dialectica, 56*(3), 261–279.
Predelli, S. (2020). *Fictional discourse*. Oxford University Press.
Predelli, S. (2021). Fictional tellers. A radical fictionalist semantics for fictional discourse. *Organon F, 28*(1), 76–106.
Priest, G. (2011). Creating non–existents. In F. Lihoreau (Ed.), *Truth in fiction* (pp. 107–118). Ontos Verlag.
Recanati, F. (2016). *Mental files in flux*. Oxford University Press.
Recanati, F. (2021). Fictional reference as simulation. In E. Maier & A. Stokke (Eds.), *The language of fiction* (pp. 17–36). Oxford University Press.
Reimer, M. (2001). The problem of empty names. *Australasian Journal of Philosophy, 79*(4), 491–506.
Reimer, M. (2005). The ellipsis account of fiction–talk. In R. Elugardo & R. Stainton (Eds.), *Ellipsis and nonsentential speech* (pp. 203–215). Springer.
Ryle, G. (1933). Imaginary objects. *Proceedings of the Aristotelian Society, Supplementary Volume, 12*, 18–43.
Sainsbury, M. (2021). Fictional names: Reference, definiteness and ontology. *Organon F, 28*(1), 44–59.
Salis, F. (2021). The meaning of fictional names. *Organon F, 28*(1), 9–43.
Salmon, N. (1998). Nonexistence. *Noûs, 32*, 277–319.
Saucelli, A. (2021). Pretending and disbelieving. *Inquiry, 67*, 1991–2004. https://doi.org/10.1080/0020174X.2021.1982403
Searle, J. (1975). The logical status of fictional discourse. *New Literary History, 6*, 319–332.
Semeijn, M., & Zalta, E. (2021). Revisiting the "wrong kind of object" problem. *Organon F, 28*(1), 168–197.
Stock, K. (2017). *Only Imagine*. Oxford University Press.
Stokke, A. (2021). Fictional names and individual concepts. *Synthese, 198*, 7829–7859.
Stokke, A. (2023). "Fictional names and Co-identification", *Philosopher's Imprint*, 23, 19
Szabó, Z. G. (2020). The goal of conversation. *Aristotelian Society Supplementary Volume, 94*(1), 57–86.
Terrone, E. (2021). Twofileness. A functionalist approach to fictional characters and mental files. *Erkenntnis, 86*, 129–147.
Thomasson, A. L. (1999). *Fiction and metaphysics*. Cambridge University Press.
Thomasson, A. L. (2003). Speaking of fictional characters. *Dialectica, 57*, 205–223.
Thomasson, A. L. (2021). What do easy inferences get us. *Philosophy and Phenomenological Research, 102*(3), 736–744.
van Inwagen, P. (1977). Creatures of fiction. *American Philosophical Quarterly, 14*, 299–308.
van Inwagen, P. (1983). Fiction and metaphysics. *Philosophy and Literature, 7*(1), 67–77.
van Inwagen, P. (2003). Existence, ontological commitment, and fictional entities. In M. Loux & D. Zimmerman (Eds.), *Oxford handbook of metaphysics* (pp. 131–157). Oxford University Press.
von Solodkoff, T. (2022). Demoting fictional names – A critical note to *Fictional discourse: A radical fictionalist semantics*. *British Journal of Aesthetics, 62*(2), 223–230.
von Solodkoff, T., & Woodward, R. (2017). To have and to hold. *Philosophical Issues, 27*, 407–427.
Walton, K. (1983). Fiction, fiction–making, and styles of fictionality. *Philosophy and Literature, 7*(1), 78–88.
Walton, K. (1990). *Mimesis and make–believe*. Harvard University Press.

Wolterstorff, N. (1980). *Works and worlds of art*. Clarendon Press.
Yablo, S. (2014). Carnap's paradox and easy ontology. *Journal of Philosophy, CXI*(9/10), 470–501.
Yablo, S. (2020). Nonexistence and aboutness: The bandersnatches of Dubuque. *Crítica, 52*(165), 77–100.
Zemach, E. (1998). Tom Sawyer and the beige unicorn. *British Journal of Aesthetics, 38*(2), 167–179.
Zucchi, S. (2021). On the generation of content. In E. Maier & A. Stokke (Eds.), *The language of fiction* (pp. 107–130). Oxford University Press.

Part V
Themes on Metasemantics and Semantics

Chapter 14
Ontology and Top-Down Metasemantics

Matti Eklund

Abstract In this paper, I explore the implications for ontology of what Jared Warren has dubbed a top-down approach to metasemantics, explaining semantic features of subsentential expressions in terms of semantic features of whole sentences. Among authors who have adopted such an approach to metasemantics are Gottlob Frege, Michael Dummett, Crispin Wright, Donald Davidson, W.V.O. Quine, Agustín Rayo, Oystein Linnebo, and Warren himself, and many of these authors have also taken this top-down approach to have ontological implications. I discuss what a top-down approach amounts to, the extent to which such an approach is plausible, and what its implications for ontology are, if any. Often a top-down approach is assumed to have broadly deflationary implications, and to provide a separate path to defending the view that abstract objects exist.

Keywords Top-down metasemantics · Context principle · Ontology · Epistemology · Metaphysical · Linnebo · Warren · The syntactic priority thesis · Abstracta · Correct assertion · Superassertibility · Truth · Truth-first · World cooperation · Metadeflationism · Inferentialism

14.1 Introduction

A number of authors discussing language have defended and stressed the thesis that the meanings of sentences are in an important sense primary and subsentential constituents are in an important sense secondary. I will refer to this general kind of view as *top-down metasemantics*. A thumbnail sketch of this view is that meaning is determined by use and it is an important fact that it is only whole sentences we use for speech acts—so the meaning of a subsentential constituent is fully determined

M. Eklund (✉)
Uppsala University, Uppsala, Sweden
e-mail: matti.eklund@filosofi.uu.se

X. de Donato-Rodríguez et al. (eds.), *Deflationist Conceptions of Abstract Objects*, Synthese Library 516, https://doi.org/10.1007/978-3-031-98717-5_14

by what it contributes to whole sentences. This view on language is encapsulated for example in Frege's so-called context principle, which in one formulation says "never ask for the meaning of a word in isolation, but only in the context of a sentence".[1] There are good questions about what exactly this means. But somehow it is a claim about the relationship between sentence meaning and word meaning and it seems somehow to amount to sentence meaning being primary and word meaning is secondary (insofar as we can and should speak of word meaning at all). In a similar vein, Quine (1981) speaks of what he describes as an "insight" regarding "the semantic primacy of sentences" which he traces back to Bentham:

> It is occasion sentences, not terms, that are to be seen as conditioned to stimulations. Even our primordial objects, bodies, are already theoretical [...] It is occasion sentences that report the observations on which science rests. The scientific output is likewise sentential: true sentences, we hope, truths about nature. The objects, or values of variables, serve merely as indices along the way, and we may permute or supplant them as we please as long as the sentence-to-sentence structure is preserved. The scientific system, ontology and all, is a conceptual bridge of our own making, linking sensory stimulations to sensory stimulations.[2]

Other theorists have subscribed to similar kinds of views.[3]

There is of course much to say about top-down metasemantics, both about what exactly it might come to and about whether any form of it is defensible. But my main focus here will not be this view on language in its own right. What I will focus on are instead the possible consequences of this kind of view as far as other philosophical issues are concerned. Specifically, I will be concerned with purported consequences for ontology.[4]

The idea that top-down metasemantics has such consequences is prominent in some parts of the literature. Here is Crispin Wright (1983), where the neo-Fregean program in philosophy of mathematics was inaugurated:

> The lynch-pin of Frege's platonism, according to our interpretation, is the syntactic priority thesis: the category of objects...is to be explained as comprising everything which might be referred to by a singular term, where it is understood that possession of reference is imposed on a singular term by its occurrence in true statements of an appropriate type.[5]

---

[1] Frege (1884, p. x).

[2] Quine (1981, p. 20).

[3] See Davidson (1979), Hale (1987), Hale and Wright (2001), Linnebo (2018), Rayo (2013), Warren (2020), and Wright (1983). Hirsch—see e.g. the essays collected in his (2011)—arguably also relies on a top-down approach but is so different from other theorists that discussion of his views is best reserved for elsewhere. See my (2011) and (2016) for some relevant discussion.

[4] Virtually all the theorists mentioned in the previous footnote take top-down metasemantics to have consequences for ontology. I have discussed Wright's and Hale's views elsewhere (see Eklund (2006) and (2016). Below I will discuss Linnebo and Warren, and I will also briefly remark on a view found in Rayo. Davidson and Quine are more complex cases. Some of their formulations suggest ontological views—I am thinking primarily of Quine's "structuralism". In my (2007a), I discuss relevant aspects of Quine in some depth.

[5] Wright (1983, p. 53).

This "syntactic priority thesis" is presented as implied by Frege's context principle. "Frege's Platonism" is of course an ontological thesis: it is a thesis about the existence of mathematical objects. So the context principle is held to be, or entail, the "lynch-pin" of an ontological thesis.

Oystein Linnebo (2018) says, in the course of arguing that there are "thin objects", where a thin object is an object whose "existence does not make a substantial demand on the world", that

> The most influential strategy for articulating the idea of thin objects is arguably Frege's. At the heart of his strategy is the context principle, which urges us "never to ask for the meaning of a word in isolation, but only in the context of a sentence"".[6]

My aim here is to question the idea that top-down metasemantics can have ontological implications of the kind at issue.

It is hard to say exactly what the ontological implications are. But as indicated by the passages from Wright and Linnebo, top-down metasemantics is used in defense of existence claims, and specifically in defense of the view that it is somehow easier for entities to exist than it may otherwise be thought. And as we will see later, some remarks in Quine suggest that Quine takes a top-down metasemantics to motivate a form of structuralism, and this is also how some commentators have interpreted Quine. (I will distance myself from this interpretation.)

Top-down metasemantics can naturally be held to have *epistemological* implications. It is a natural for anyone who adopts a top-down metasemantics to think that our cognitive and epistemic access to the world is via whole sentences, or correlates thereof. When I see or understand or know or perceive something, I in the first instance see or understand or know or perceive *that* such-and-such. I am not directly cognitively related to the object itself as opposed to something's being the case.

This epistemological view, like top-down semantics itself, can of course be questioned. What does it come to? Can some version of it be accepted? But again, it will not be my focus. My focus will be ontology. Of course, the epistemological view outlined can be of relevance for ontology. For example, it can be held to be of relevance to some issues faced by some ontological views. Suppose I hold that there are Xs but it is objected that if there were, we couldn't know that there are. I may well want to be able to say something in response to this objection and turn to some epistemological view in the course of doing so. The epistemological view outlined may be a natural candidate: I can say that the problem is defused if we attend to my use of whole sentences wherein Xs are referred to or quantified over. But be that as it may: the epistemological view itself is still at best of indirect relevance for ontological matters.

In Frege (1884), the metasemantic thesis is clearly supposed to provide an answer to the question: "How, then, are the numbers to be given to us, if we cannot have any ideas or intuitions of them?". This is about our cognitive relation to numbers. It is closely linked to epistemology. How can we have knowledge of numbers, if we cannot have any ideas or intuitions of them? And at least intuitively, it is clear how the

[6] Linnebo (2018, p. 107).

metasemantic thesis can form an important part of the answer. The reason the question of how we can have knowledge of numbers seems hard is because we are wondering about we can connect individual words to abstract objects. But if we instead focus on whole sentences, we can tell a story about the use of them, without bringing in anything about how individual subsentential expressions relate to their worldly correlates.

And then there is the ontology. I made some remarks above about roughly what kind of ontological theses are at issue. But is at least more unclear exactly what the ontological upshot is supposed to be. In the next section, I will say something about what some theorists inspired by Frege have said—and I will focus mostly on such theorists also in most of my later discussion. Only toward the end will I return to Quine. As I will argue, it is unclear to what extent Quine wants to draw ontological, as opposed to epistemological, conclusions.

As mentioned, what I will do in this paper is to raise some problems for the idea that top-down metasemantics has ontological implications of the kind that it is often taken to have. I will both bring up general issues and, in the later parts of the paper, discuss specific recent proposals due to Jared Warren (2020) and Oystein Linnebo (2018).

Here is the plan in more detail. In Sect. 14.2, I will briefly say something more about the supposed role of the context principle—a paradigm of a top-down metasemantics—in ontology, as well as discuss what may be the most obvious suggested use in metaphysics of a top-down approach: the legitimization of contextual definitions such as abstraction principles. I will argue that even if a top-down approach has such uses there are problems, including that a general top-down approach seems unnecessarily powerful for the job. In Sect. 14.3, I will discuss two general kinds of possible suggestions regarding how a top-down metasemantics could have metaphysical implications. I will raise some obvious concerns about both of them. While I will not attempt to pin the ideas discussed in that section on any particular theorist, I believe that the—rather obvious—problems with those ideas are rather instructive. In Sects. 14.4 and 14.5 I will look in more detail at some discussions where top-down approaches are used for the purposes of metaphysics: the discussions found in Warren (2020) and Linnebo (2018). In both cases my conclusion will be that a top-down approach does not play any real role, despite what some of the rhetoric may suggest. In Sect. 14.6, I will briefly discuss the later Quine.

14.2 The Context Principle in Ontology

Consider again the above quoted passage from Wright:

> The lynch-pin of Frege's platonism, according to our interpretation, is the syntactic priority thesis: the category of objects…is to be explained as comprising everything which might be referred to by a singular term, where it is understood that possession of reference is imposed on a singular term by its occurrence in true statements of an appropriate type.[7]

[7] Wright (1983, p. 53).

An obvious concern about the view sketched is that it seems from the formulations that already by Wright's own lights, the context principle does not help with the truth of the sentences themselves, so it is open to the anti-platonist to just deny that the sentences are true. My suspicion is that Wright must think that the context principle helps with the truth claim as well. The reason is that if not, it is unclear how it could be so central to Frege's platonism as according to Wright it is. But *how* might it be helpful there?[8]

One may think that my question has a straightforward answer and one very well-known from the literature—and that the answer is prominent in the work of Crispin Wright and Bob Hale in defending neo-Fregeanism. Let me explain.

If we can regard principles like

The direction of line a = the direction of line b iff lines a and b are parallel

and

The number of Fs = the number of Gs iff the Fs and the Gs are equinumerous.

as providing definitions of the left hand sides in terms of the right hand sides then we have arguably demystified both the metaphysics and the epistemology of certain kinds of abstracta, directions and numbers. Contextual definitions like these promise to provide nice payoffs. If the principle concerning numbers—*Hume's principle*, to use the common label—is indeed a definition, then a case can be made that arithmetic has been reduced to (second-order) logic plus definitions, despite its ontological commitments.[9]

But from a certain perspective, these purported definitions are problematic: how *can* we think of them as definitions when their truth seems to depend on the existence of directions and numbers? One may think: it can happen to be the case that there just are no such entities as directions and numbers and then the right hand sides can be true without the left hand sides being so.

Top-down metasemantics can be seen to provide a response to this otherwise seemingly crucial objection. The objection presupposes that there is an in some sense independent question of whether there are entities such as directions, but given top-down metasemantics, we have provided all we need to provide regarding the meaning of direction- and number-talk when we have provided whole sentences containing direction- and number-terms with truth conditions, as when biconditionals like the above one are used as contextual definitions.[10]

One concern about the mentioned use of contextual definitions, going back to Dummett, is that even if some true sentences containing by what syntactic criteria are singular terms can get their meanings via contextual definitions like this, there is no guarantee that the singular terms are not semantically idle, and more strongly the singular terms seem positively to be semantically idle. There is a dilemma that

[8] For further discussion of this, see Eklund (2006) and (2016).

[9] For elaboration, see Wright (1983) and Hale and Wright (2001).

[10] Of course there are independent complications. For example, are the proposed definitions sufficient to endow the expressions introduced with relatively determinate meanings. See Hale and Wright (2001) for discussion of some concerns on this score.

can be presented. Either the sentences introduced by the contextual definitions mean the same as the sentences that serve as definiens, or they do not. In the first case, Dummett's objection does seem to get purchase. In the second case, the sentences demand different things of the world. Whether or not the dilemma is in the end serious, a top-down metasemantics does not seem to help. It can at most help provide the result that the introduced sentences mean the same as the sentences that serve as definiens, but that would just impale us on the first horn of the dilemma.[11]

But suppose that objections like this can be satisfactorily dealt with. And suppose moreover that contextual definitions of the kind mentioned are legitimate. Needless to say this is to suppose a lot. But I will not pause on this. There are significant questions that can be raised even granted all this.

If what has been described is the full extent of the usefulness of a top-down metasemantics for ontological purposes, then it is of help only regarding Fs such that Fs can be seen as characterized by definitions like the contextual definitions mentioned. Other ontological issues remain untouched. For example, relating specifically to abstracta: this is a viable philosophy of abstracta only to the extent that reductive definitions like these can be found for all abstracta we wish to countenance. Someone might be inclined to say in response: abstracta just are the entities characterized by definitions of these—they are entities generated by a process of abstraction. But this seems to amount to saying: all non-spatial, non-causal etc. entities are like that. And what is supposed to justify such a restrictive ontological claim? Going beyond abstracta, singular terms for concrete entities may also be introduced by contextual definition.[12] The top-down approach can help vindicate also such appeal to contextual definition. But again there is the same limitation. What about the existence of entities not referred to by terms introduced by such definitions?

So there is the concern that appeal to top-down metasemantics does not provide a fully general approach to ontological issues. This may be important to keep in mind, but obviously does not in principle need to be that bad. When I illustrated the point that a top-down metasemantics was thought to have ontological implications, I related to what Wright and Linnebo have said. And I only quoted Wright on a particular approach to the question of the existence of numbers, and Linnebo on an approach to thin objects. Nothing in what I quoted Wright and Linnebo as saying in itself uses or needs a top-down approach to motivate a fully general view on ontology.

But at least some of Wright's claims indicate that he wants to say something fully general about ontology. Appealing to the context principle he concludes that "there can be no philosophical science of ontology, no well-founded attempt to see past our categories of expression and glimpse the way in which the world is truly

[11] See Dummett (1981a, pp. 494–505), (1981b, pp. 424–5), (1991, pp. 189–99) and (1998, pp. 384–6). For discussion see Assadian (2019) and Ebert (2015).

[12] Linnebo (2018, section 2.3).

furnished".[13] This is a more general claim than merely that some particular existence claims can be defended and explained by appeal to the context principle.

And Linnebo too seems committed to something more general. First, he holds that "reference to abstract objects is constituted by more basic facts that do not themselves involve the relevant abstract objects".[14] This by itself does not amount to a general ontological claim. One can subscribe to this claim while adding that there may be many other abstract objects besides: it is only that these other abstract objects cannot be referred to. However, Linnebo centrally characterizes objects as possible referents of singular terms.[15] This rules out that some objects could not be referred to; and Linnebo's claim about reference to abstract objects must concern abstract objects generally.

But what might be a more general and significant view on ontology motivated by a top-down metasemantics?

14.3 Two Suggestions

In this section I will turn to some possible suggestions, perhaps not plausibly pinned on any specific authors. The aim is to highlight some general issues.

Consider first the following metaphysical view, which parallels top-down metasemantics *individual entities* are grounded-ish in *truths* (/propositions/facts). The truths are thereby more fundamental than individual entities. (It is not standard to take individual objects to be relata of the grounding relation—hence I speak not of grounding but of something grounding-ish. But from now I will drop the "ish".)

Someone defending both top-down metasemantics and this corresponding *truths-first* view, as we may call it, might hold that the worldly correlates of sentences are metaphysically prior to the worldly correlates of subsentential expressions in the way that a top-down metasemantics can be thought to yield that meanings of sentences are prior to meanings of words.

There are some good questions about the details of the truths-first view. What about truths like *there are Fs*? I assume that like the Fs themselves they are always grounded on the view in question. The view would not be very helpful if such truths were themselves sometimes among the ultimate grounds. What is gained by banning individual entities from the ultimate grounds if these existence truths are among the grounds? But banishing these existential truths from the ultimate grounds gives rise to new questions. What is the criterion for what truths should by the same token be consigned to the class of the grounded?

Perhaps such questions about the details can be satisfactorily dealt with. A more general complaint would still remain. What we are investigating is just one

[13] Wright (1983, p. 52).

[14] Linnebo (2018, p. 128).

[15] Linnebo (2018, p. 22).

metaphysical thesis among others. What is more, the metaphysical thesis, whatever its fate, seems wholly independent of metasemantics, even if in some ways it parallels the metasemantic view. There is no reason why the metasemantic thesis should lend any plausibility to the metaphysical thesis.

A different ontological view which might be thought to go along with top-down metasemantics is this. Any sentence that meets criteria for *correct assertion* is thereby guaranteed to be true. In particular, a correctly assertible sentence committed to the existence of Fs can't be untrue *simply* on the basis that it could turn out that there are no Fs. The underlying thought is that we can make use of a notion of correct assertion, separate from truth, and use that to frame theses about truth. The reason this could be thought to be a way of making use of a top-down metasemantics would be that what we assert are whole propositions, the correlates of sentences, and so tying truth to a suitable notion of assertibility would be a way to respect the supposed priority of sentences.

Again there are good questions about the details. Most obviously there is the issue of how to characterize the relevant notion of correct assertion. If "correct assertion" just is a different name for truth then appeal to correct assertion hardly constitutes progress. And a correct assertion in the relevant sense is not simply a warranted assertion, since some warranted assertions may turn out to be false for straightforward reasons. (I assert that p on the basis of warranted belief, but p is false.) A more plausible suggestion is that a correct assertion is an assertion whose content is superassertible, where superassertibility is the following property:

> A statement is superassertible…if and only if it is, or can be, warranted and some warrant for it would survive arbitrarily close scrutiny of its pedigree and arbitrarily extensive increments to or other forms of improvement of our information.[16]

Crispin Wright introduced superassertibility as being a candidate for being a notion of truth.[17] An immediate consequence of taking truth to consist in superassertibility is that then no statement that cannot be warranted can be true. If the suggestion at issue had the consequence that there cannot be truths for which a thinker cannot have warrant that might be a problem. But the suggestion at issue here is just that superassertibility is a *sufficient* condition for truth. By itself, this is compatible with there being true statements that are not superassertible.

But whatever in the end to say about the details, there is, again, a more general complaint. Where does the supposed cooperativeness of the world come from? *Why* would correct assertibility, in whatever exact sense is at issue, suffice for truth? The claim that it does so suffice appears to be a separate metaphysical thesis, scarcely justified on the basis of metasemantics.

What might be said in response to this complaint? Here is what seems to me to be the correct response for the friend of the conception under investigation: Given top-down metasemantics, the only constraints on the meanings and truth-conditions of sentences come from facts having to do with the use of whole sentences, in such

[16] Wright (1992, p. 48).

[17] See the discussion in Wright (1992).

a way that the very idea that sentences can be correctly used but still fail to be true because of some mismatch between what the sentence demands at the subsentential level and what the world actually provides is simply unintelligible. It is not strictly that the world is cooperative enough to meet the relevant demands; it is rather that the extra demands cannot coherently be imposed. I believe that what I have just outlined does illustrate how a top-down metasemantics in principle can be used for ontological ends. What I do not see is how to go on to justify the claims made. Also, it deserves emphasis how radical the claim is. Just to illustrate the radical nature of what is suggested, compare a more modest idea.[18] According to the modest idea, attention to actual language use shows that we often use ontologically committing sentences in such a way that they do not demand anything more from the world in order to be true than do sentences without the specific commitments in question. It is compatible with this modest idea that certain ontologically committing sentences can have more demanding meanings. What we are considering now involves ruling this out.

14.4 Warren on the Consequences of Top-Down Approaches

After this general discussion of how a top-down approach could be relevant, let me now discuss in more detail what some specific authors have said. In this section, I will discuss Jared Warren's recent work. In the next section, I will discuss Oystein Linnebo. While Warren's and Linnebo's respective approaches are very different, the point I will make about them both is at bottom the same: while they give the impression of relying on top-down approaches to metasemantics, such reliance does no real work. Note that while I will be critical of Warren and Linnebo, all that I will be concerned to show is that a top-down approach does no real work for them. It is compatible with my arguments that their general outlooks should be accepted. My points will only concern the (lack of a) role of the top-down metasemantics. (The stone soup may be very good, but it is no thanks to the stone.)

I will start by discussing Warren. Warren defends a conventionalism about logical and mathematical truths: "According to mathematical inferentialism/conventionalism, the existence of numbers is not a substantive and controversial fact about our world, but is instead a *trivial byproduct* of our using our arithmetical language in the way that we do".[19] Generally, Warren's view is that rules can be meaning-constitutive for concepts, and among the meaning-constitutive rules may

[18] What I am calling the modest idea is prominently at issue in Rayo (2013), ch. 1. He argues that we actually speak languages where sentences with different structure (like sentences flanking the biconditional in one of the contextual definitions above) but we could speak a language where the structure of a sentence matters in such a way that only atomic sentences whose subsentential structure matches the world's metaphysical structure can be true.

[19] Warren (2020, p. 213).

be and are ones that allow us to conclude existential claims just by appeal to these rules.[20]

An objection to this view—an objection central to Warren's discussion of these issues—is that if one can establish the existence of some things, like numbers, this way, can one not also establish, say, the existence of God this way—can one not, for example, use this kind of conventionalism to vindicate the ontological argument for the existence of God?[21]

In response to this objection, Warren launches into a general metaontological discussion: "The standoff can't be broken at the merely ontological level. To break this deadlock, we need go meta. Finding the right way to think about metaontology is the only way to advance matters".[22] Warren's metaontological discussion is centered around the following claim:

> Quantifier Role Priority: Subsentential expression Q in a language L is an (unrestricted) type i quantifier expression just in case Q plays, in L, the inferential role of an (unrestricted) type i quantifier expression..[23]

More informally, the thought behind Quantifier Role Priority is that

> if there is a subsentential expression in a given language that plays the (structural) inferential role of an unrestricted existential quantifier, then that expression is, really and truly, an unrestricted existential quantifier, *semantically* speaking.[24]

Warren relies on inferentialism, the view that the meaning of an expression is determined by certain meaning-constitutive inferences in which it partakes. This is for example on display through his appeal to inferential role in the statement of Quantifier Role Priority. He says that his approach "is to see all semantic categories as deriving from syntactic/inferential categories, against the backdrop of a top-down approach to metasemantics".[25] He thus explicitly adopts a top-down approach, and this is tied to his inferentialism and the relevant kind of priority thesis.

Warren goes on to introduce the theses Name Role Priority and Predicate Role Priority, supposed to be analogous to Quantifier Role Priority:

> Predicate Role Priority: Subsentential expression Δ in a language L is an n-place predicate expression just in case Δ plays, in L, the inferential role of an n-place predicate expression.

---

[20] A complication regarding Warren's idea is that different meaning-constitutive principles may be inconsistent with each other. For discussion see, e.g., Eklund (2002) and (2007b).

[21] Warren (2020, pp. 202f and 214).

[22] Warren (2020, p. 214).

[23] Warren (2020, p. 216).

[24] Warren (2020, p. 217).

[25] Warren (2020, p. 219).

Name Role Priority: Subsentential expression α in a language L is a name expression just in case α plays, in L, the inferential role of a name expression.[26]

So far this is all just philosophy of language, and there is no obvious connection to ontology—although one might perhaps surmise that what Warren says about quantification could be of significance for ontology. When tying this to ontology, this is how Warren does it. First, relying on a top-down approach to metasemantics together with role priority theses, Warren characterizes different forms of what he calls "metadeflationism".

Top Down + Quantifier Role Priority = Quantifier Metadeflationism
Top Down + Predicate Role Priority = Predicate Metadeflationism
Top Down + Name Role Priority = Name Metadeflationism.[27]

"Top Down" is here, Warren, says, a "placeholder" for "any top-down metasemantics".[28]

Warren then argues that metadeflationism leads to *pluralism* about existence and reference, of the following kind:

the ontologically relevant kind of pluralism involves the possibility of different languages each having (for example) unrestricted existential quantifier expressions that *do not mean the same thing* and so cannot be translated into each other. This isn't at all trivial: it is not the mere claim that the words "there is" or "exists" could have meant something different in some alternative language. That much is entirely trivial and is secured by the conventionality of linguistic signs. [...] Nor is it the claim from above that the word "exists" or the like could have meant something similar while obeying slightly different logical rules of inference. Instead the claim is that the logical rules can be exactly the same for two expressions in different languages, and so both expressions can, by Quantifier Metadeflationism, be unrestricted existential quantifier expressions, *while still differing in meaning*.[29]

Here is one natural way to gloss the idea—where, importantly, the gloss due to me, not Warren. Assuming inferentialism—again, this is Warren's own favored top-down approach—then (roughly) there are different possible expressions that are unrestricted existential quantifiers. This is pluralism. It will be (relatively) trivial that if it is as much coherent that Ks exist, then there will be an existential quantifier $\exists_i$ such that "$\exists_i$xKx" is true. One problem with this gloss (which, again to stress, is mine) is this: why think "K" means the same in each of the languages at issue? And if it doesn't then even if there is a language where "$\exists_i$xKx" we cannot conclude that

[26]Warren (2020, p. 217). There are differences between the role priority theses. The Predicate Role Priority thesis and the Name Role Priority theses are general characterizations of the semantic categories *predicate* and *name*. But the Quantifier Role Priority thesis as he states it is not a general characterization of a corresponding notion of a quantifier. Instead it is a characterization of the notion of an existential quantifier. From some perspectives the slide is not innocuous. One may, for example, in principle accept the Name Role Priority thesis but think that what makes the specific name "Joe Biden" the name that it is, is not something characterizable in terms of inferential role alone.

[27]Warren (2020, p. 219).

[28]Warren (2020, p. 219).

[29]Warren (2020, p. 222).

it is a language such that Ks "exist" in the sense of the existence concept of that language.

Anyway, to sum up what the view is: Warren, relying on a top-down approach, and more specifically inferentialism, defends what he calls metadeflationist theses. The metadeflationism in turn leads to pluralism. This is how a top-down approach is supposed to be ontologically relevant. I have some concerns about this.

First, the role of the metadeflationism, and specifically the supposed deflationary aspect of metadeflationism, is somewhat unclear. Warren does not pause on it, except to say that it leads to pluralism. The "metadeflationism" is deflationary to the extent that the pluralism is. But why should the pluralism be thought to be deflationary? It would be in an intuitive sense deflationary if the possible gloss I mentioned above was correct. But given the problems with that gloss, the "pluralism" does not go beyond the claim that there are different unrestricted quantifier expressions—and why exactly is that deflationary? In general, there is nothing deflationary or otherwise philosophically significant about there being several expressions of a given kind.

Second, why does not the thesis of Quantifier Role Priority suffice *by itself* for the pluralist consequence—why the need to add Top Down? Already given the Quantifier Role Priority thesis, one can argue that in different languages there are different expressions with the inferential role of an (unrestricted) type i quantifier expression. Whether or not such an argument is persuasive, it seems no less persuasive than it would be without the "Top Down" part, which appears idle. Now, as stated Quantifier Role Priority as stated arguably relies on inferentialism—given its appeal to the inferential roles of quantifiers, and given that inference is an operation on sentences, Quantifier Role Priority is in effect Top Down. (However, it is common to hold that logical expressions are characterized by their inferential roles without accepting inferentialism across the board, so there remains the question of whether a general inferentialism is required.)

Third, Quantifier Role Priority is simply a thesis about what it takes to be a quantifier of type i. It is a purely classificatory thesis. How can a thesis of that kind be of any significance? it is easy to formulate "bottom-up" theses which have similar pluralist consequences. Consider again the formulation of Quantifier Role Priority:

> Quantifier Role Priority: Subsentential expression Q in a language L is an (unrestricted) type i quantifier expression just in case Q plays, in L, the inferential role of an (unrestricted) type i quantifier expression.

A bottom-up thesis that promises to have the same kind of consequence is

> Quantifier Reference Priority: Subsentential expression Q in a language L is an (unrestricted) type i quantifier expression just in case Q has, in L, a semantic value/referent of the kind that an (unrestricted) type i quantifier expression has.

This thesis has pluralist consequences just in case there are different semantic values of the kind in question. Quantifier Role Priority may well be more *plausible* than Quantifier Reference Priority, and therefore pluralism built around the former is more plausible than pluralism built around the latter. But that is a different matter and irrelevant to the point I am making. The point is just that both theses are purely classificatory: how liberal should we be about calling something a type i quantifier expression?

I have criticized the general metaontological remarks Warren makes, where he supposedly makes use of a top-down approach to metasemantics. Back to conventionalism and God—the issue that in Warren serves to motivate the foray into metadeflationism and pluralism. Having discussed the role of top-down approaches, Warren returns to the objection concerning the existence of God, and addresses it as follows:

> Critics might still wonder why this approach doesn't allow us to just as easily prove the existence of God? Why, they ask, can't we inferentialists just add the following inference rule to our language:
>
> $$(God)\ \frac{}{\exists x\left(x = \mathrm{God}\right)}$$
>
> There is no in principle barrier to our following this rule, so if we added this as a basic rule of our language, unrestricted inferentialists like me would have to admit that it is valid, and thus that the rule's conclusion is true. But, the objection continues, this conclusion entails that God exists and that the term "God" refers (to God)! So isn't this just to admit the validity of a simplified ontological argument for the existence of God? What has gone wrong here?
>
> The simplest response to this is that *nothing* has gone wrong. [...] The result the critic is pointing to is exactly what we should expect and want. The problem is only that the critic is badly misdescribing the result. The possibility of a linguistic practice like ours except with the additional rule *(God)* added, *does not show that God exists*. The term "God" means something in that language that it does not mean in our language. Even though the sentence "God exists" is true in that language, why should we think the same sentence is true in our language, on *that* basis?[30]

I want to make two points about what Warren says here. First, I think that what Warren is saying is completely and straightforwardly *right*. He effectively dismisses the objection. But second, this is straightforward also in another way: there is absolutely no appeal, or need to appeal, to the apparatus of top-down approaches, priority theses and metadeflationism. There is no such appeal in what Warren says in response to the objection; nor does what Warren says in any way depend on it. Warren's actual response to the God objection is simple: an expression "God" governed by the rule *(God)* does not mean what the actual expression "God" means.

[30] Warren (2020, p. 232 f).

14.5 Linnebo on the Context Principle

After this discussion of Warren, let me now turn to Oystein Linnebo's (2018) discussion of the role of the context principle. Linnebo begins his discussion of this by saying (what I quoted already early on),

> The most influential strategy for articulating the idea of thin objects is arguably Frege's. At the heart of his strategy is the context principle, which urges us "never to ask for the meaning of a word in isolation, but only in the context of a sentence".[31]

He goes on to note that this idea may seem unhelpful:

> When we explain the meaning of a sentence, must we not also explain the reference of every singular term involved in the sentence? If so, Frege will only have transformed one hard problem into another that is at least as hard.[32]

And he could have gone on to note that the situation is actually worse than this. It is not merely that we are left with an *equally hard* problem: the problem seems to be the *very same* as that we started out with—explaining the meanings of the subsentential expressions.

Linnebo goes on to describe Frege's response to this predicament:

> Frege [...] responds that, when we explain the meaning of each of the relevant sentences, "we must reproduce the content of this sentence in other terms, avoiding the use of" the singular terms whose reference we are trying to explain (§62).[33]

and sums up the strategy as follows:

> Frege's strategy in the Grundlagen for explaining how numbers are "given to us" [...] relies on two main ingredients: the context principle and the idea of "reproducing" meaning in other, and less problematic, terms. These ingredients are supposed to work together in tandem.[34]

The reproduction idea is what is at issue in principles like the contextual definitions discussed above, in Sect. 14.2. The idea is that the right hand side reproduces the meaning of the left hand side.

Linnebo says that the strategy relies on both the context principle and the reproduction idea. Apparently they are supposed to be separate ingredients. However, given the reproduction idea, what is the separate role of the context principle? Does not the reproduction idea—the idea that a sentence's content may be reproduced in other terms—suffice by itself?

It could still be that the context principle is somehow needed for the reproduction idea. But this does not seem to be the case either. What the reproduction idea needs is simply that it is not the case that the contents of two sentences are different

---

[31] Linnebo (2018, p. 107).

[32] Linnebo (2018, p. 109).

[33] Linnebo (2018, p. 109).

[34] Linnebo (2018, p. 116).

already if one has a constituent with a given meaning such that the other sentence has no constituent with the same meaning.

Linnebo distinguishes between two different versions of the general strategy outlined. The first version is this:

> The first explanatory strategy combines the context principle with the "ultra- thin" conception of reference discussed [...] According to this conception, an expression refers provided that, first, it has all the appropriate syntactic and inferential characteristics of a singular term, and second, the expression figures in appropriate true sentences. Since the first requirement is spelled out in entirely non-semantic terms, the only semantic requirement is the second one, which concerns whole sentences. The result is an extremely undemanding conception of reference..[35]

Against this Linnebo brings up a version of the complaint due to Dummett that I mentioned already earlier: on this view, singular terms of the extended language are "semantically idle".[36] This brings us to the second strategy, centered around the following added semantic constraint:

> ...that the identity predicate be given its ordinary semantic interpretation of standing for the relation of identity. For this semantic interpretation to work, the singular terms that flank the identity predicate must obviously be interpreted as referring to objects; otherwise, the identity relation would have no relata.[37]

Linnebo's concern about the second strategy is that it "imposes incompatible requirements on the notion of meaning":

> On the one hand, the meanings in question must be coarse-grained enough for two sentences to share the same meaning although they differ in syntactic-semantic structure. Let us call this *the requirement of recarving*. On the other hand, the meanings must be fine-grained enough to enable the reference of the singular terms occurring in a sentence to be determined by the sentence's meaning and its syntactic-semantic structure. Let us call this *the requirement of reference determination*.[38]

Having dismissed these two strategies, Linnebo turns to his own positive (and as he puts it, "speculative") suggestion regarding the context principle—appeal to metasemantics:

> The context principle is concerned with a metasemantic question, namely how reference is constituted. It is not a primitive metaphysical fact that an expression refers to some object. There are some more basic facts in virtue of which the relation of reference obtains. *Perhaps the more basic facts in virtue of which an abstraction term refers to an abstract object need not mention or otherwise involve this abstract object.*[39]

Somehow the thought must be that the supposed fact that the more basic facts do not involve the abstract object provides Frege with what he needs. But how exactly is this supposed to work?

---

35 Linnebo (2018, p. 123 f).

36 Linnebo (2018, p. 124).

37 Linnebo (2018, p. 125).

38 Linnebo (2018, p. 125).

39 Linnebo (2018, p. 128).

How, to begin with, can metasemantics be relevant at all? It is clear from Linnebo's discussion that he is talking about sentences. It is meta*semantics*. But if propositions are the fundamental truth bearers, a more basic question concerns what it takes for a *proposition* about abstracta to be true. Metasemantics, concerned with the question of in virtue of what *sentences* mean what they do, simply does not speak to that issue.

Relatedly, and more specifically, it appears that going metasemantic does not get around the dilemma that Linnebo calls attention to. Whatever the metasemantic story behind the meanings of the sentences on the left and right hand sides of a contextual definition of the kind at issue may be, those sentences are either determined to have the same meaning (this is the first horn of the dilemma) or different meanings (in which case they demand different things of the world—and this is the second horn of Linnebo's dilemma). So if the dilemma is genuine in the first place—something I am not weighing in on—the metasemantic strategy that Linnebo proposes does not get around it.

More generally, consider the question of in virtue of what a given sentence is true. This question splits into two. There is (a) the question of in virtue of what the sentence expresses what it does and (b) the question of in virtue of what that in turn is true. As Paul Boghossian states it, a sentence S is true iff for some p, S means that p, and p. Issues having to do with what the sentence means only speaks to the first conjunct, what the sentence means. It does not speak to the truth of p.[40]

A natural conclusion to draw from this kind of consideration is that metasemantics can only ever speak to one aspect of what makes a sentence true (whether it means that p, for some given p) and not to the other aspect (whether p)—and this spells doom for a strategy such as Linnebo's.

However, one view on (b) is that the answer sometimes is: *nothing*. Some propositions are *vacuously* true: true but not by virtue of any way the world is. This is a minority view, but with some pedigree. In cases where that is the answer to (b), the answer to (a) amounts to a full answer to the question of in virtue of what the sentence is true. So the friend of the metasemantic strategy that Linnebo describes might feel helped by this view (b).

But this view on (b) of course stands in need of defense. And whatever a plausible defense might be, it is not clear why a top-down approach would be involved.

I conclude that Linnebo fails to show that a top-down approach plays a significant role in ontology. To stress, this is a limited criticism of Linnebo. All I have argued is that insofar as Linnebo draws significant and justified ontological and metaontological conclusions, a top-down approach to metasemantics does not play a significant role.

[40] Boghossian (1996, p. 364).

14.6 Quine on "Structuralism"

Before closing, let me say just a few words on the role of top-down metasemantics in Quine.[41] Some of Quine's formulations may make it sound as if he draws strong (meta)ontological conclusions based on top-down metasemantics. Here is Quine (1992a):

> Reference and ontology recede thus to the status of mere auxiliaries. True sentences, observational and theoretical, are the alpha and the omega of the scientific enterprise. They are related by structure, and objects figure as mere nodes in the structure. What particular objects there may be is indifferent to the truth of observation sentences, indifferent to the support they lend to theoretical sentences, indifferent to the success of the theory in its predictions...[42]

There is a focus on whole sentences and their truth conditions—a top-down metasemantics (here restricted to scientific language)—and the claim that related to this, reference and ontology are "mere auxiliaries". There are questions about how this works. The thought, of course, is that permutations of what objects are assigned as referents of terms and values of variables do not matter to the truth of whole sentences, these permutations are a matter of indifference. But if this is the thought, a natural response is that all of the objects assigned as referents of terms and values of variables must exist in order for the truth-values of sentences to remain unaffected.

But however it may be with Quine's reasoning, Quine does not consistently flirt with the idea of drawing (meta)ontological conclusions. Compare an article from the same year, Quine (1992b). Quine there does say,

> The conclusion is that there can be no evidence for one ontology as over against another, so long anyway as we can express a one-to-one correlation between them. Save the structure and you save all . . . This global structuralism may seem abruptly at odds with realism. let alone naturalism . . .

and so far it may sound like Quine is using a top-down approach to draw conclusions about ontology. In the first sentence, he reiterates the conclusion of his indeterminacy argument, which in turn relies on a top-down approach. In the last sentence, he calls the conclusion "global structuralism", and says that it is at odds with realism. But then Quine goes on to take back what is suggested. He asks rhetorically, "Are . . . [light] rays, molecules and nerve endings themselves disqualified now as mere figments of an empty structure?" and answers the question in the negative: adverting to naturalism: "Naturalism looks only to natural science, however fallible, for an account of what there is and what what there is does. Science ventures its tentative answers in man-made concepts, perforce, couched in man-made language, but we can ask no better". And he goes on to conclude, "My global

[41] I discuss Quine in more depth in my (2007a).

[42] Quine (1992a, p. 31).

structuralism should not, therefore, be seen as a structuralist ontology . . . My global structuralism is a naturalistic thesis about human activity.[43]

Quine's reasoning here sounds strange. The view seems to be that the naturalism commits us to taking the claims made by science at face value. But there is the obvious possibility of accepting that he claims made by science are true—there are light rays, molecules and nerve endings—but use the kind of philosophy Quine engages in, itself part of total science, to investigate what it means for it to be true that there are light rays, molecules and nerve endings But while I do not think that this is completely clear, something like this seems to be going on. Although Quine does speak of a "structuralism", the structuralism at issue seems turns out not to be either a thesis directly concerned with ontology. Naturalism provides the ontology. The "structuralism" has no implications for what kind of ontology to accept but is only a broadly epistemological claim having to do with our cognitive relation to ontological claims. Quine then does not seem to use a top-down metasemantics for ontological purposes in the way other theorists I have discussed here do.

14.7 Concluding Remarks

I have here been concerned with the possible ontological significance of top-down metasemantics. Different authors have claimed that top-down metasemantics is of ontological significance. I have here discussed problems with such claims. Again to stress, my aims are in a certain respect modest. I have not sought to argue against the ontological views defended by the theorists discussed. Instead, I have only been concerned to emphasize that top-down metasemantics can at most play a limited role.

References

Assadian, B. (2019). Abstractionism and mathematical singular reference. *Philosophia Mathematica, 27*, 177–198.

Boghossian, P. (1996). Analyticity reconsidered. *Noûs, 30*, 360–391.

Davidson, D. (1979). The inscrutability of reference. *The Southwest Journal of Philosophy, 10*, 7–19.

Dummett, M. (1981a). *Frege: Philosophy of language* (2nd ed.). Harvard University Press.

Dummett, M. (1981b). *The interpretation of Frege's philosophy*. Harvard University Press.

Dummett, M. (1991). *Frege: Philosophy of mathematics*. Duckworth.

Dummett, M. (1998). Neo-Fregeans: In bad company? In M. Schirn (Ed.), *Philosophy of mathematics today* (pp. 369–387). Clarendon Press.

Ebert, P. (2015). Dummett's criticism of the context principle. *Grazer Philosophische Studien, 92*, 23–50.

[43] Quine (1992b, pp. 8–9).

Eklund, M. (2002). Inconsistent languages. *Philosophy and Phenomenological Research, 64*, 251–275.

Eklund, M. (2006). Neo-Fregean ontology. In J. Hawthorne (Ed.), *Philosophical perspectives, 20, metaphysics*. Cornell University.

Eklund, M. (2007a). The ontological significance of inscrutability. *Philosophical Topics, 35*, 115–134.

Eklund, M. (2007b). Meaning-constitutivity. *Inquiry, 50*, 559–574.

Eklund, M. (2011). Review of Eli Hirsch, quantifier variance and realism. *Notre Dame Philosophical Reviews*. Available at https://ndpr.nd.edu/reviews/quantifier-variance-and-realism-essays-in-metaontology/

Eklund, M. (2016). Hale and Wright on the metaontology of neo-fregeanism. In P. Ebert & M. Rossberg (Eds.), *Abstractionism*. Oxford University Press.

Frege, G. (1884). *Die Grundlagen der Arithmetik, Wilhelm Koebner, Breslau* (English translation by J. L. Austin). The Foundations of Arithmetic, Blackwell.

Hale, B. (1987). *Abstract objects*. Blackwell.

Hale, B., & Wright, C. (2001). *The reason's proper study*. Oxford University Press.

Hirsch, E. (2011). *Quantifier variance and realism*. Oxford University Press.

Linnebo, O. (2018). *Thin objects: An abstractionist account*. Oxford University Press.

Quine, W. V. O. (1981). Things and their place in theories. In *Theories and things* (pp. 1–23). Harvard University Press.

Quine, W. V. O. (1992a). *Pursuit of truth*. Harvard University Press.

Quine, W. V. O. (1992b). Structure and nature. *Journal of Philosophy, 89*, 5–9.

Rayo, A. (2013). *The construction of logical space*. Oxford University Press.

Warren, J. (2020). *Shadows of syntax: Revitalizing logical and mathematical conventionalism*. Oxford University Press.

Wright, C. (1983). *Frege's conception of numbers as objects*. Aberdeen University Press.

Wright, C. (1992). *Truth and objectivity*. Harvard University Press.

Chapter 15
Sortal Combat. Semantic Inferentialism and Ontological Deflationism About Abstract Objects

James Henry Collin

Abstract This paper sets out an inferentialist approach to abstract objects—arguing that it provides a deflationary account of abstracta—and draws attention to some distinctive and attractive features of the account. In brief, a consequence of the kind of semantic inferentialism developed by Brandom is that quantifiers are essentially sortally restricted. As a result, there is a plurality of particular quantifiers, a plurality of senses of 'exists', and a plurality of kinds of existence. The inferentialist litmus test for which senses of 'exists' are 'ontologically robust' or 'weightless', I argue, is whether the standards of correctness and incorrectness for existence claims are internal to the linguistic practice itself, or are, in some part, external to the linguistic practice. In the former case, existence claims are responsible only to the internal rules of the linguistic practice and are ontologically weightless. In the latter case, existence claims are responsible to worldly facts and are ontologically robust. On this account, ordinary empirical claims (for example) are ontologically robust, whereas claims about mathematical objects (for example) are ontologically weightless.

Keywords Inferentialist · Abstract objects · Sortal · Plurality of quantifiers · Plurality of existence

15.1 Introduction

Pragmatism—more specifically, the semantic inferentialist pragmatism associated with Robert Brandom—underwrites a deflationary conception of abstract objects. It does so by providing a metasemantic framework in which it is intelligible that there are different kinds of existence, and that has the expressive resources to distinguish

J. H. Collin (✉)
College of Arts and Sciences, VinUniversity, Hanoi, Vietnam
e-mail: james.c@vinuni.edu.vn

X. de Donato-Rodríguez et al. (eds.), *Deflationist Conceptions of Abstract Objects*, Synthese Library 516, https://doi.org/10.1007/978-3-031-98717-5_15

vocabularies whose existence claims are ontologically heavyweight and vocabularies whose existence claims are ontologically weightless. These are points I'll draw out initially by making a few, somewhat impressionistic, remarks about the differences between pragmatist and representationalist approaches to metasemantics. This will set the stage for some more specific observations about the inferentialist treatment of singular terms, intentionality and their bearing on the metaphysics of abstract objects (in this case, natural numbers). In outline: for the inferentialist, using an expression as a singular term is only intelligible in the context of the ability to make identity judgements using that term. The practice of making identity judgements is, in turn, only intelligible in the context of the ability to apply sortals, such that any singular term or quantifier can only be introduced in connection with a sortal. The inferentialist view of quantifiers as essentially sortally restricted means that, for the inferentialist, there is a plurality of particular quantifiers, a plurality of senses of 'exists', and a plurality of kinds of existence. The inferentialist litmus test for which senses of 'exists' are 'ontologically heavyweight' or 'weightless', I argue, is whether the standards of correctness and incorrectness for existence claims are internal to the linguistic practice itself, or are, in some part, external to the linguistic practice. In the former case, when existence claims are responsible only to the internal rules of the linguistic practice, they are ontologically weightless. In the latter case, when existence claims are responsible to worldly facts, they are ontologically robust. On this account, ordinary empirical claims (for example) are ontologically robust, whereas claims about mathematical objects like natural numbers (for example) are ontologically weightless.

15.2 Representationalism and Pragmatism

Let's start by contrasting representationalism with pragmatism, with respect to metasemantics. Here I'll think of representationalism as consisting of two claims—or, really, postures that could be precicified into more determinate claims—about semantic meaning. The first has to do with the explanation of semantic meaning. The representationalist treats representational notions as (in some theoretically important sense and in some theoretically important ways) more fundamental than practical ones in understanding how linguistic items come to acquire semantic contents. Representationalism, in this sense, begins with the notion of representational content—or with consanguineous representational notions like *denotation*, *signification*, *satisfaction*, and so on—and explains other features of language, such as the goodness or badness of inferences, in terms of those representational notions.

It is fair to say that mainstream model-theoretic semantics reflects the representationalist order of explanation. Take some language $\mathcal{L}$ with constant symbols c_1, c_2, c_3, ..., relation symbols R_1, R_2, R_3, ..., function symbols $f_1, f_2, f_3 \ldots$, and logical connectives $\neg$, $\rightarrow$. A denotation function δ is defined from $\mathcal{L}$ to a domain M, with an object $c^{\mathcal{M}}$ for each constant c of $\mathcal{L}$, an n-tuple $c^{\mathcal{M}_1}, c^{\mathcal{M}_2}, \ldots c^{\mathcal{M}_n} \subseteq M^n$ for each n-place relation symbol R, and a function $f^{\mathcal{M}} : M^n \rightarrow M$ for each function symbol

f of $\mathcal{L}$. δ is here defined as a function from the set of elements of the object language *itself* to the *domain* of that language—the collection of *things* that language is about (given the interpretation). One can then understand the truth of atomic sentences in terms of satisfaction by n-tuples of elements of M, the truth of complex sentences truth-functionally, as depending on the truth-values of the atomic sentences composing them, and define a consequence relation $\models$ in the usual way, such that a sentence ϕ is a consequence of a set of sentences Γ just in case any interpretation that satisfies Γ also satisfies ϕ. In this way, proprieties of reasoning in a given language are explanatorily downstream of the language's broadly representational properties. The representionalist gloss on this is that the *semantic* explanatory order of explanation exhibited by formal languages like these provides a *metasemantic* picture of how languages in general come to be meaningful and how semantic meanings come to be grasped. Meaning is, at base, instituted by denotation relations between words and objects, so that coming to grasp the meaning of a language is, at base, coming to learn the denotation relations of the language, and so to learn the truth conditions of sentences of that language.

This first claim or posture is often taken to support the second claim that I'm including under the label 'representationalism'. This is the claim that language is more or less uniformly in the business of representation. For if one is required to appeal to representational notions to explain, say, semantic meaning or logical consequence, then any meaningful vocabulary or any vocabulary closed under logical consequence has to be understood as representing the world as thus and so. A representationalist order of explanation then commits one to a uniform, or near uniform, view of language as in the business of representation.[1]

The two pillars of representationalism support a particular framing of metaphysical debates. This is a defeasible, but still default, assumption that any given vocabulary has the role of representing the world as thus and so, so that assertions made in that vocabulary—at least those made using atomic sentences of the vocabulary—depict the world as containing such and such objects standing in such and such relations, and so carry with them ontological commitments. The assumption is defeasible because one can argue that particular vocabularies ought to be given special treatment which would defuse their apparent ontological commitments, either by (i) giving a "non-standard semantics" for that vocabulary—for instance manifested in various forms of expressivism—(ii) arguing that the vocabulary can be reduced to another whose ontological commitments are more perspicuous—for instance manifested in reductionist programmes—or (iii) arguing that we can

[1] Often one finds that this background is not made explicit, but that it is informing metaphysical debates no less for that. These considerations are however brought, influentially, to the surface in Benacerraf (1973). They are also given clear expression by (Soames, 2010, p. 1): 'The central fact about language is its representational character. Exceptional cases aside, a meaningful declarative sentence S represents the world as being a certain way. To sincerely accept, or assertively utter, S is to believe, or assert, that the world is the way S represents it to be. [...] For S to be meaningful is for it to represent the world as being a certain way, which is to impose conditions the world must satisfy, if it is to be the way S represents it.' See also (Soames, 2014, p. 7).

understand the atomic sentences of the vocabulary as false, without thereby coming into conflict with any of our other commitments—for instance manifested in fictionalism or other error theories.[2]

Now, pragmatist views of language, as I will understand them, are *antirepresentationalist*, though not in the sense that they see no role for representational idioms in understanding language, nor in the sense that they deny that language can (at least in some cases) be used to represent how things stand in the world. Rather, the pragmatist rejects the *order of explanation* that characterises representationalism. She sees the order of explanation as running in the other direction. The pragmatist begins with practices of treating some inferences as proper and others as improper—that is, with notions pertaining to language *use*—and explains other features of language, for instance representational or denotational purport, in terms of those practical notions. Some pragmatists (Rorty is their avatar) have held the additional radical view that representational notions aren't worth recovering. But this is not required of the pragmatist *qua* pragmatist; inferentialists such as Brandom explain representational notions in terms of practices of deontic scorekeeping (Brandom, 1994, chs.5, 8; 2019, chs. 2, 12)—something we will revisit later. The pragmatist then can still see vocabularies as being used to represent the world as thus and so. This, however, is something that comes into focus further down the explanatory horizon, and there is no guarantee, at the outset of inquiry, that every vocabulary should be seen as in the business of representation.[3]

The pragmatist order of explanation is also operative in the pragmatist's understanding of formal semantics. We noted a moment ago that mainstream model-theoretic semantics exhibits the representationalist order of explanation, that the representationalist gloss on mainstream model-theoretic semantics is of mainstream model-theoretic semantics as picturing the relationship between language and the world—with interpretation functions as picturing representational relationships between singular terms and things, predicates and sets, and so on—and that this is sometimes taken to be a consideration in favour of the second pillar of representationalism: that language is more or less uniformly in the business of representing the world. The point I want to make here, to clear the path for a more specific inferentialist and deflationist account of abstract objects, is that the pragmatist should be wholly unmoved by this kind of consideration. The pragmatist has a very different gloss on the philosophical significance of mainstream model-theoretic semantics than that of the representationalist. The pragmatist begins her order of explanation with an account of what one is *doing* when engaging in formal semantic theorising. And, of course, what one is *doing* in assigning an interpretation δ to an object language $\mathcal{L}$ is taking sentences of $\mathcal{L}$ and sentences of a metalanguage, and linking them

---

[2] The tripartite division is due to (Price, 2013, ch. 2).

[3] This is a claim we will need to nuance later. A preview: for the specifically *inferentialist* pragmatist, there is *one sense* in which every autonomous discursive practice—that is, a discursive practice that can be engaged in without engaging in any other discursive practice—should be seen as in the business of "representation" (see Brandom (2011, 2013)). But the sense, I will argue, is a purely deflationary one.

with biconditionals. So in providing truth conditions for the sentences of $\mathcal{L}$ one does so by using sentences of a metalanguage. (How else could it be done?)[4]

What is pertinent here is that formal semantics *per se* does not come packaged with representationalist commitments. The metalanguage can only inject representationalist juice into the object language through the biconditionals connecting the sentences of each, if one *already presupposes* that the role of the metalanguage is to represent or picture the world. But this is a philosophical *add-on*, not something somehow internal to the practice of model-theoretic semantics. It is possible, as the pragmatist may, to assert the biconditionals that make up any given interpretation, *without* thereby undertaking any commitments *vis-à-vis* the representational or non-representational nature of a given vocabulary.[5] We will see shortly the implications a specifically *inferentialist* understanding of language has for the metaphysics of abstract objects. But what is already clear is that a broadly pragmatist approach to language, eschewing, as it does, a representationalist order of explanation, does not share the metasemantic assumptions informing much of the contemporary debate on abstract objects.

15.3 The Role of Singular Terms

Having set out this pragmatist background, we can now move to considering a specifically *inferentialist*, pragmatist conception of meaning and its implications for the metaphysics of abstract objects. The guiding thought behind semantic inferentialism, is that inferential relationships between claims are constitutive of the meaning of those claims, and that understanding a claim involves grasping the inferential relationships it bears to other claims. Unpacking this a bit: on the inferentialist picture, claims stand in deontic relationships to one another. Commitment to a claim deontically precludes commitment to certain other claims—that is, any given claim stands in the *incompatibility* relationship to certain other claims. Additionally,

[4] This point was recognised by (Rorty, 2018, pp. 261–62): 'The point of constructing a "truth theory of English" is not to enable philosophical problems to be put in a formal mode of speech, nor to explain the relationship between words and the world, but simply to lay out perspicuously the relation between parts of a social practice (the use of certain sentences) and other parts (the use of other sentences).'

[5] There is, note, no barrier to the pragmatist, not just endorsing (as true) the biconditionals specifying an interpretation, but also in endorsing (as informative in various ways) the practice of mainstream model-theoretic semantics. One might, for instance, understand mainstream model-theoretic semantics as akin to model-building in the empirical sciences, and illuminating in the way analogous to the way that an idealised model of a physical system can be illuminating. (See Peregrin & Svoboda, 2021). That said, for the pragmatist there is a special kind of illumination to be had by (non-mainstream) formal semantics that reflect the pragmatist order of explanation. Robert Brandom and Alp Aker (2008) make inroads into formal semantics that do exactly this. More recently, Ulf Hlobil and Robert Brandom (2024) have developed a formal semantics of this kind in far greater detail.

commitment to a claim deontically requires commitment to certain other claims—that is, any given claim stands in a committive *commitment-preservation* relationship to certain other claims. Finally entitlement to a claim can license entitlement to certain other claims—that is, any given claim stands in a permissive *entitlement-preservation* relationship to certain other claims. For instance, a commitment to 'Ursula is a grizzly bear' deontically precludes commitment to 'Ursula has no spatial extension', deontically requires commitment to 'Ursula is a mammal', and if one is entitled to 'Ursula is a grizzly bear' one is also entitled to 'Ursula is a mammal'.[6] To grasp the meaning of a claim is to be able to navigate, in practice, those proprieties of inference. Being a language-user essentially involves keeping track of one's commitments and entitlements—the commitments and entitlements one *acknowledges* in the first instance, and the commitments and entitlements that follow from them—as well as the commitments and entitlements of others. This practice Brandom calls *deontic scorekeeping*.[7]

Inferential relationships between claims, with claims expressible as *sentences*, are fundamental insofar as claims are the minimal unit one can attach pragmatic force to, undertake responsibility for, or use to make a deontic move in a language game.[8] But the meaning of *sub*sentential expressions are given by their *contribution* to the inferential role of claims. Two subsentential expressions have the same meaning if substituting one for another in a sentence makes no difference to the proprieties of inference associated with that sentence. The subsentential expressions we are interested here are *singular terms*: those expressions that have singular referential purport (i.e., those expressions that aim to pick out exactly one thing). What kind of inferential contribution is required of a subsentential expression for it to function as a singular term? Consider two singular terms 'the shortest serving British Prime Minister of the twentieth century' and 'Bonar Law', and the inference from:

Bonar Law was of Scottish descent

to

---

[6] Strictly speaking, relationships of incompatibility, commitment-preservation, and entitlement-preservation should be understood as holding between *sets* of claims, in order to accommodate nonmonotonicity. The pairs of single claims I use to illustrate the idea is an idealization (though a harmless one, since the examples are monotonic).

[7] This is elaborated in Brandom (1994, chs. 2,3) and (2000, chs. 1, 6). Note that these inferences are not good in virtue of being logically valid—for this they would need to be supplemented with conditional premises—but rather in the sense of being *materially* good, i.e., inferences it is correct to make in virtue of the proprieties of inference constitutive of the meanings of the claims involved in those relationships. For the inferentialist, logical vocabulary like conditionals are used to make explicit what inferences are *already* good in virtue of the meanings of the non-logical vocabulary they contain. As Brandom develops the view, the structure of rationality is articulated in terms of the deontic relationships I've just mentioned, but there are natural ways of extending the account to accommodate probabilistic relationships between claims. See González de Prado Salas, de Donato Rodríguez, and Zamora Bonilla (2021).

[8] See Brandom (1994, pp. 79–85) and (2000, pp. 12–15).

The shortest serving British Prime Minister of the twentieth century was of Scottish descent

If we endorse this inference as materially good *in virtue of the contribution of the singular term* in each, then we are thereby committing ourselves to a battery of other inferences as being materially good. The first thing to notice is that the inferences made good by singular terms are *symmetrical*. In accepting the inference from:

Bonar Law was of Scottish descent

to

The shortest serving British Prime Minister of the twentieth century was of Scottish descent

in virtue of the contribution of the singular terms 'the shortest serving British Prime Minister of the twentieth century' and 'Bonar Law', we are committed to also accepting the reverse inference from:

The shortest serving British Prime Minister of the twentieth century was of Scottish descent

to

Bonar Law was of Scottish descent

In this way the contribution of singular terms to the proprieties of inference governing a claim differs from the contribution of predicates. For while predicates *may* make good *some* symmetrical inferences, this will not *exclusively* be the case; their contribution will always make good some asymmetrical inferences. For instance, the inference from

Bonar Law was of Scottish descent

to

Bonar Law was of British descent

is good in virtue of the contribution of 'of Scottish descent' and 'of British descent', but the reverse inference is not good. (Being of Scottish descent guarantees being of British descent, but not vice versa.) The inferences licensed by the contribution of singular terms, unlike predicates, are *de jure* symmetrical.[9] We can also notice that by accepting one such inference, we are committed to many more. Because it is the *singular term* that is materially involved in our commitment to the goodness of the inference (and not the predicate), we are committed generally to the

[9] Aren't the inferences licensed by predicates *de jure* symmetrical when those predicates are equivalent? No. The inferential contribution of, e.g., 'Triangular' and 'trilaterial with straight sides' may be such that, for any singular term t, one is licensed to infer 't is triangular' from 't is trilateral with straight sides' and vice versa, but these predicates will also license non-symmetrical inferences; from 't is triangular' one can infer 't is a polygon', but the reverse inference is bad.

goodness of other inferences involving the substitution of ‘Bonar Law’ for ‘The shortest serving British Prime Minister’, and vice versa, *no matter which predicates are used.* For example, a commitment to, ‘Bonar Law played chess’ now deontically requires commitment to ‘The shortest serving British Prime Minister played chess’, and vice versa, ‘Bonar Law invented Mr Whippy ice cream’ deontically requires commitment to ‘The shortest serving British Prime Minister invented Mr Whippy ice cream’, and vice versa. So in undertaking commitment to a single inference in virtue of the contribution of two singular terms, one undertakes commitment to many more symmetrical inferences involving the substitution of those singular terms.

These inferential relations are made explicit by informative identity claims, characteristically of the form ‘(the x)[Dx] = t’ such as ‘The shortest serving prime minister of the twentieth century is Bonar Law’ or ‘The smallest number expressible as the sum of two cubes in two different ways is 1729’.[10] Brandom calls these *recognition statements.* The reason that it is specifically identity claims that make explicit the inferential contribution of singular terms is, at least, twofold. Firstly, using a set of expressions as singular terms necessarily involves being able to make judgements about whether the things picked out by those terms are the *same* or *distinct.* (An object is necessarily the sort of thing that can be distinct from other things.) A practice of using singular terms is only intelligible then in the context of practices that manifest that ability. The locution that makes explicit judgements about when the things picked out by two given singular terms are the same or distinct is the identity claim. So if we have two terms with singular referential purport such as ‘The shortest serving British PM’ and ‘Bonar Law’ then, if we are to manifest the ability to use them as singular terms, we must be able to make judgements about whether those terms are coreferential or not. And the claim that makes this explicit is ‘Bonar Law = the shortest serving British PM’ or ‘Bonar Law ≠ the shortest serving British PM’.

Secondly, the proprieties of inference that arise in virtue of the contribution of singular terms are not just *de jure* symmetrical, but also *de jure* reflexive and transitive. Singular terms then are grouped into equivalence classes that provide sufficient conditions for when one is committed to *applying* a singular term, and the commitments that are incurred by applying the term, that are indifferent to the predicates

[10] This is why it isn’t in general the case, on Brandom’s account, that inferences from a claim of the form ‘Pt’ to a claim of the form ‘Pu’ are *de jure* symmetrical, when p ≠ u. There is a *sense* in which we sometimes regard inferences of the form ‘Pt, therefore Pu’ as good “in virtue of the contribution of the singular terms ‘t’ and ‘u’” when p ≠ u, but this is when we take (perhaps implicitly) certain predicates to apply to the objects denoted by ‘p’ and ‘u’, and regard the inference as good in virtue of the application of those predicates. One kind of case like this is when the names pick out positions in a structure, or pick out objects that necessarily bear particular relations to other objects, such as integers. (Consider, in this regard, ‘n is a positive integer, therefore 2n is not prime’.) Here predicates are involved in the material goodness of the inference, and the inference will not be *de jure* symmetrical. Material inferences that are good in virtue of the contribution of singular terms are only *de jure* symmetrical when one regards those singular terms as co-referential. We can say further, part what it *is* to regard two singular terms as co-referential is to treat material inferences that are good in virtue of the contribution of those singular terms as *de jure* symmetrical.

appearing in the sentences into which those singular terms can be substituted. Identity locutions make explicit this kind of open-ended, symmetrical, reflexive, and transitive intersubtitution licence. In short: the ability to use singular terms is only intelligible in the context of the ability to make identity judgements.

Identity claims, however, are intelligible only with respect to a sortal.[11] Consider the predicate *is large*, in contrast to the (sortal) predicate *is a dog*; having learned that a is a dog and that b is a dog, it makes sense to ask whether a is the *same* dog as b, but having learned that a is large and that b is large, it makes no sense to ask whether a is the same large as b. So treating a subsentential expression as a singular term—as picking out something that can be identical or non-identical to something picked out by another term—essentially involves treating that term as falling under a sortal concept. Without bringing a term under a sortal, and so using it in practice as a singular term, counting is impossible.[12]

As a consequence, a grasp of sortal concepts is a prerequisite for the intelligibility both of the kind of inferential commitments made explicit by identity claims (and hence of singular terms), and of our ability to count objects. This has immediate implications for quantifiers, for quantifiers *quantify*, and, as we have just seen, quantifying is unintelligible unless indexed to a sortal concept. Sortally *un*restricted quantifiers such as 'thing' or 'object', then, are pseudosortals; though syntactically like sortals, there is no fact of the matter as to how many *things* there are in the room, though there is a fact of the matter as to how many books, pens, molecules, shadows, spatial regions of such and such a shape, there are in the room. As a result, there is no generic (unsortalized) existential quantifier, but a plurality of sortally-restricted quantifiers—even when that sortal restriction is implicit, or messily disjunctive.

For the inferentialist, the meanings of quantifiers are given in substitutional terms—in the way familiar from proof systems—by their inferential role with respect to the singular terms that fall under their respective sortals. For instance, entitlement to a claim of the form '$\exists xPx$' can be gained via entitlement to there being (or to our being able to introduce) a *vindicating substituend*: a singular term *a* such that *Pa*. Though quantifiers all play the same structural *role*—and for that reason qualify as quantifiers—different quantifiers have different *specific* proprieties of inference. Since any given quantifier is sortally restricted, the only singular terms that qualify as legitimate vindicating substituends are those that fall under the same sortal concept. (One cannot vindicate a commitment to 'There is a *book* in the room' by referentially picking out a *molecule* in the room.) So each quantifier is bound up in different sets of inferential commitments. The set of non-quantificational claims that commit one to the quantificational claims made using a specific

[11] Brandom (1994, pp. 436–40; 2015, p. 217).

[12] Hence Frege: 'If I place a pile of playing cards in [someone's] hands with the words: Find the number of these, this does not tell him whether I wish to know the number of cards, or of complete packs of cards, or even say of honours cards at skat. To have given him the pile in his hands is not yet to have given him completely the object he is to investigate; I must add some further word—cards, or packs, or honours' (Frege, 1884, sec. 22).

quantifier, as well as the set of non-quantificational claims that one becomes committed to in virtue of asserting quantificational claims using a specific quantifier, will be different for each quantifier. In the inferentialist framework, every quantifier has a different meaning.

Free logics are available to the inferentialist; there is nothing incoherent about a practice in which singular terms which do not pick out anything can act as vindicating substituends for commitments couched using the existential quantifier. As such, inferentialism allows for a distinction between existentially non-committing particular quantification and fully-fledged existential quantification. The latter is a kind of particular quantification, but one in which the vindicating substituends are restricted to a special subset of singular terms, which Brandom calls *canonical designators*. Roughly speaking, these are the terms that, within a linguistic practice, are treated by default (though perhaps defeasibly) as picking out something, and are individuated such that they cannot pick out more than one thing. Canonical designators, in other words, are those terms one is entitled to take to have *successful singular referential purport*. Although all existential quantification shares this common structure, different discourses have different sets of canonical designators.[13]

15.4 Abstract Objects

We are now in a position to see how this paves the way for a deflationary treatment of abstract objects. A tempting thought as this juncture would be to see the potential for a deflationary treatment of abstract objects as residing in the distinction between existentially non-committing particular quantification and fully-fledged existential quantification. But that tempting thought is a non-starter, since, as we will see, quantification over arithmetical objects—paradigmatic abstract objects—is fully-fledged existential quantification. In fact, inferentialism opens up space for a deflationary treatment of abstract objects by providing the conceptual resources to distinguish between existential quantification that is ontologically heavyweight and existential quantification that is ontologically weightless.

Note first that how one gains entitlement to treat something as a canonical designator will differ significantly from vocabulary to vocabulary. In a physical vocabulary, a canonical designator may be something in the ballpark of 'The horse occupying the spatial region <x,y,z>'. The practices that underwrite entitlement to that definite description succeeding in picking out one, and exactly one, thing are very different to the practices underwriting entitlement to holding that 'the number 2' or 'the only even prime number' succeed in picking out one, and exactly one, thing. How do we gain entitlement to treating *numerals* as canonical designators? Here is Brandom:

[13] The inferentialist account of singular terms is elaborated in Brandom (1994, ch. 6), (2000, ch. 4), and (2002).

> In the paradigmatic case of natural numbers, numerals such as '121' can serve as canonical designators because they are systematic abbreviations of successor numerals: elements of the sequence
>
> 0, 0', 0'', 0''', 0'''' ...
>
> Peano's axiomatization ensures that numerals of this form cannot fail to refer to numbers, and further that their lexical distinctness (marked by the number of successor marks they bear) suffices for the distinctness of the numbers they refer to. Claiming that some noncanonical numerical expression succeeds in referring to a number is just claiming that it is intersubstitutable (saving discursive commitments) with some element in the canonical sequence of successor numerals. Existential commitment with respect to this sort of object, natural numbers, is a disjunctive substitutional commitment linking the candidate numerical expression to some canonical substituend. Saying which number a numerical expression refers to is producing the canonical designator that is intersubstitutable with it. (Brandom 1994, p. 442)

That numerals are systematic abbreviations of successor numerals, combined with the guarantee provided by the Dedekind-Peano axioms that there is a number picked out by 0 and a successor for every number, ensures that numerals are always successful in picking out exactly one number. As a result, numerals function as canonical designators; we are *entitled* to take them as succeeding in picking out an object.[14]

As well as functioning as canonical designators, numerals are used as meaningful singular terms. Recall that for the inferentialist, using a subsentential expression (such as a numeral) as a singular term is only intelligible in the context of being able to make *identity* judgements using that term. Trivial identities such as

0''' = the successor of 0''
0''' = the successor of the successor of 0',

do not express substantive inferential commitments, as they merely inferentially link two singular terms that are notational variants of each other. What is required to grasp a numeral *as* a singular term, is the ability to make *substantive* identity judgements involving that term. Our inferential grasp of arithmetic, however, allows us to do this; we make substantive identity judgements, such as 'The smallest number expressible as the sum of two cubes in two different ways is 1729' or '2 is the only even prime number'—the kind of substantive identity judgements constitutive of our ability to grasp the inferential role of numerals, and so to understand them and use them as singular terms. As meaningful singular terms that function as canonical designators, numerals play the role of vindicating substituends for quantifier claims about natural numbers.

Let's take stock. An upshot of the foregoing, is that, on the inferentialist picture, there are different sorts of existence. Numerical existence is not the same as physical existence, for example. In the first place, as we have already seen, a natural number quantifier has a different meaning from, for instance, various physical quantifiers, since the sets of singular terms that are candidate vindicating substituends

[14] In going along with Brandom that the Dedekind-Peano axioms guarantee the referential success of numerals, I'm bracketing, for the time being, familiar epistemic issues surrounding claims about the existence of mathematical objects.

are different for each quantifier (falling, as they do, under different sortal concepts). In the second place, there is a different set of canonical designators for physical things and natural numbers. So, and for broadly the same reason that different quantifiers have different meanings, 'exists' when applied to physical objects has a different meaning than 'exists' when applied to numbers, and both pick out different sorts of existence. Numerical existence and physical existence are different. Quite generally, for the inferentialist, different sortally-restricted quantifiers have different inferential-subtitutional roles relating different sets of canonical designators, and so different senses of 'exists' expressing different kinds of existence.[15]

15.5 Deflationism About Abstract Objects

I think the inferentialist understanding of abstract objects is best understood as, in some sense, *deflationary*. In the first place, the pragmatist is inoculated against two widespread causes of inflationary flareup, the first due to Benacerraf (1973), and the second to Quine (1986). We started out by noting that the pragmatist rejects representationalist orders of explanation, and, with them, the default (if defeasible) assumption that language is uniformly in the business of depicting how things stand in the world. She understands mainstream model-theoretic semantics as, fundamentally, making explicit the inferential relationships between an object language and a metalanguage, but she is not compelled to add the representationalist gloss that this, in every instance, provides a picture of the relationship between language and world.

Seen in this light, the sorts of considerations Benacerraf (1973) raises have no inflationary import. Benacerraf defends a Tarskian account of mathematical truth. The thought is that this will imply truth conditions for mathematical sentences that are ontologically inflationary. But if we have not *already* given a representationalist gloss on formal semantics there are no such inflationary consequences. When we supply truth conditions for a mathematical language in the Tarskian way, we give those truth conditions in a metalanguage. But, as we noted, once we have dropped the representationalist claim that standard model theoretic semantics gives a metasemantic picture of how language must relate to the world, the truth conditions given this way will only be inflationary if we already take the metalanguage to be ontologically heavyweight. Only then can claims made in the metalanguage inject

[15]There is still a sense in which 'exists' is univocal, since each quantifier plays the same structural role, though in each case indexed to a different set of canonical designators: 'what these different sorts of existential commitments share—what makes them all varieties of *existential* commitments—is the way in which their pragmatic significance is determined by a set of expressions playing the role of canonical designators. The manifest differences between them are consequences of the very different sorts of expressions that serve as canonical designators in each case.' (Brandom, 1994, p. 440) Different quantifiers have different proprieties of inference and, so, different meanings. But at a higher level of abstraction, all quantifiers play the same inferential role. We will reserve the term 'strictly univocal' for views holding that there is only one, or one fundamental, quantifier.

ontological juice, through the biconditionals specifying the interpretation function, into the object language.

Understanding quantifiers as essentially sortally restricted also innoculates against Quinian arguments for ontologically inflationary views of abstract objects. Quine defends the claim that the existential quantifier expresses existence on the grounds that its meaning is given by the English phrase 'there is an object x such that...' and that this phrase expresses existence (Quine, 1986, p. 89). If one takes 'existence' to be univocally ontologically heavyweight, then we have a quick route to ontological inflation, via the truth of claims quantifying over abstract objects. But for the inferentialist, existential quantifiers are not conceptually prior to the sortals and singular terms characteristic of a given vocabulary for which a quantifier can be introduced. As such, one cannot *first* determine the meaning of "the" existential quantifier, and then use that antecedently specified meaning to draw conclusions about the meaning of, e.g., mathematical claimables and their ontological commitments. This is to get things backwards. Singular terms are 'properly introduced only in connection with some at least implicit sortal' (Brandom, 1994, p. 439), so the meaning of a quantifier (and of any existence claims) will depend on the prior meaning of the vocabulary in which it is embedded. The proper order of inquiry, for the inferentialist, is to first determine whether, for example, arithmetical linguistic practices are such as to institute heavyweight ontological commitments, and then, on that basis, to determine the ontological significance of the existential quantifier indexed to the sortal *natural number*. For the inferentialist then, Benacerraf and Quine's arguments for ontological inflation—and the many arguments downstream of Benacerraf and Quine, sharing the same relevant presuppositions—don't gain any traction.

So what positive account can the inferentialist give of when a vocabulary should be understood in deflationary or inflationary terms? The answer, I think, lies in Brandom's account of intentionality. In fact, Brandom elaborates two distinct accounts of intentionality. The first we'll consider, *practical* intentionality, is a kind of *world-directedness* that emerges when a creature engages in a particular process of skilful interaction with the world, in which the creature's practical commitments are updated as a result of the process.[16] The second, *discursive* intentionality, is a kind of *object-directedness* that results from treating sentences as incompatible with one another.[17]

Consider first practical intentionality. Practical intentionality, as the inferentialist understands it, is instituted by skillful interactions with features of an agent's environment that exhibit a particular kind of progressively corrective structure, characterised by 'the cycle of perception, inference, action and perception' (Brandom, 1994, p. 332). Brandom elaborates:

[16] This idea is already at play in Brandom (1994, p. 529), but is further developed in Brandom (2008, ch. 6) and Brandom (2019, *passim*).

[17] This is discussed in Brandom (1994, chs. 7, 8; 2000, ch. 5; 2008, ch. 6, 2009, ch. 1; 2011, 2013).

> [A] creature's practical engagement with the world exhibits *practical intentionality* insofar as it is feedback-governed, that is, specifiable (in a sufficiently rich vocabulary) as having an algorithmic TOTE structure in which each cycle is mediated by its differential responses to the effects of its own performances. Specifying the behaviour of a system in such terms of taking or treating it as practically *directed toward* the features of its environment that play a suitable dual role in the reliably covarying causal chains of events that serve as both inputs and outputs from the system that engages in a process with this structure. (Brandom, 2008, pp. 182–83)

What this amounts to is mostly clearly seen with a toy example. Consider a community of language-users who employ the term 'steel*', the meaning of which can be characterised by two conditionals, making explicit, respectively, the circumstances of application and the consequences of application of 'steel*':

$$\forall x\left(x \text{ is grey and reflective} \Rightarrow x \text{ is steel} *\right)$$

$$\forall x\left(x \text{ is steel}* \Rightarrow x \text{ is rigid}\right)$$

such that *d is steel** is a committive material consequence of *d is grey and reflective*, *d is rigid* is a committive material consequence of *d is steel**, any proposition materially incompatible with *d is steel** is materially incompatible with *d is grey and reflective*, and any proposition materially incompatible with *d is rigid* is materially incompatible with *d is steel**.[18] In employing the term 'steel*' language users in the community are undertaking the responsibility to endorse, in their practice, these inferential commitments. Put differently, in using the term 'steel*', one is according it (as characterised by the two conditionals) *authority* over one's reasoning, such that having undertaken a commitment to *d is steel** one is *obliged* not to endorse any proposition materially incompatible with that commitment.

Nevertheless, the authority accorded to 'steel*' is not absolute. In discursive practices exhibiting practical intentionality, the language-user accords inputs, brought about by the her worldly interactions, the authority to update and correct her discursive commitments, in some circumstances. Here the language-user must be able to reliably differentially respond to features of her environment, where the response is to take on a new discursive commitment—a commitment that sometimes deontically requires the agent to revise one or more of her existing commitments. Say the language-user perceives that an item *d* is grey and reflective. She now has incurred commitments to *d is steel** and *d is rigid*. Say further, though, that through subsequent interaction with *d*, she finds that *d* is malleable, and incurs a commitment to *d is not rigid*. The language-user is now in an aporetic state, with a set of mutually incompatible commitments, and has the obligation to amend her commitments to repair the incompatibility. Here the circumstances of application could be altered so that a commitment to an item being grey and reflective no longer incurs a commitment to the same item being steel*, or the consequences of application could be altered so that a commitment to an item being steel* no longer incurs

[18] The use of the arrow with two horizontal lines marks that the conditionals here should not be understood as material conditionals, but as expressing modal force.

a commitment to the same item being rigid, or, alternatively, the language-user could simply refrain from using the term 'steel*', ceasing to accord it any authority over her reasoning.

For the pragmatist then, practical intentionality is understood diachronically, as a process or practice, and, for the inferentialist in particular, as an ongoing and unbounded sequence of feedback-governed processes that are characterised by 'a Test-Operate-Test-Exit (TOTE) cycle of perception, performance, assessment of the results of the performance, and further performance' (Brandom, 2008, p. 178). These processes, Brandom notes, are "thick" in the sense of essentially involving commerce between a language-user and bits of her environment. It is this thickness that 'engender[s] the possibility of *friction* between the world and the deployment of vocabulary in a practical cycle of perception-and-performance articulated by those material inferential and incompatibility relations.' (Brandom, 2008, pp. 185–86).

TOTE-structured interactions with the world, of the kind sketched here, are the mechanism by which one makes one's commitments *responsible* (and responsive) to the world, and thereby *world-directed*. The alethic modal or nomic properties of, for instance, steel* *itself* gain purchase on the deontic modal proprieties of inference that characterise the linguistic community's concept 'steel*'. The friction made possible through these TOTE-structured interactions, mean that the world itself is able to inform the content of our commitments and the inferential structure of our concepts. This is one of the cruces of practical intentionality, since for one's commitments to be *world-directed*—to be *about* the world—is for the world to be accorded a certain kind of *authority* over those commitments: the authority to determine whether those commitments are correct or incorrect, true or false. But to *treat* the world as authoritative over one's discursive commitments *just is* to engage in a practice in which, according to the rules of that practice, one could and would alter one's discursive commitments as a result of the features the world in fact has. Where one does not or cannot engage in a practice like that, one is not exhibiting practical intentionality, and one's commitments are not, in that instance, world-directed.

Now this can be contrasted with discursive intentionality. One can get a handle on the object-directedness of discursive intentionality by considering *de re* propositional attitude ascriptions such as *Ruairidh believes **of** the theropod **that** it is a sauropod*, or *Magnus believes **of** the herbivore **that** it is a carnivore*.[19] A function of *de re* propositional attitude ascriptions, according to the inferentialist, is to, on the one hand, *attribute* some discursive commitment to another, and to *acknowledge* a discursive commitment of one's own. In saying, for example, *Ruairidh believes **of** the theropod **that** it is a sauropod* I acknowledge my own commitment to the specification *theropod* and attribute a commitment to the specification *sauropod* to another, while, at the same time, making explicit that both my and the other's specifications pertain to the same object. My commitment falls within the scope of the 'of', while the other's commitment falls within the scope of the 'that'. This is a way

[19] This is elaborated in Brandom (1994, ch. 8; 2000, ch. 5).

of recognising that my, and the other's, inferential commitments can be different, while simultaneously recognising that, despite this, they are *about* the same *things*.

Notice that, in these cases, I treat my commitment as *incompatible* with the commitment of the other. This treatment of commitments as incompatible, this aspect of deontic scorekeeping, is a central part of how the inferentialist understands co-referential purport, since the deontic incompatibility of two or more sentences always involves co-referential purport. In the simplest case, to treat two simple atomic sentences as incompatible when attributing contrary properties is to treat the tokening of the singular term in each sentences as co-referential. 'd is a carnivore' and 'd is a herbivore' are incompatible, but 'd is a carnivore' and 'e is a herbivore' (when $d \neq e$) are not. Incompatible properties, such as *being a herbivore* and *being a carnivore*, are incompatible in the sense that they cannot be simultaneously co-instantiated in a single object.[20] So to treat two sentences predicating *being a carnivore* and *being a herbivore* as incompatible, is to regard the singular term in each sentences as purporting to refer to the same thing. Co-referential purport is also required in more complex cases of incompatibility. Consider for example an incoherent triad: 'd is the mother of e', 'd is a theropod', 'e is a sauropod'. A similar kind of triangulation determining co-referential purport takes place here; the incoherence only exists because each tokening of 'd' purports to refer to the same object and each tokening of 'e' purports to refer to the same object. So to treat the the triad as incoherent is to regard each tokening of 'd' to purport to refer to the same object and each tokening of 'e' to purport to refer to the same object. In this way, *object-directedness* is built into linguistic deontic scorekeeping practices in which some sentences are treated as incompatible with others—that is to say, all linguistic deontic scorekeeping practices.[21]

So the inferentialist framework allows one to distinguish between two phenomena, that are quite distinct, but which both fall under the rubric 'intentionality' insofar as they both pick out a kind of directedness.[22] Discursive intentionality, when it is not paired with practical intentionality, shares with disquotational,[23] minimalist,[24]

[20] 'The incompatibilities among properties are essential to the individuation of objects, since objects (particulars) are units of account for incompatibilities of properties (universals). If $a = b$, then it cannot be that Pa and Qb for incompatible properties P and Q. But if $a \neq b$, it is possible that Pa and Qb. To be an object (a single object) is to exclude incompatible properties.' (Brandom, 2015, p. 69) See also Brandom (2008, pp. 187–89), (2009, pp. 42–45), and (2019, ch. 5).

[21] At a more fundamental level this requires treating distinct tokenings of singular terms as part of a single anaphoric chain. See Brandom (1994, ch. 7) and Collin (2023).

[22] In fact, Brandom distinguishes between *that*-intentionality—the kind of 'intentionality' afforded to a linguistic item in virtue of it expressing a proposition—and *of*-intentionality—the kind of 'intentionality' afforded to a linguistic item in virtue of it being about an object or objects. So we can distinguish at least three phenomena that fall under the rubric 'intentionality': propositional contentfulness, object-directedness, and world-directedness. Brandom (2013, p. 108) mentions, in addition, practices of mapping and tracking as kinds of representation. What matters here is the distinction between object-directedness and world-directedness.

[23] See e.g., Quine (1990) and Field (1994).

[24] See e.g., Horwich (1998).

and prosentential[25] accounts of truth the fact that it can be understood in terms of purely "horizontal" relationships between linguistic items. For the same reasons that these accounts of truth should be regarded as deflationary, then, we should also think of discursive intentionality as a deflationary kind of intentionality—something comprised of and explicable in terms of the relationships that linguistic expressions bear to one another. Practical intentionality, in contrast, is irreducibly *in*flationary, and cannot be understood except in reference to 'vertical' relationships between linguistic items and worldly objects.

Now, a goal of Brandom's is to explain discursive intentionality in terms of practical intentionality.[26] Here Brandom is surely right that specifically linguistic world-directedness must be explained at least partly in terms of practical intentionality. What I want to draw attention to however, is that there is a degree of latitude or flexibility in the sorts of linguistic practices we can engage in, such that there can be linguistic practices that exhibit both discursive intentionality and practical intentionality, but also linguistic practices that exhibit only discursive intentionality. Consider the idealised but coherent example of a community of language users with no prior mathematical vocabularies, who, subsequently, introduce arithmetical terms. In treating some arithmetical sentences as incompatible, they engage in the kind of triangulation required to treat them as co-referential, just sketched above. It is clear then that the newly introduced arithematical vocabulary exhibits deflationary discursive intentionality. But it is equally clear that it does not exhibit inflationary practical intentionality, as there are no TOTE-structured cycles that can serve to modify the inferential proprieties of arithmetical claims. On the contrary, the standards of correctness and incorrectness for numerical claims are fixed in a way 'internal' to language, without appeal to thick, world-directed practices. Arithmetical assertions are neither responsive to nor responsible to the world.

This is closely related to the point, emphasised by Field (1989, 2016, ch. 1), that mathematics must be *conservative* in order to be good, where a mathematical theory M is conservative just in case it is consistent with every internally consistent theory about the physical world (or any other non-mathematical domain). It is more or less trivial that pure mathematics is conservative, since consistent pure mathematical theories do not entail anything about the physical world. Things are different with impure mathematical theories, containing predicates and function symbols that relate non-mathematical objects or relate non-mathematical objects to mathematical objects. As Field points out (1989, pp. 56–57), a variation of ZF (call it ZFU_{FIN}), allowing for sets of non-sets, with specification and replacement schemas some instances of which range over non-sets and contain non-mathematical predicates, and including the negation of the axiom of infinity, is not conservative. From the union of ZFU_{FIN} and a theory implying the existence of a infinite discrete linear ordering of non-mathematical entities, it is possible to define a function mapping those entities onto an infinite set of integers—a result that is inconsistent with

[25] See e.g., Grover (2014) and (Brandom, 1994, ch.5).

[26] Brandom (2008, p. 179).

ZFU_{FIN}. The very fact, however, that ZFU_{FIN} implies substantive facts about the physical domain makes it unacceptable.[27] Good mathematical theories are conservative. In contrast, it is a requirement of good physical theories that they be non-conservative; if some theory T is consistent with every theory about the physical domain, T cannot contain any information about the physical domain. Similarly, for T to be empirically testable, it must be possible to derive consequences about the physical world from T, and this requires non-conservativeness.

One consequence of the conservativeness of acceptable mathematics is that acceptable mathematical theories cannot be subject to the process of deriving consequences, exercising reliable differential responses to features of the environment to test those consequences, and correcting the theory, that characterises TOTE-cycles. As Field notes, the observation that, while empirical theories *cannot* be conservative if they are to be good, mathematical theories *must* be conservative if they are to be good 'helps undercut the influential idea of mathematics as quite continuous with science' (Field 1989, p. 61). The inferentialist can add that it marks another very fundamental discontinuity between empirical vocabularies and mathematical vocabularies: that the former, but not the latter, are world-directed. Arithmetical vocabularies are not taken up in the practices required for practical intentionality, and so lack the inflationary kind of representational purport.[28]

---

[27] Field puts the point this way: '[I]f it were to be discovered that standard mathematics implied that there are at least 106 non-mathematical objects in the universe, or that the Paris Commune was defeated … all but the most unregenerate rationalists would take this as showing that standard mathematics needed revision. Good mathematics is conservative; a discovery that accepted mathematics isn't conservative would be a discovery that it isn't good.' (Field 1989, p. 13)

[28] *Applied* mathematics does involve TOTE-structured interactions with the world. Does then the claim that mathematical vocabularies, as they are used, should not be regarded as exhibiting practical intentionality become untenable once applied mathematics is brought into the picture? Not if we keep clear-headed about what one is doing in applying mathematical vocabularies to empirical domains. Here I have in mind the account of Leng (2012, 2020, 2021), though see also Bueno and French (2018) and Ketland (2021). How, roughly, does this go? Take (i) some axioms Γ in a language $\mathcal{L}$ with an interpretation δ^{M}, such that Γ characterises a structured domain $\langle D^{M}, R^{M}\rangle$ of mathematical objects, and (ii) some sentence ϕ of $\mathcal{L}$ such that $\Gamma \vDash \phi$. Any structured domain that satisfies Γ also satisfies ϕ, Let then δ^{N} be an interpretation of $\mathcal{L}$ such that Γ now characterises a structured domain $\langle D^{N}, R^{N}\rangle$ of physical objects. Then $\Gamma \vDash \phi$, *where* ϕ *is now interpreted as a claim about a target physical system*. In this way we can use mathematical languages to show that, given certain physical assumptions, certain physical facts must obtain. Typically the structures characterisable this way will be highly idealised counterparts of real physical systems. When there is no characterisation of the real system in these terms, it is impossible of course to define a formal structure-preserving mapping between the model and the real system, but a gamut of informal methods—appeal to resemblances etc.—will be appealed to to establish the relevance of the model to the real system, which allow one to reasonably draw conclusions about the real system on the basis of inferences about the idealised structure. The point to emphasise here, is that, even though mathematical modelling practices undoubtedly involve the kind of TOTE-cycles described by Brandom, mathematical theories are never revised as a result. What *does* get revised are our claims about which physical systems satisfy which axiom sets or which physical systems relevantly resemble models of which axiom sets. At no point do our mathematical commitments or mathematical vocabularies exhibit inflationary, world-directed, practical intentionality.

We can now draw two strands of thought together. Firstly, numerical existence is different from physical existence. Secondly, numerical existence is ontologically weightless. The first is a consequence of the conceptual space the inferentialist framework makes for a plurality of sortally-restricted quantifiers, and, corresponding to these, a plurality of different kinds of existence. The second is a consequence of understanding the world-directedness of a vocabulary as instituted by TOTE-structured cycles of interaction and correction, and recognising that arithmetical vocabularies are not bound up in cycles of that kind. Because arithematical vocabularies do not exhibit practical intentionality, they are not, for the inferentialist, world-directed. And because they are not world-directed, they ask nothing of the world. On the view described here, taking arithmetical claims to make demands on the world is to engage in a kind of category error. So although we can talk about natural numbers—in the deflationary sense of 'about'—and natural numbers can correctly be said to exist—in the sense of 'exist' appropriate to natural numbers—the existence of natural numbers places no demands on reality. This is an ontologically weightless kind of existence. Here we have an inferentialist way of cashing out the suggestive Sellarsian metaphor (Sellars, 1974), echoed some decades later by Wright (1992, p. 181), that abstract objects 'are no more than shadows cast by the syntax of our discourse'.

15.6 Inferentialism by the Method of Differences

I have been arguing that, if one accepts an inferentialist understanding of singular terms and of intentionality, one can also endorse a deflationary conception of abstract objects, or, at least, one paradigmatic instance of abstract objects: natural numbers. I'll end as I began, with some comments positioning the pragmatist inferentialist position in relation to more familiar positions in the philosophical landscape. As before, this will just be a sketch, but it should suffice to make visible some of the key similarities and differences between inferentialism and these other positions.

One of the principle advantages of the inferentialist account, I take it, is its provision of a framework in which it is intelligible that there be multiple distinct existential quantifiers, some of which are ontologically heavyweight and some of which are ontologically weightless. In this regard, the inferentialist view of abstract objects is further from the fictionalism espoused by Leng (2010) and Field (2016)—that treats mathematical claims as, strictly speaking, false—and closer to the nominalism of Azzouni (2004, 2007, 2010, 2017), the noneism of Priest (2016)—that both distinguish between quantifier commitments and (heavyweight) ontological commitments—and the 'subtle Platonism' of Rayo (2013, 2015, forthcoming)—that gives mathematical claims trivialist truth conditions, making no requirements on the world. It differs from both Azzouni and Priest however, who both hold that quantifiers are ontologically neutral, in regarding *some* quantifiers as ontologically heavyweight. And Rayo's view is more liberal than the inferentialist's, insofar as it permits the introduction of referring singular terms that do not feature in substantive identity judgements. With Linnebo's (2018) 'thin' conception of abstract objects, the inferentialist draws an ontological distinction between mathematical objects and concreta. But the thin conception sees abstract objects as making a small demand on the world, while for the inferentialist arithmetical claims are not world-directed, and so the truth of those claims cannot make *any* demands on the world. (In this respect, the inferentialist is closer to Rayo's 'ultra thin' conception of objecthood.) The inferentialist shares with Thomasson (2014) the broad methodological commitment of appealing to features of language *use* to motivate a form of ontological deflationism (as well as an understanding of the most fundamental forms of quantification as being sortally restricted), but differs insofar as Thomasson motivates her deflationism within the boundaries of a strictly univocal conception of quantification and existence. Like Hirsch (2011) and Hofweber (2016), the inferentialist opposes strictly univocal conceptions of the existential quantifier. Unlike Hirsch, however, there is nothing in the inferentialist account that immediately implies the doctrine of *quantifier variance*, so understood as to include the claim that there 'is no uniquely best ontological language with which to describe the world' (Hirsch, 2011, p. xii). And, unlike Hofwebber, who argues that numerical singular terms function syntactically like names, but do not purport to refer to mathematical objects, the inferentialist regards numerals as referring to mathematical objects, but goes on to understand this in deflationary terms.[29]

So inferentialist deflationism has many affinities with other forms of deflationism, while remaining unique in a number of key respects, and adds to the landscape is its own distinctive motivations for making the distinction between ontologically heavyweight and ontologically weightless commitments and for taking mathematical commitments to fall into the weightless camp. The semantic inferentialist also has her own resources for addressing epistemic issues in the philosophy of mathematics. Traditional platonist views face epistemic objections arising out of the conceptual gap they create between referential purport and referential success. The world must cooperate with our belief-forming processes if our beliefs are to be true. Given the peculiar attributes of mathematical objects, one can then argue that there are in-principle difficulties in explaining the reliability of our mathematical beliefs (Field 1989, p. 26), which may constitute a defeater for those beliefs, or that there are in-principle reasons to doubt that our mathematical beliefs could meet the modal conditions required for knowledge (Collin, 2018). Now whether epistemic objections to platonism are ultimately successful is an open question. But the inferentialist deflationist position appears to address these epistemic issues at the root. Because arithmetical beliefs are not, for the inferentialist, world-directed, there is no question of the world failing to cooperate with those beliefs, so the epistemic issues surrounding platonism cannot arise in the first place. Of course, a good deal more needs to be said to motivate that train of thought, but it suggests that another theoretical advantage of the inferentialist view is that it undermines the intelligibility of a gap between referential purport and referential success in the case of arithmetic.

[29] See Collin (2022) for a more detailed survey of these views.

References

Azzouni, J. (2004). *Deflating existential consequence*. Oxford University Press.
Azzouni, J. (2007). Ontological commitment in the vernacular." *Noûs* (Detroit, Mich.) *41*(2), 204–226. https://doi.org/10.1111/j.1468-0068.2007.00644.x
Azzouni, J. (2010). Ontology and the word 'exist': Uneasy relations. *Philosophia Mathematica, 18*(1), 74–101. http://philmat.oxfordjournals.org/cgi/doi/10.1093/philmat/nkp011
Azzouni, J. (2017). *Ontology without borders*. Oxford University Press.
Benacerraf, P. (1973). Mathematical truth. *The Journal of Philosophy, 70*(19), 661–679.
Brandom, R. (1994). *Making it explicit: Reasoning, representing, and discursive commitment*. Harvard University Press.
Brandom, R. (2000). *Articulating reasons: An introduction to inferentialism*. Harvard University Press.
Brandom, R. (2002). The significance of complex numbers for Frege's philosophy of mathematics. In *Tales of the mighty dead: Historical essays in the metaphysics of intentionality* (pp. 277–297). Harvard University Press.
Brandom, R. (2008). *Between Saying and doing*. Oxford University Press.
Brandom, R. (2009). *Reason in philosophy*. Harvard University Press.
Brandom, R. (2011). Pragmatism, expressivism, and anti-representationalism: Local and global possibilities. In *Perspectives on pragmatism: Classical, recent and contemporary*. Harvard University Press.
Brandom, R. (2013). Global anti-representationalism? In *Expressivism, pragmatism and representationalism* (pp. 85–111). Cambridge University Press.
Brandom, R. (2015). *From empiricism to expressivism: Brandom reads sellars*. Harvard University Press.
Brandom, R. (2019). *A Spirit of trust*. Harvard University Press.
Brandom, R., & Aker, A. (2008). Appendices. In *Between saying and doing* (pp. 141–175). Oxford University Press.
Bueno, O., & French, S. (2018). *Applying mathematics: Immersion, inference, interpretation*. Oxford University Press.
Collin, J. H. (2018). Towards an account of epistemic luck for necessary truths. *Acta Analytica, 33*, 483–504.
Collin, J. H. (2022). Mathematical nominalism. In J. Fieser & B. Dowden (Eds.), *The internet encyclopedia of philosophy*. https://iep.utm.edu/mathematical-nominalism/
Collin, J. H. (2023). Of marriage and mathematics: Inferentialism and social ontology. *Topoi, 42*, 247–257.
de Prado, G., Salas, J., de Donato, X., & Rodríguez, and Jesús Zamora Bonilla. (2021). Inferentialism, degrees of commitment, and ampliative reasoning. *Synthese, 198*(4), 909–927. https://doi.org/10.1007/s11229-017-1579-5
Field, H. (1989). *Realism, mathematics and modality*. Wiley-Blackwell.
Field, H. (1994). Deflationist views of meaning and content. *Mind, 103*(411), 249–285.
Field, H. (2016). *Science without numbers* (2nd ed.). Oxford University Press.
Frege, G. (1884). *Die Grundlagen Der Arithmetic: Eine Logisch-Mathematische Untersuchung Über Den Begriff Der Zahl*. Verlage Wilhelm Koebner.
Grover, D. (2014). *A Prosentential theory of truth* (Vol. 194). Princeton University Press.
Hirsch, E. (2011). *Quantifier variance and realism: Essays in Metaontology*. Oxford University Press.
Hlobil, U., & Brandom, R. (2024). *Reasons for logic, logic for reasons: Pragmatics, semantics, and conceptual roles*. Routledge.
Hofweber, T. (2016). *Ontology and the ambitions of metaphysics*. Oxford University Press.
Horwich, P. (1998). *Truth*. Clarendon Press.
Ketland, J. (2021). Foundations of applied mathematics I. *Synthese. An International Journal for Epistemology, Methodology and Philosophy of Science, 199*, 4151–4193.

Leng, M. (2010). *Mathematics and reality*. Oxford University Press.
Leng, M. (2012). Taking it easy: A response to Colyvan. *Mind; A Quarterly Review of Psychology and Philosophy, 121*(484), 983–995.
Leng, M. (2020). Mathematical explanation doesn't require mathematical truth. In *Current controversies in philosophy of science* (pp. 51–59). Routledge.
Leng, M. (2021). Models, structures, and the explanatory role of mathematics in empirical science. *Synthese, 199*(3), 10415–10440.
Linnebo, Ø. (2018). *Thin objects: An abstractionist account*. Oxford University Press.
Peregrin, J., & Svoboda, V. (2021). Moderate anti-exceptionalism and earthborn logic. *Synthese, 199*(3), 8781–8806.
Price, H. (2013). *Expressivism, pragmatism and representationalism*. Cambridge University Press.
Priest, G. (2016). *Towards non-being: The logic and metaphysics of intentionality* (2nd ed.). Oxford University Press.
Quine, W. V. O. (1986). *Philosophy of logic* (2nd ed.). Harvard University Press.
Quine, W. V. O. (1990). Truth. In *Pursuit of truth*. Harvard University Press.
Rayo, A. (2013). *The construction of logical space*. Oxford University Press.
Rayo, A. (2015). Nominalism, Trivialism, Logicism. *Philosophia Mathematica, 23*(1), 65–86.
Rayo, A. (forthcoming). The ultra-thin conception of objecthood. *Inquiry*, forthcoming.
Rorty, R. (2018). *Philosophy and the mirror of nature* (30th Anniversary ed.). Princeton University Press.
Sellars, W. (1974). Empiricism and abstract entities. In *Essays in philosophy and its history* (pp. 245–286). Springer.
Soames, S. (2010). *Philosophy of language. Princeton foundations of contemporary philosophy*. Princeton University Press.
Soames, S. (2014). *The analytic tradition in philosophy*. Princeton University Press.
Thomasson, A. (2014). *Ontology made easy*. Oxford University Press.
Wright, C. (1992). *Truth and objectivity*. Harvard University Press.

Chapter 16
The Implicature View of Ontological Commitments and Denials

Jody Azzouni

Abstract It is shown that we express ontological commitments in English purely by means of Gricean implicatures. That is, when we commit ourselves to the existence of bears as we do in conversation when we assert "There are bears," or "Bears exist," this is not because of the meaning (the semantics) of the phrase "there is" nor of the word "exist." Rather it is because the sentences in which these words appear often exhibit implicatures that speakers and their audience experience as the assertion of ontological commitments. Alternative views of the linguistic phenomena, that these phrases are ambiguous or polysemous, are shown to face serious counterexamples. Positions that take phrases such as "real," "taking place," "obtains," and so on to involve ontological commitment by virtue of their meanings are also undercut by the use of counterexamples.

Keywords "Exist (s)," generics · Gricean implicatures · Meinongianism · Ontological commitment · "Real," "there is"

My thanks to John Collins for invaluable "push back" (in several emails, late May, early June, 2019) which led to substantial changes throughout the paper, but especially to Sects. 16.5 and 16.6. My thanks to the organizers of the *Abstract objects and circularity conference* (at MCMP, July 6th and 7th, 2019) for inviting me (and consequently, for bringing this paper into existence), and to the audience for their comments and responses that yielded additional changes.

J. Azzouni (✉)
Department of Philosophy, Tufts University, Medford, MA, USA
e-mail: jody.azzouni@tufts.edu

X. de Donato-Rodríguez et al. (eds.), *Deflationist Conceptions of Abstract Objects*, Synthese Library 516, https://doi.org/10.1007/978-3-031-98717-5_16

16.1 The Implicature View of Ontological Commitments and Denials

People think that some things exist and that some things don't; people argue and disagree about this. More dramatically, wars are fought, friendships and love affairs shattered … it's a venerable tradition to torture (and execute) people because of their beliefs or disbeliefs in various things—mostly supernatural things, I regretfully note. Even today (which is more like yesterday than I ever thought possible) atheists are still widely seen in the United States as a kind of human so depraved that just about anyone else (perhaps just about *anything* else) is regarded as more suitable for presidential office.

"Having ontological commitments" is the official jargon among philosophers for a person's holding, taking, believing, etc., things to exist. Quine (1953) notably focused attention on ontological commitment *itself* (rather than on ontological commitments *themselves*)—thus the jargon is his. He was the first philosopher to do so, both because he discussed ontological commitment explicitly and because he gave a theory of it. (Before Quine, I think, no one considered the possibility of—or the need for—a "theory of ontological commitment.") Using van Inwagen's phrase, Quine, therefore, was the first "metaontologist." His view of ontological commitment, however, is both particularly simple and compelling, and yet (from the perspective of ordinary usage) pretty bizarre. That's because he takes people's ontological commitments to be derivatively had, by virtue of sets of sentences they hold true (or believe). In turn, any such set of sentences D has among its commitments the entities in the extension of a first-order predicate P, if, relative to a transcription of D into D* (a set of interpreted first-order sentences), $(\exists x)\mathrm{P}x$ is an implication of D*. This is Quine's "criterion" for ontological commitment—perhaps the first official ontological criterion philosophers have ever had.[1]

Variations on Quine's criterion emerged soon after the criterion did—after all, his audience was mostly philosophers, who can't leave anything alone. A popular move is to broaden the criterion so it applies directly to languages other than

[1] Caveat. Quine claimed his criterion is ecumenical: applicable to any set of sentences, whether one believes them or not. That is, the criterion supposedly yields commitments without requiring the criterion-user to share those commitments. Quine always characterized his criterion semantically: philosophers quickly realized—Cartwright (1954) among the earliest—that this is puzzling because semantic characterizations (Tarskian ones) *acquire* the ontological commitments of the targeted sentences, unless those commitments are trapped within an intensional operator in the language of the semantics. Quine's recommendation of metaontological intensionality belies his first-order extensionalist commitments. (It's especially odd since Quine writes as if one can characterize one's *own* commitments using his criterion.) The purely syntactic criterion I've attributed to Quine avoids this; it's essentially the "alternative" of Church (1958). In the first-order case (which is the only one Quine takes seriously) the two criteria have similar yields because of completeness. Although Quine subsequently alludes to Church's article, he never discusses Church's proposal, indicates if he thinks it's an alternative, or otherwise compares it, to his own. This is puzzling. Craig (2017, p. 104) distinguishes between "Quine's criterion" and "Church's criterion." He may be right to do so.

first-order ones, thus obviating transcriptions—instead, singling out a hodgepodge of expressions (ones that *philosophers* often call "quantifier-expressions"), "there be," "some," "exist," "real," "most"—even, "occur," "obtain," etc.—as sources of ontological commitment by virtue of their semantic properties.

When Quine's criterion is understood as already holding of natural language (instead of being a regimentation or piece of conceptual engineering applied *to* natural language), it becomes a linguistically-testable hypothesis about the expression of ontological commitments and denials in the vernacular. And that's the form I'll challenge it in: instead of ontological commitments and denials arising from the semantics of certain words, I say it arises from and *only from* implicatures.[2] Consequently, our concept of EXISTENCE (the metaphysical notion) occurs not straightforwardly among the meanings of our words (or phrases) but among the implicatures of uttered sentences.

Let me introduce some relevant (Gricean) jargon.[3] Call "what is said" the linguistically-mandated (literal) meaning of an expression. Call an "implicature" what's conveyed by an utterance of a sentence that goes beyond what is said. The content of a "conventional" implicature goes beyond what is said but is determined purely by that.[4] A *particular* conversational implicature, instead, conveys additional content via other factors besides the meaning of the uttered sentence, e.g., what's said/implicated earlier in the conversation, context, perceived speaker-intentions, etc. Listeners are taken to gain access to particular implicated content, more or less consciously, by reasoning their way to it using Grice's maxims. Particular conversational implicatures, thus, because they depend on specific communicative set ups, are often sentence-wide: they usually don't depend on the meanings of specific words, nor are they experienced as due to specific words. They are also (again generally) experienced as different from what's asserted.[5]

---

[2] My claim is only being evaluated in this paper for English. I suspect, but will provide no evidence, that all natural languages are the same in this respect.

[3] I'm largely using the apparatus from Grice (1987). Nearly all of it is contested (and generalized along various dimensions) in an enormous literature in both philosophy and linguistics. That *something* recognizably Gricean, with recognizably Gricean categories for various speech acts, is true of how we speak is all that everyone in this literature seems to agree on. I'm hoping that my application of these tools to ontological commitments and denials sidesteps controversy insofar as what I claim translates easily to these competing alternatives.

[4] Grice's (1987, p. 25) example is "He is an Englishman; he is, therefore, brave," where it is *not said* that his bravery follows from, *in particular*, his being an Englishman. ("Therefore," only literally conveys that his being an Englishman follows from *something*, it doesn't say what.) Nevertheless, the phrase, when used as in Grice's example, is usually understood this way. Importantly, it's heard as *asserting* this. "Asserted," as I use it here, is an *experiential* category. See what follows.

[5] Grice's (1987, p. 32) examples: A: *I am out of petrol.* B: *There is a garage round the corner.* A: *Smith doesn't seem to have a girlfriend these days.* B: *He has been paying a lot of visits to New York lately.* In both cases, the implicatures (the garage is open and has petrol; Smith has or may have a girlfriend in New York) are recognizably not part of the content of what B says (of what is said). They are also recognizably not *asserted.*

Lastly, as Grice (1987, p. 37) puts it, "... the use of a certain form of words in an utterance would normally (in the absence of special circumstances) carry such-and-such an implicature or type of implicature": these are "generalized conversational implicatures." Unlike particular conversational implicatures, one often *does* experience a particular word or phrase to carry the implicature. Further, one often experiences these implicatures as *asserted* even though standard Gricean tests tell us that, nevertheless, they're implicatures (e.g., because they can be "canceled"; see below).[6] Because the usual experience is to hear generalized conversational implicatures as asserted, Grice's maxims are used in calculations to explain the *absence* of implicatures as opposed to their presence, as with particular conversational implicatures. (I'll give examples later.)

There are several good usage tests that operationalize the Gricean distinction between semantic content and implicature. One is to give two contexts in which the same expression nevertheless conveys something different. A widely-used example:

1. John went to the edge of the cliff and jumped.

It's supplemental that John jumped *off* the cliff—but it's the default interpretation (at least among people who don't study sentences like these for a living). We recognize, however, that the jumping-off reading is supplementary because the sentence can be read this way instead:

2. John went to the edge of the cliff and jumped (up and down at its edge),

where John isn't studiously attempting suicide, but just being reckless and stupid. We can make explicit that the jumping-off reading is an implicature and not part of the meaning of "went to the edge of the cliff and jumped" by using Gricean cancellation. Implicatures can (often) be explicitly denied after uttering phrases that would otherwise carry them, without the result sounding contradictory—as it would sound if we try to take back something that's perceived as part of the semantic content of a phrase. Consider: "John went to the edge of the cliff and jumped—but not *off* the cliff." This "cancels" the jumping-off reading of "John went to the edge of the cliff and jumped."

---

[6]Grice's (1987, p. 37) example (nearly enough): "John went into a house yesterday and found a tortoise inside the front door." This implicates the house isn't John's. But (most) people experience this as *asserting* the house doesn't belong to John. Imagine the following conversation after B learns that the house *does* belong to John. B: *But you said it wasn't his house*. A: *No I didn't; I said* ***a*** *house*; ***his*** *house is* ***a*** *house, after all.* B: *That's what it* ***means*** *to say* ***a*** *house—not* ***his****; you're being dishonest.* B, strictly speaking is wrong; *but* B (and other normal people) hear what A is saying as *asserting* it isn't John's house. Compare this to the flap over Trump's recorded remarks about Meghan Markle: "I didn't know that she was nasty." Democrats: *Trump said she was nasty and then denied he said it and we have it on* ***camera****!* Republicans: *He never* ***said*** *she was nasty.*

I will say (but not worry too much about it) that the distinction between generalized conversational implicatures and conventional implicatures isn't transparent to me. In Sect. 16.7, I'll make certain claims about generalized conversational implicatures—but perhaps these are actually claims about conventional implicatures. This detail doesn't matter.

16.2 Usage Evidence Against the Ambiguity of Phrases Used for Ontological commitment and Denial

I'm arguing for an implicature-specification of an earlier view of mine (Azzouni, 2004), that ontological commitment in the vernacular is *solely* a matter of pragmatics; no words or phrases convey ontological commitment/denial by virtue of their meaning alone. This view can only be established on the basis of usage. Unfortunately (for me) one important usage-fact about ontological commitments and denials supports both the pragmatic-commitment position as well as competing ambiguity views. This is that we don't *always* ontologically commit ourselves when we use "there be," "exist," etc.; we often assert (and are up to) something else—not ontological commitment or denial. Every ordinary-language locution philosophers treat as conveying ontological commitment or denial is regularly used without thoughts of ontology:

3. All the orcs in these books are evil
4. Three-fourths of the deities in the Indian pantheon are female.
5. No Roman god except Janus has two faces
6. The average star has 2.4 planets.
7. Although no one thinks Disney would allow this, sexually-depraved cartoon characters exist.
8. You're just wrong about that. There really are good orcs in Tolkien's fiction.
9. Things (or: Objects) that don't exist puzzle philosophers.
10. There are as many fictional witches as warlocks.[7]

Many contemporary metaphysicians think that the above illustrations and others show that these words are *ambiguous*—these different meanings are sometimes described as ontological "heavyweight" and "lightweight" ones. This is a linguistic claim, however; and so it faces long-standing linguistic tests.[8] None are definitive, of course: it's *patterns* of such tests (and facts about usage) that indicate whether we're facing pragmatic phenomena (as I claim) or instead semantic ambiguity. The conjunction reduction test, for example, shows genuinely ambiguous words resist combination in one use. A common illustration: we can say,

[7] It's widely thought "fictional" induces an intensional context within the scope of which reference occurs without incurring ontological commitment; alternatively, that the word implicates a make-believe context—see, e.g., Voltolini (2018, p. 101). The same has been claimed about "cartoon character." "Fictional" may *sometimes* do this, but any such *generalization* is counterexampled by the acceptability of: "Oh no!—all sorts of fictional beings have come to life and are wandering around New York causing trouble." ("Cartoon character" works too.) "Fictional" is an adjective attributable to something because it originates in a genre of writing/film. When something originates that way it *usually* (maybe always—in *our* world) doesn't exist; but that's not a semantic requirement of "fictional." Further, "fictional," isn't metaphorical when describing someone who has come to life: "Few fictional characters have ever come to life." So too: "Only some fictional characters exist."

[8] See, e.g., Zwicky and Sadock (1975). For a nice overview of the topic, see Sennet (2016).

11. The feathers are light and the colors are light,

although we can't (comfortably) say,

12. The feathers and colors are light.

Applying the test to (1) "John went to the edge of the cliff and jumped," we *can* say:

13. Both John and Peter went to the edge of the cliff and jumped, but only John jumped off the cliff.

This indicates that "jump" isn't ambiguous, with two meanings, one which we need to jump off the cliff to satisfy and another where we needn't: we can cancel the implicature with respect to Peter but not John, using the same token of the word. We can similarly show that, "there be," "exists," etc., aren't ambiguous. First, "there be." We *can* say, for example,

14. There are cartoon characters and ordinary people who look like one another,

where the speaker (like most of us) thinks ordinary people exist but cartoon characters don't. Next, "exist." I'm a nominalist; nevertheless, *I* can say the following (indeed, I think it's *true*):

15. Numbers and people exist who are equally famous—π and Einstein are two examples.

Examples like this can be multipled easily.

Next, Gricean cancellations; consider:

16. There are cartoon characters and ordinary people who look like one another, although only ordinary people exist.[9]
17. Numbers and people exist who are equally famous—π and Einstein are two examples—although numbers, of course, don't really exist.[10]
18. There are more Greek gods than goddesses—not that there *are* supernatural beings.

[9] I discuss this phenomenon in Azzouni (2010a, p. 83), although I don't characterize it there in Gricean-cancellation terms, speaking instead of a "contrastive use." I'm thus now showing parallels between ontological commitments/denials and standard cases of implicatures, and using these parallels against published purported counterexamples to my view that ontological commitments and denials are pragmatic (not semantic)—I haven't tried this before. (Hofweber, 2016, p. 55, writes of my view that, "an articulation of an ontological question requires the help of a secret pragmatic ingredient." Implicatures aren't a well-kept secret, I'd like to think.)

[10] Some—e.g., Voltolini (2018, p. 101)—follow Evans (1982) and Walton (e.g., 1990), claiming that "really" is a device for indicating the simple truth of something uttered within a pretense. "Really," is *really* complicated, but despite that, there's little reason to think it regularly involves pretense—as what I've just written illustrates. Its effect in (17) looks similar to its effect on verbs generally. Consider: "Is John running on the track the way the coach required him to?" Response: "Yes, he's running…but he's not *really* running." What's indicated is his obeying the letter of the demand (he's on the track, he's moving quickly; he's not walking or sitting), but we wouldn't normally call *that* (skipping, say, or jogging) running. Pretense isn't involved.

What follows the hyphen in (18) is naturally said as soon as the speaker detects (or fears) someone's taking her as ontologically committing herself by what she's saying.

Cancelling an implicature of ontological commitment is complicated because the only way to do it is *with* a word or phrase that's normally used in constructions in which an ontological commitment implicature *isn't* cancelled. Nevertheless, as (16)–(18) show, it's natural, and, as (18) particularly shows, an ontological implication can be cancelled by using the same phrase "there are" *twice*, the second time negated. "Exist," however, is often the word of choice to cancel ontological implicatures, as when it's said,

19. There are hobbits (over there) that I'm hallucinating and that don't exist.[11]

Ambiguity views allow, at least in principle, that the order in a sentence of the same words or phrases bearing different meanings doesn't matter. On a view on which Gricean cancellations of ontological implicatures are involved, order *does* matter—in particular, the words or phrases used to cancel an implicature have to occur *after* the word or phrase with the implicature (to be cancelled) occurs. Although one *can* say,

20. There are numbers that do not exist,

in the right context (for example, where one is denying that complex numbers, say, exist), one *can't* say (in any obvious context I can pretty much think up),

21. Numbers do not exist, and there are numbers.[12]

In (20), the ontological implicature of "there are numbers" is cancelled by "do not exist." In (21), there is no cancellation. According to an ambiguity view on which "there be" and "exist" can be either lightweight or heavyweight, (21) is semantically admissible. On an ontological implicature view it's not admissible because no ontological implicature is cancelled in the first clause of (21), and so the statement looks simply contradictory. (Although ambiguists and polysemists can

[11] "That don't exist" isn't redundant (despite an earlier reference to hallucinations). I could, after all, hallucinate something that *does* exist, a friend I haven't seen for a while, for example.

[12] Variants of (21) don't seem to work either: "Numbers don't exist, although there are numbers" and "Numbers don't exist, but there are numbers" are bad; "There are numbers, although they don't exist," and "There are numbers, but they don't exist," are fine. (21) is due to Schaffer (2009, p. 358).

Here's an example, however, thanks to Hanoch Ben-Yami (email, July 16, 2019)–I've modified his example a little:

Pagan Child: Dad, I've read all these people believe in all these gods. Some believe in green gods, some in purple gods, and some in both. And our teacher told us that some of these gods really exist and some don't. Can you help me with this?

Pagan Dad: Sure, let's start with the Bughee panthcon. So, in the Bughee pantheon, there are green gods, and they really exist, but the Bughee have no purple gods. Next, the Mughee pantheon. Green Mughee gods don't exist, although there are green Mughee gods, in fact plenty of them; but they're sheer nonsense, forget about them. However, the purple Mughee gods do exist, you should respect them!

help themselves to Focus to explain some asymmetries—see Hofweber (2016) for discussion and references–Focus doesn't help with the cases I give here.)

This said, and despite (18), we generally can't use a phrase containing a word to cancel an ontological implicature that an antecedent phrase carries (in which an earlier appearance of that word appears). This isn't surprising since the same phrase twice in a sentence to both convey and not convey an implicature is usually confusing. "Penguins exist, although they don't exist" is awkward, as Asay (2010, p. 305) points out. Indeed, as he also points out, "Penguins exist, although they don't really exist" sounds awkward too. Asay (2010, p. 303) writes: "… it is unclear what it could possibly mean." That's false: context is crucial with cancellations of implicatures. Imagine a climate-conspiracy theorist who thinks penguins don't exist—they're, he claims, a (typical) invocation of nonexistent animals to stoke groundless extinction-event fears. He *could* say, also expressing intense aesthetic reactions to both real and (what he takes to be) unreal animals, "Famous animals that aren't cute exist—naked mole rats, penguins and Sphynx cats—although, contrary to what people think, penguins don't really exist." (Why "don't really exist"? The conspiracy theorist might think there's a common kind of fish that people pretend is a bird—a so-called "penguin." With a Trumpy belief like this, it's quite natural to say: "Penguins don't *really* exist.") Cancellations of implicatures are sensitive to context; we can't always comfortably cancel an implicature. This holds of ontological implicatures as much as any other sort of implicature. (I'll add that the story I tell about bare existentials in Sect. 16.6—that they're generalized conversational implicatures—applies to phrases like "Penguins exist," and "Penguins are real." This is why these phrases sound straightforwardly ontologically committing.)[13]

It's striking that phrases involving negations of "exist," "there be," etc., retain ontological implicatures so that they can cancel ontological implicatures of earlier uses of "there be" or "exist." Satisfyingly, this arises from a more general implicature phenomena: Straightforward negations often negate not the literal meaning of an expression but its implicated meaning.[14] Consider "John brushed his teeth and got out of bed." There is an implicature of temporal order: John brushed his teeth (first), and then got out of bed. But this can be cancelled:

22. John brushed his teeth and got out of bed, but not in that order.

---

[13] One result of the foregoing, however, is that we must *carefully* apply Grice's "detachability" implicature test. It looks like an implicature of "there are vampires," that vampires exist, because we can use "exist" to cancel: "There are vampires, but they don't exist." If we take "there are As" and "As exist" to mean the same, we have a problem because it's not an implicature of "Vampires exist" that vampires exist—Consider:? "Vampires exist, but they don't exist." (I owe this concern to John Collins, email June 10, 2019.) The response is: it *is* an implicature of "vampires exist" that they EXIST (ontological implicature retained in "exist"); we can use "don't exist" to cancel the ontological implicature of "there are vampires" because the ontological implicature remains in "don't exist"; but we can't use "don't exist" with "vampires exist" because two tokens of the same word "exist" runs interference—that's why we use "really exist."

[14] Theories about the interaction of negation with implicatures are controversial. That said, there are many examples, like the forthcoming (23), where what's negated isn't literal meaning.

The following negation of (22), however, is admissible:

23. John didn't brush his teeth *and* get out of bed—because he got out of bed *first*.

(23) negates the implicature of "John brushed his teeth and got out of bed," not what it says. Something similar happens with denials of ontological commitments. Consider,[15]

24. There are no blue unicorns.

This is ontologically dismissive. It's not, that is, an ontologically-neutral remark; and it's hard to see how to cancel this ontologically-dismissive implicature—if an implicature is what it is. Asay (2010, p. 303) concludes, instead, that one "can express negative ontological commitment in the vernacular…" where he means that "There are no…" denies an ontological commitment by virtue of its *semantics*. But, as (22) and (23) show, this doesn't follow. In a wide range of cases, negations negate implicatures, not literal meanings. That's why, "There are no blue unicorns," "Blue unicorns don't exist," "and "Blue unicorns aren't real," all read as ontologically dismissing. That's also why, as I indicated earlier, "doesn't exist" can be used to cancel an ontological implicature while retaining its own ontologically-dismissive implicature.

Not all the usage data distinguishes the implicature view successfully from its ambiguous and polysemous competitors. Consider:[16]

25. There are vampires in stories, but bats really exist.

Speaking to a child, I'll likely mean to be both ontologically committing to bats and ontologically *dismissive* of vampires. An implicature must be involved (on *anyone's* view) because no phrase appears in the sentence that conveys the ontological dismissing of vampires. According to ontological implicaturists, the ontological implication of "there are vampires" isn't explicitly canceled; rather, the canceling is implicated by "bats really exist" being contrasted with vampires. (Bats really exist in contrast to vampires in stories which really don't.) Can ambiguists say the same? It seems so: they're allowed implicatures; and in this case, the ontologically-committing reading of "there are vampires" is ruled out by an implicated contrast (because of "really exists") between bats and vampires—no contrast, however, can be computed if "there is" is read committingly. This raises a general empirical worry: Different semantic/pragmatic machineries can predict the same hearer-speaker intuitions. Although ambiguity views, I believe, are definitively sunk by conjunction reduction tests, polysemy views survive them. Ruling out polysemy calls for a more delicate analysis.[17]

---

[15] This example is due to Asay (2010).

[16] I owe this example to Eli Hirsch, given during his comments on *Ontology without borders*, Author meets critics, January 8, 2019, APA, New York.

[17] See Sect.16.7.

I've previously described my view as one in which the phrases "exist," "there is," etc., are "semantically neutral"—open to being interpreted (or not) as ontologically committing. I've lately realized this misleads because it sounds like there are semantic *slots* for serious ontology in these words and phrases (it sounds like semantic underspecification). Ontological implicaturists deny this. Consider the earlier mentioned explanation of the perceived role of tense in sentences like "John brushed his teeth and got out of bed," as an implicature and not as part of the semantics of "and" (as originally suggested by Grice (1987)). It's misleading to say "and" is *neutral* with respect to tense. Tense isn't involved in the semantics of "and" *at all*. No more is ontological commitment involved with words and phrases like "exist" and "there is," or (I claim) with *any* words or phrases of the vernacular.

16.3 Meinongianism

Meinongianism is the *metaphysical* position that some things don't exist, where the so-called things, nevertheless, have metaphysical presence. I'm using "metaphysical presence" to cover several possibilities ("Meinongianism" is a slippery family of positions). Some Meinongians distinguish levels of "being": things that don't exist nevertheless have being or subsistence, etc. Other Meinongians, accepting the dichotomy, exists/doesn't-exist, nevertheless require nonexistent beings to have properties of some sort. Importantly, Meinongian semantic theories use Meinongian objects as truth-makers for sentences about those objects.[18]

Those like myself who (in contrast) repudiate all such beings along *all* dimensions—i.e., who claim that something that doesn't exist can't have properties and can't be (genuinely) referred to—avoid semantic theories like this. To render sentences "about" nonexistents true (ones which "attribute" properties to nonexistents, e.g., "Mickey Mouse has black ears"), semantic theories that are interpreted ontologically neutrally must instead be relied on. These are, in effect, noncompositional—at least with respect to items that don't exist—although this doesn't prevent the use of truth-conditional semantic theories.[19]

Moltmann (2018 and elsewhere) claims: "There is considerable support for the Meinongian view from natural language," where nonexistent objects are "semantic

---

[18] E.g., Parsons (1980), Routley (1980), or Zalta (1983).

[19] See Azzouni (2010b), chapter 5, Azzouni (2017, appendix to the general introduction), or Azzouni (2012a). Those like myself—who think objects either exist or are in no way at all, unlike Meinongians—can't even *refer* to nonexistent objects because, on our view, there's nothing to refer *to*. See Azzouni (2013a) and Azzouni (2012b) for (i) discussion of this, (ii) for how one can nevertheless use "terms that refer to what doesn't exist," e.g., "Mickey Mouse," in true sentences, without, nevertheless, referring to anything, and (iii) for how one can finesse the ordinary use of "refer" which allows a "relation" to what doesn't exist. See Collins (2017, section 5) for a discussion of truth conditions and truth making broadly along parallel lines.

values" of words and phrases like "the king of France," "Vulcan," "there are," "some," "the building," as in (Moltmann's examples):

26. There are objects of thought that do not exist.
27. The king of France does not exist.
28. Vulcan does not exist.
29. Some buildings the guide mentions do not exist.
30. The building built there last year no longer exists.

This evidence *doesn't* show "natural language ontology" supports a Meinongian reading for these words and phrases. The problem is that the cancellation role, of "do-not/does-not/no-longer exist(s)" in (26)–(30), can also be played by "not that there *are*," as in (18) "There are more Greek gods than goddesses—not that there *are* supernatural beings"; and by "aren't real":

31. Some buildings the guide mentions aren't real.

Furthermore, as (7) indicates ("Although no one thinks Disney would allow this, sexually-depraved cartoon characters exist"), "exist" itself often occurs in ontologically non-committing utterances. The phenomenon, that is, that Moltmann explains by pressing "exist" into a semantic ontological-committing role (and "there be" into a semantic Meinongian non-existents role) is broader than this: other words have these roles too, and so *all* the relevant usage is better explained by an ontologically-committing implicature view in which *any* of the phrases, "exist," "real," "there be," etc., can occur in utterances that implicate existence, although none of them carry ontological commitment semantically. Further, her view that "there be" conveys a Meinongian status (that contrasts with existence) is belied by the cancellation role of "exist" in (26)–(30) also being played by phrases designating any Meinongian status *whatsoever*. Consider:

26* There are objects of thought with no properties,
26** There are objects of thought that don't subsist.
26*** There are objects of thought that don't exist, subsist, or have any sort of being at all.
26**** There are objects of thought that aren't the semantic values of any phrases of English.[20]

Similar astericked statements for (27)–(30) can be designed equally easily. The upshot: A presupposition of Meinongianism in "natural language ontology," specifically with respect to "there be," can't be established on the basis of usage patterns.

[20] (26****) is "paradoxical" *only given* the presupposition that the truth value of (26****) requires a truth maker targeted by "there is" for (26****) to be true—that is, that there *must be* a specific semantic target for "objects of thought" for (26****) to be rendered true (as opposed to a non-compositional, sentence-wide, truth condition for this); neutralist interpretations of truth-conditional theories show this is wrong (see, e.g., Azzouni (2010b), chapter 5).

16.4 When Do We Actually Ontologically Commit or Deny?

Let's be honest: we *pretty rarely* ontologically commit ourselves to things, or ontologically deny them. Proponents of Quine's criterion applied to ordinary language—and those who treat ontological commitment as arising from natural-language expressions (with certain meanings)—grossly overstate the occurrence of ontological commitments and denials among us. There *are* ontological debates, of course, where implicatures of ontological commitments *do* occur. God *exists*; God doesn't *exist*. In just those words. *There are* extraterrestrials among us. *There aren't* extraterrestrials among us; indeed, *there aren't* extraterrestrials *at all*. In just those words.

Needed is a distinction between (i) *implicating* ontological commitments or denials and (ii) saying something true about something we *are* (or *aren't*) ontologically committed to, but implicating nothing one way or the other about this. I think cars exist. I also think cars were invented by Karl Benz. I don't think Mickey Mouse exists. Nevertheless, I *do* think Mickey Mouse was invented by Walt Disney. (This is a truth about Mickey Mouse; it's not a truth about something else—Mickey-Mouse cartoons or Mickey-Mouse thoughts.)

If we can accept (hold, believe) truths about things that don't exist—or that we would deny exist (if we were asked)—that creates space between implicating ontological commitments to something and saying something true about something to which one *has* an ontological commitment (but implicating nothing about this); similarly, there is space between implicating an ontological denial and saying something true about something that one thinks doesn't exist. To say (for example) that As were invented by B isn't, by saying *that*, to implicate an ontological commitment to As (or to B).

This is a natural ("intuitive," as philosophers say) distinction. *How* natural is disguised somewhat by talk of implicature because, as I indicated earlier, some implicatures are heard as asserted and some aren't. (I hypothesize that generalized conversational implicatures are usually heard as asserted but particular conversational ones usually aren't.) Here's how the distinction plays out with ontological commitments and denials. "God *exists*," I say (with an emphasis or in a context—during a certain kind of lecture, say—that indicates I mean this *ontologically*). I can also say similarly that "extraterrestrials *don't exist*," and thus deny an ontological commitment. These are *heard* as asserting ontological commitments or denials. But consider:

32. Some eggs are in the refrigerator
33. The raccoons around here will eat garbage if it's left out,
34. A cow is in the barn.

These *don't allow* a stress that enables an *easy* implicature of commitment; similarly, these aren't used to strike positions in ontological debates. We don't (normally) hear them as asserting—ontologically—*anything*. I can certainly stress that *some* eggs are in the refrigerator, that *the raccoons* around *here* will eat garbage if it's left out, or that *a cow* is in the barn. But these aren't the kinds of stress that convey ontological commitments or denials: "There *are* unicorns," "Vampires *aren't* real," "Some superheroes *are* real," and "Shrees aren't *things*" clearly implicate ontological commitments (and denials). And they're *heard* as asserting ontological commitments (and denials).

Indeed, it looks like no genuine natural-language quantifiers[21] can be easily used to implicate (and thus stake out) ontological commitments or denials. As preliminary evidence, consider two teenagers debating the existence of extraterrestrial aliens while sitting in a livingroom (looking out onto a lawn); suppose during this debate a rocketship lands on the lawn (We've all seen movies like *this*.) The proponent can slyly respond to the naysayer, rolling her eyes, "Extraterrestrials are on the lawn *again*." Although this conversationally implicates an ontological commitment to extraterrestrials—it's not like "There are extraterrestrials." It sounds like an insinuation of commitment and not an assertion of it, although it sounds like an ontological assertion to say (with stress) "There *are* extraterrestrials."[22]

Here's more evidence. Consider,

35. Most ways of getting around this are awkward.
36. A futile attempt is a sad thing to see.
37. Vampires are frightening.
38. Bears are in the yard again.

We must distinguish ontological debates from other debates ("bears *are* in the yard again—look over *there*") that aren't ontological debates because ontological commitments or denials aren't implicated. Because ontological commitment is only conveyed by implicatures—which are subtle—it's easy to *not* realize if an ontological implicature is in play. Nevertheless, the distinction is real and easily grasped with a little care; it's also an important empirical data point.

Some locutions in which "exist," "there is," and "real," occur implicate ontological commitments or ontological denials either straightforwardly or with stress. Can other words occur in utterances that implicate ontological commitments or denials in this way? Consider, again, (35)–(38). None of these (normally) implicate ontological commitments or ontological denials. We can't stamp a foot, and say: "Vampires are frightening," or,

39. Most bears are frightening,

and in this way make an *ontological* claim. Most speakers *do* presuppose there are no vampires and that there are bears; but neither stance is generally conversationally implicated, or otherwise conveyed, by these sentences. Notice that,

40. Most vampires are frightening,

---

[21] "Real," "there is," and "things" aren't examples of natural-language quantifiers; "a few dogs," "most cats," "seven cows" and "penguins" *are* examples.

[22] Again: I hypothesize that "There are extraterrestrials" carries a *generalized* conversational ontological implicature, and "There are extraterrestrials on the lawn again," in this context, carries a particular conversational ontological implicature. I discuss this further in Sect. 16.6.

is as acceptable to say as (39)—no cancellation of the existence of vampires is needed—that's because neither (39) nor (40) implicates ontological commitments to begin with. Consider,

41. Some bears are frightening.

This, for the same reason, doesn't *convey* ontological commitment either. We can say, as well,

42. Some vampires are frightening.

Similar examples show that none of natural-language quantifiers, "A few As," "some Bs," "a C," "most Ds," etc., can be used to straightforwardly implicate ontological commitment. "Exist," "real," or "there is," are needed.

Proper names are the same. Just as we can say,

43. Mars is a planet,

we can say,

44. Juno is a goddess,

or

45. Dracula is frightening.

In none of these cases is ontological commitment implicated. But consider,

46. Mars is a planet, but it doesn't exist.[23]

Something is wrong. Balaguer suggests that there is a strong intuition of inconsistency (personal communication), but that's wrong. We *do* sense something off, and we know we presuppose Mars exists—that invites Balaguer's diagnosis that the presupposition that Mars exists contradicts the second clause. But since (44) and (45) are fine, this *can't* be the explanation. *We* know that Mars exists, but that isn't a presupposition of the *sentence*, "Mars is a planet."

Actually, that (46) "sounds bad" strikes me as simply lack of imagination. "But" demands a contrast; without it the utterance sounds weird—regardless whether what's talked about exists or not. After all,

47. Juno is a goddess, but she doesn't exist,

sounds off too. On the other hand,

48. Mars is a planet that doesn't exist,

is fine. (We *disagree* with (48); but it's not "weird" the way (46) is.)

Provide (46) with an appropriate background that gives "but" something to do, however, and it's fine. Imagine a debate over the kinds of things that people mistakenly think exist. B thinks magical beings and artifacts are examples, but she

[23] This example is due to Balaguer (2018).

(ignorantly) claims no one thinks this about planets. Many of us would say: "Vulcan is a planet people believed in, but it doesn't exist." A very confused/ignorant person, however (who, say, thinks he has recently discovered something about Mars that no one else knows), could respond to B—enthusiastically thinking he's refuting her: "*Mars* is a *planet*—but it doesn't exist."[24]

The foregoing usage evidence shows very few turns of phrase can be used to ontologically commit and ontologically dismiss: "there are Ps," "Ps are real," "Ps exist," and "there aren't Ps," "Ps aren't real," and "Ps exist." This, in turn, supports these phrases carrying *generalized* conversational ontological implicatures. Indeed, even tensed modifications of "exist" and "there is," e.g., "now exists," "did exist," "there were," etc., *don't* carry generalized conversational implicatures. Consider:

49. A state of emergency now exists, but it doesn't really exist.

This reads as bizarre because we can't cancel an ontological-commitment implicature if there's no implicature there to be cancelled. Of course, the speaker doesn't presuppose that states of emergency (really) exist or states of emergency (really) don't exist. Rather, the question of the ontological status of a state of emergency (or states, generally) wouldn't naturally occur to *anyone* (unless he had to consider it because, say, he was surrounded by aggressive philosophers). The same phenomenon is exhibited by.

50. Tornadoes did once exist in Alabama, but of course tornadoes don't really exist.
51. There once were tornadoes in Alabama but of course tornadoes don't really exist.

---

[24] To my ear, "Dracula is frightening, but he doesn't exist," is *fine*. That's because it's easy to imagine this being said to reassure a listener by stressing something he can be taken to know, but (because of irrational fears) nevertheless needs reassurance about. Imagining backgrounds carefully can dispel impressions of infelicity (although *not* always).

Marianna Antonutti has pointed out to me (July 7, 2019) that "The present King of France doesn't exist" is okay, but "Jacques is the present King of France, and he doesn't exist," isn't ok. Similarly, "Mars doesn't exist," is okay; even "Mars is a planet that doesn't exist," is okay (although we think these are false). "Juno is a goddess that doesn't exist," however, sounds weird. Antonutti suggests that our uses of certain words "planet," "goddess"—because of background assumptions—triggers expectations that what we're talking about is real (or isn't). These, however, aren't *ontological implicatures* of the sentences we use these words in. If we say, "Juno is a goddess," the expectation is that Juno isn't real (but that's not an implicature). Thus, to then add that Juno doesn't exist sounds redundant (violating a Gricean maxim). That's not true with "Mars is a planet"—to go on to claim *it* doesn't exist isn't similarly redundant (because the background assumption is that it *does* exist). This shows (again) that what makes Balaguer's example weird isn't the sheer denial of Mars' existence but the use of the word "but." This sets up an expectation that "Mars is a planet" is about to be implicitly hedged or qualified in some way—but the nonexistence of Mars isn't a hedging or qualification of its being a *planet*. That's why we need to construct a complex background (involving people who are weirdly ignorant or with strange beliefs) against which these utterances are appropriate.

(50) and (51) read bizarrely (unlike (18)). A tensed "there is" functions like a tensed "exist"—it doesn't (naturally) help to implicate ontological commitments in these kinds of constructions.[25]

52. There once were tornadoes in Alabama, but tornadoes don't exist there any longer,

is fine. There are no ontological implicatures in (52), and so, none are being cancelled.

Now consider,

53. The demonstration is taking place.
54. The accident we were all fearing happened.
55. This law is valid.
56. The accident John mentioned did not occur.
57. The state of emergency that Trump declared only obtains in Kentucky.

Moltmann (2018) claims that "obtain," "occur," "happen," are "existence predicates," and consequently, that (53)–(57) are ontological commitments—although to particular sorts of (Meinongian) entities.[26] Given the implicature role of "exist" and "there be" have already been established, the important question is whether sentences containing "obtain," "occur," and "happen," implicate ontological commitments or denials as "there be" or "exist" do. If not, Moltmann's grounds for claiming natural language (English, anyway) treats states of emergency as a kind of Meinongian entity that "obtains" (but doesn't "exist") are slender. Consider:

58. The state of emergency still obtains.
59. The accident occurred last year.

and now consider these attempted Gricean cancellations:

60. The state of emergency still obtains, but it doesn't exist.
61. The accident occurred last year, but it doesn't exist.
62. The state of emergency still obtains but it never really obtained.
63. The accident occurred last year but it didn't obtain.

[25] Consider, for example, "There were objects of thought that don't exist." This reads badly. Even "There were objects of thought that didn't exist," reads badly. We don't tense generalized conversational ontological commitments and denials.

[26] Moltmann (2018) stipulates that "existence predicates differ from ordinary predicates in that with a non-referring subject (or rather a subject term not standing for an actual object), they generally yield true sentences if they are in the scope of negation … and false sentences if they are not in the scope of negation." Just as "There are objects of thought that do not exist," and "The book John is thinking about does not exist," mark "exist" as an existence predicate, according to her, so too, "John's speech never happened," "Trump's state of emergency doesn't obtain," "This law is not valid," and "Hillary Clinton's victory speech never occurred," mark "happen," "valid," "occur," and "obtain," as "existence predicates." According to Moltmann, corresponding to these predicates are various kinds of "entities" with being but not existence. I've already argued against using this contrastive property of "exist" as a mark of ontological meaning, as Moltmann does, because it fails to distinguish "exist" from "there be" or "real."

None are felicitious. Consider these other attempts at Gricean cancellations:

64. The law is valid but isn't real.
65. The law is valid but doesn't obtain.

I can multiply failures *ad nauseam*, but perhaps it's better to give up now. Conclusion: Even though usage shows that there are selection restrictions on verbs like "exist," "happen," etc., so that the nounphrases that take some of these verbs don't take others, there is no reason to think this provides any evidence that these words are used to convey metaphysical commitments—to entities of any sort. Our only evidence of ontological implicature occurs with certain sentences containing a very narrow swathe of words, "there be," "exist," and "real."

16.5 What Is Said by "There are Greek goddesses"

Consider this worry (due to John Collins, email, June 19, 2019): Speakers use what is said, along with background and other factors, to "compute" implicatures. Presumably, "John brushed his teeth and got out of bed," truth-functionally joins brushing teeth and getting out of bed as its linguistically mandated content—the temporal order is implicated. Now consider,

66. There are many Greek goddesses,

and let's take it, as the implicature view prefers, that (in the right contexts) this implicates ontological commitment that isn't part of its content. Question: What *is* its linguistically-mandated content? What *is* being said?[27]

This is hard to make out. Let's note first that linguistically unsophisticated speaker-hearers badly fumble demands to express what is said. B says (66). If C says, in response, "Oh wow! You're a pagan? That is *so* cool," B is likely to say something like, "Of course not. I'm talking about myths." Now C presses further, "Well, what did you mean when you said, "*There are* many Greek godesses? You did say *there are* these things, right?" B won't respond by telling C *what is said* by "There are many Greek goddesses." B instead will *backtrack*, try to say what she said *without* the phrase "there are."[28]

Why isn't what is said easily accessible to speaker-hearers in these cases? Relative inaccessibility of even major aspects of meaning *is* generally recognized as

[27] The concern can (and has been) raised directly. Asay (2010, p. 308) writes: "Azzouni leaves us with a puzzle about what bare existential judgments like 'There are Arctic penguins' could possibly mean, and how they could be true."

[28] I've noted in Azzouni (2007, p. 216, 2nd paragraph of footnote 38) that this backtracking occurs; I've diagnosed it as a shift in context—as I would put it now, a shift from a context where the implicature was canceled (by background considerations) to one where it's not—a shift that the wily B has trapped C into not seeing has occurred.

typical (despite these often being easily elicited by prompting).[29] *What is said*, is, therefore, no different. In this case, speakers clearly know what *isn't* being said; their verbal behavior—under pressure—indicates that *what is said* doesn't involve ontological commitment or denial.

But what *is* being said? (26*)–(26****), and other sentences formed with "exist" and "real," like "Objects of thought exist that don't subsist," "Some unreal objects of thought don't subsist," etc., show there isn't some ontological status of things that's conveyed by what is said; nor that what is said is that such things are in "a domain of discourse." (The latter invites the thought that *there are* things in a domain of discourse.) Instead, a linguistic condition is being given about a certain sentence. Consider, "There are gods." What is said is that "There are gods," belongs to the (vaguely designated) discourse being undertaken—we can now use "gods" in the discourse in a positive way, e.g., "there are gods," is *true*. This is how I want to interpret B's protest above—that she's talking *about* myths (and saying something true "about" them).

The formulation I've just given is *metalinguistic*. It's about the presence of a term in a discourse, and how (so far) it can be used; and so this can't *literally* be what is said either. What is said by *there are many Greek goddesses*, where no ontological status of any sort is given to Greek goddesses, is the *object-language correlate* of the metalanguage remark: "The sentence, 'there are many Greek goddesses', is true." I put what is said metalinguistically, in cases like this, because *we have no other way of saying what is said in natural languages*. This is one reason why what is said by these phrases is elusive to speakers (both philosophers and others), and why speakers unsuccessfully attempt saying something different (try to "paraphrase") when they're challenged. They *know* what they *weren't* saying: they weren't saying anything ontologically committing, about any sort of ontological status at all—they weren't even thinking of anything like that. (And their grasp of meaning in this case is sufficient for them to understand how the ontological implicature contrasts with what is said.) But this *doesn't* mean speakers can formulate, in a positive way, what they *are* saying—other than by repeating the remark again; and *that's* ruled out because they're now in a context where ontological commitments are being implicated by these phrases. The same phenomenon occurs, I claim, with "exist," "real," and other words within sentences that implicate ontological commitments and denials.[30]

[29] My favorite example is "house." Chomsky (2000, pp. 35–6) describes many aspects of "house" that nonspecialists overlook, including, "that a brown house has a brown exterior, not interior … if I see the house, I see its exterior surface, seeing the interior surface does not suffice … If Peter and Mary are equidistant from the surface—Peter inside and Mary outside—Peter is not near the house, but Mary might be … the interior [of a house] is abstractly conceived; it is the same house if I fill it with cheese or move the walls—though if I clean the house I may interact only with things in the interior space, and I [may be] referring only to these when I say that the house is a mess or needs to be redecorated …. If my house used to be in Philadelphia but is now in Boston, then a physical object was moved … in contrast [to] my home …."

[30] Three points. First, there is *more* than an analogy with a view Carnap held at one time about "pseudo-object sentences." See Carnap (1937, § 74). (My thanks to Benjamin Marschall for this

One may worry that my gloss on what is said is (unhelpfully) circular. I've given "e is true" as a characterization of the meaning of e. (And, so, I can imagine some philosophers saying: "*Really*?") But, as I noted in footnote 19, what's called for with nonreferring expressions is a semantics that doesn't *require* truth-making individuals, such as Santa Claus, and the like. Instead, truth-inducing plays a role (although not in the *semantics*—which is exhausted, in this case, by the Tarski biconditionals). The sentence, "Sherlock Holmes is smarter than Daffy Duck," is truth-induced by *several* fictional discourse-practices—ways that we tell stories, cartoons, etc. The same is true of "existence" statements: truth-inducing relates to discourses, even though the truth-conditional semantics of such statements doesn't. Speakers understand this much about such statements—enough to say, for example: "I was talking about *myths*."

16.6 Bare There-Existentials

Collins (forthcoming) notes that there-existentials "are neutral with respect to fictive/mythic and non-fictive/mythic discourse." In particular, the following are natural utterances for disbelievers in such things (what follows are all Collins' examples):

67. There are several ghosts in Poe.
68. There are many Roman gods who are female.
69. There is a Greek god with wings.

But, he points out, "bare *there*-existentials—those that lack modifiers or codas—are much less happy with fictive/mythic pivots given accepted ontological commitments":

70. There are orcs.
71. There are several ghosts.
72. There are many Roman gods.
73. There is a Greek god.

These are usually unacceptable to utter if one doesn't believe in orcs, ghosts or (pagan) gods.[31]

reference, July 8, 2019.) Second, one thing "There are many Greek goddesses," *doesn't say*, despite being metalinguistic, is that there are many specific *names* N such that "N is a Greek goddess," is true. That's true in this particular case, but not, say, of "there are many orcs in Tolkien's fiction." (My thanks to Roy Cook, July 7, 2019, for pressing me on this.) Third, the content of ontologically-*committing* expressions is (pretty much) equally inaccessible to nonprofessional speaker-hearers. "What do you mean by 'exist'?" when posed directly to someone will get answers in the neighborhood of: "You know, like *out there*."

[31] We can, perhaps, cook up specific backgrounds where utterances of some of these are acceptable. Someone memorizing Roman gods, for example, for a test, after many hours of studying, could burst out with: "There are (*way too*) *many* Roman gods." (Thanks to Hanoch Ben-Yami, July 8, 2019, here, for a—slightly different—version of this example.)

Collins also notes that the following inferences seem to fail:

74. There are several ghosts in Poe; so, there are ghosts.
75. There are many Roman gods who are female; so there are Roman gods.
76. There are holes in the dress; so, there are holes.
77. There is a way to avoid the problem; so, there is a way.

These are unhappy if read (as they naturally are) with ontologically-committing consequents.[32]

Collins (forthcoming) explains all this in a way that fits either polysemy or implicature views. He writes:

> The reason the bare existential is not good for talking about fiction or myth or make believe, etc. is precisely because it is bare, and so, out of the blue, fails to provide any fictive/mythic background to localize the claims. If such a background is presupposed, then the bare existential is fine. In other words, the distinction is not really a linguistic one, but a pragmatic one, concerning the background assumptions required to differentiate between talking about what there is in some *punkt*[33] sense, from what there is relative to some discourse. Language does not encode this difference itself; it merely provides the means to express the *punkt* commitment free of any background assumptions, precisely because no such assumptions are required.

Similarly, with the inferences (74)–(77), "the bare existential default [in the consequent] asserts a global ontological commitment" but this isn't true of the antecedents.

I tell this story with conversational implicatures; thus I focus on background contextual assumptions as Collins does. As is well-known, implicatures don't have to be explicitly cancelled if the context does it for us. I don't have to cancel the (generalized) conversational implicature of (1) that John jumped *off* the cliff if there's a background discourse assumption that we're discussing dangerous jumping exercises at edges of cliffs. So too, just as implicatures are often retained when embedded within negation, they are also often retained when embedded within conditionals—and for the same reason (because they're heard as asserted).

The typical calculations using Grice's maxims seem at work as well. In Grice's example (footnote 6), where it's said, "John went into a house yesterday and found a tortoise inside the front the door," the thought is, "if it had been John's house she would have said so." Similarly, for "There are vampires," one thinks, "if he believed in vampires, he would have said so."

Furthermore, Grice's distinction between generalized conversational implicatures and particular conversational implicatures bears directly on these cases. (70)–(73) are all generalized conversational implicatures; the expression is heard, in Grice's (1987, p. 37) language, "normally (in the absence of special circumstances)

---

[32] Hofweber (2016, p. 12) writes of the same inference form with respect to properties, numbers and propositions (e.g., "There is a number of moons of Jupiter, so there is a number," "There is a property that Fido has, so there is a property," and "There is a proposition that 'Fido is a dog', so there is a proposition") that, "…let's be honest: are there really any easier and more compelling arguments anywhere in philosophy?" My ears beg to differ.

[33] This is German for "point." So I read Collins as saying, "in some pointed sense."

[to] carry such-and-such an implicature or type of implicature." That's why bare existentials are unhappy with fictive/mythic pivots; they need special circumstances to cancel the ontological implicatures they'd otherwise carry. On the other hand, "There are cartoon characters and ordinary people who look like one another," "Numbers and people exist who are equally famous," and "There are more Greek gods than goddesses," are *not* unhappy with fictive/mythic pivots the way bare existentials are; that's because they don't carry generalized conversational implicatures.[34]

16.7 Polysemy (More of an Empirical Challenge?)

Call "polysemy" (following Collins, 2017, pp. 678–679) the possession of ontologically-distinct but otherwise intimatedly related meanings—"intimately related" in the sense that the meaning usages in question *fail* the standard tests for ambiguity.[35] One can have a polysemy view of "exist" and "there be," think that "exist" and "there be," respectively have two (or more) of the same related meanings, and that attempts to distinguish them fail the standard tests for ambiguity.[36] Empirically comparing polysemy and implicature views about "exist" and "there be" is correspondingly hard. I'll, nevertheless, try to do so here—to the detriment of polysemy.

I'll start by summarizing the distinctive features (and resources) of the implicature view. First, it treats "there be," "exist," and "real," barring grammatical restrictions, as on a par. They all can be used to implicate ontological commitment; none of them carry commitment semantically. Second, I take on-board the standard

---

[34] Several years ago, Jeff McConnell argued (in conversation with me) that "Apples are real," is hard to hear as not ontologically committing. He took that to be evidence that 'real' has an ontologically-committing reading. Rather, that kind of utterance also carries a generalized conversational ontological implicature.

[35] An example is "London is large, polluted and unfriendly," where London *qua* geographic area, London *qua* atmosphere, and London *qua* many *many* unfriendly people are nevertheless copredicated. As Collins (2017, p. 689) notes, "copredication is an under-investigated phenomenon." It's also somewhat controversial since some examples, e.g., names of cities and persons, are ones where opponents claim there aren't "ontologically-distinct" meanings in play. I'm avoiding these details—see Collins (2017) for a good discussion of the issues.

[36] Hofweber (2016) expressly commits himself to a polysemy view of natural-language quantifiers, where they have a domain-conditions reading and a inference-role/substitutional reading. I'm not sure he's entitled to describe his view as a polysemy one. He never mentions ambiguity tests (and his favorite example of polysemy (63), "Before I get home I have to get some beer to get drunk" flunks the conjunction reduction test). He characterizes the distinction between ambiguity and polysemy in terms of word identity ("bank," as used for financial institutions or sides of rivers, is two words; "get" as in his example, is one). But this tool is useless for drawing a distinction in actual cases. Lastly, the meanings of quantifiers in terms of the presence of something in a domain or in terms of an inference to another sentence are different enough for it to seem that the suggestion is ambiguity rather than polysemy.

Gricean distinction between generalized conversational implicatures and particular conversational implicatures.[37]

I do, however, introduce a fresh wrinkle by distinguishing between what is said (the linguistically-mandated content of an expression) and what is *heard* as asserted. The latter is a matter of how we *experience* expressions.[38] My hypothesis (which has been satisfied—as far as the examples in this paper are concerned) is that this distinction explains why negating expressions sometimes negate implicatures and sometimes what is said: we always hear the negation of an expression as negating what is asserted. The utterance of an expression carrying a generalized conversational implicature is generally heard as asserting the implicature, and so a negation of that expression is heard as negating the implicature. Uttering an expression carrying a particularized conversational implicature isn't heard as asserting the implicature (it sounds, instead, like an insinuation or a "implication"), and correspondingly, negating the expression isn't heard as negating the implicature.

This neat package explains several puzzles about "exist" and "there be." It explains:

(i) why negations of "exist," "there be," and "real" easily cancel ontological implicatures usually carried by "exist," "there be," and "real." Simple sentences starting with phrases like "there are Ps …," "Ps exist …," and "Ps are real …," all carry general conversational implicatures; "don't exist," "aren't real," and "there aren't," are all naturally heard as negating ontological implicatures. This is the mechanism behind ontological implicature cancellation.

(ii) Why, nevertheless, is "doesn't exist" the phrase of choice for cancellations? Correspondingly, why is "exist"—to the native ear—more closely associated with the ontological notion than "there is"? The answer is purely grammatical. Although (as (18) "There are more Greek gods than goddesses—not that there *are* supernatural beings,") shows, "not that there *are*" can be used to cancel an ontological implicature, it's hard, grammatically, to use it to cancel an ontological implicature carried by "exist." Consider (20), "There are numbers that do not exist." It's *grammatically awkward* to flip this so that a negated use of "there be" cancels the ontological implicature carried by "exist." This is because cancellations have to *follow* phrases with the implicatures to be cancelled. In addition, in this case, we need to put "exist" into adjectival form. The result is the hideous, "Existing numbers, there aren't."[39]

(iii) As I showed in Sect. 16.4, we hear a distinction between actually ontologically committing or denying, and cases where we're doing something else, and this

[37] With a caveat, see footnote 6

[38] See Azzouni (2013b, especially chapter 3) for further discussion. The literature tends to assimilate the theoretical notion, *what is said*, to what we hear (or see) as *asserted*. That these come apart is important.

[39] Which, to my ear, *almost* works; it *almost* sounds like an assertion of nominalism!

distinction seems not semantic but pragmatic.[40] One wants, however, to explain why certain phrases, e.g., bare existentials, seem, by default, to carry ontological commitment, while other phrases, e.g., "many Ps," "A few Ps," etc., don't. The distinction between generalized and particular conversational implicatures is what's needed.

The foregoing (discussed in Sect. 16.4) undercuts views that take natural-language *quantifiers* to be polysemous. This is because there is *no* reading—other than with heavy background assumptions that allow particular Gricean implicatures—on which these expressions have ontologically-committing readings (recall the extraterrestrials-on-the-lawn case). Still left standing is a narrowed polysemous view of "there be," "exist," and "real."

Can so-limited polysemy views manage the usage data that the implicature view handles? The primary problem for polysemy are *multiple readings* that we *don't* hear (this includes utterances where we don't hear a reading *at all*—even though polysemy views predict we should). To manage these embarrassing expressive riches, polysemy views need additional resources. I'll illustrate the issue in what follows.[41] Consider, here repeated,

19. There are hobbits (over there) that I'm hallucinating and that don't exist,
20. There are numbers that don't exist.
14. There are cartoon characters and ordinary people who look like one another.

Consider (19) and (20) first. Suppose a polysemy view that takes there to be *two* closely related meaning, for "there be" and "exist" (it's worse for views that take there to be more than two). *Prima facie*, we therefore have *two* (non-contradictory) readings for each of (19) and (20) and not just the one (for each) that are heard. The implicature view rules out the extra reading because the occurrences of "don't exist" have to be read as carrying the ontological implicature cancelled in, respectively, "there are hobbits" and "there are numbers." What prevents, on the polysemy view, the reading of (20) on which "there are" has the ontologically-committing reading, and "don't exist" carries the non-committing reading?[42] One can invoke the sugges-

[40] Ironically, Quine's coinage "ontological commitment" sounds like a label for a *speech act*: surely *Quine* had nothing like that in mind.

[41] I raise this objection to polysemy (and ambiguity) in Azzouni (2017, section 3.4), although my discussion of examples here is different. My thanks to Thomas Hofweber for spirited conversation on July 7, 2019, that forced me to revisit the issue. He remains utterly unconvinced. (Philosophy doesn't seem to be about convincing people of anything.)

[42] One maneuver open to the proponent of polysemy is to link the related meanings, so that the ontologically-committing meaning of existence/there-be implies the non-committing meaning. This is, generally, not how these meanings have been understood. Hofweber's (2016) view, presumes, in general that the non-committing inferential-role meaning, and the committing domain-conditions meaning aren't linked this way. Although, the latter *could* be treated as implying the former, it's not typical to think that, in general, everything in a possible domain of discourse can be individually named or described. So too, on more ordinary Meinongian views, it's not (generally) thought that items that exist also have being or subsist. These are seen as exclusively disjunctive.

tion ((ii) above) that "exist" is more closely associated with the ontologically-committing reading than "there is";[43] but as (18) indicates, this reading isn't mandatory; and so the other reading should be *hearable*—certainly if prompted. I say it isn't.

(14) raises a different—although related—problem. This is that an ontological-committing reading isn't heard *at all*. On polysemy views, *four* possible readings are hearable, but (I suggest) only the reading in which *there is no ontological commitment whatever* can be heard. On the implicature view, this makes sense because although this is an expression ("there are ...") that carries a general ontological implicature, it's cancelled by our background knowledge that speakers don't believe in cartoon characters.[44] In general, ontological commitments are heard only in certain cases; polysemy views make them hearable far more often; indeed, the view seems to suggest that we should be able to hear them all the time—even with sentences like (33), The raccoons around here will eat garbage if it's left out," and (34), A cow is in the barn, where (barring peculiar circumstances that can trigger particular conversational implicatures), we don't hear them at all. So, neither the general quantifier version of the polysemy view nor the narrowed polysemy version survive scrutiny.

16.8 A Loose End: Fine on Generics

Regimentation ("conceptual engineering") comes in both Carnapian and Quinean flavors. Carnapians replace natural language with a language artifact—in fact, *many* of them. Quineans remain in natural languages although they change (some of) their syntactic/semantic/logical properties. In both cases—at least in practice—there seems to be an implicit assumption that pragmatics, specifically implicatures, can be dropped: there is no good reason to construct or modify a language, and then use the resulting sentences nonliterally.

Assuming this is right (although I have doubts which I'll not discuss now), when we characterize ontological commitments and denials in a formal setting, one approach keeps the quantifiers non-ontological, but introduces an "existence" predicate E that explicitly (semantically) conveys ontological commitments and denials (as I suggest in Azzouni (2004)). Doing things this way (although it isn't the only way) is tempting because—in the vernacular—the word of choice by far for cancelling ontological implicatures is "exist." Be that as it may, Fine (2009) has challenged the Quinean approach to formalization. I now show that his challenge

[43] Suggested to me by Thomas Hofweber, July, 7, 2019.

[44] Similarly, we wouldn't hear the implicature, in (1), "John went to the edge of the cliff and jumped," if we didn't think that anyone ever jumped *off* of cliffs, but instead there was a widespread practice of jumping up and down at cliff edges. It wouldn't occur to us to hear it as suicidal, even if we did hear that reading with variants like "John went to the edge of the roof of the building and jumped," where there was no such weird-exercise practice on roof-edges.

amounts only to the charge that the Quinean refuses to distinguish between "There be" and "exists" in the vernacular. The Quinean *does* refuse distinguish these, and responds: Why should I? We have to now introduce considerations that Fine doesn't mention.

Consider the sentence,

78. Integers exist.

This, I've argued, ontologically implicates integers. How should it be rendered in a first-order formalism in which natural-language ontological implicatures have been semantically formalized? Fine suggests that the English statement (78) entails

79. Natural numbers exist.

But rendering (78) as $(\exists x)\mathrm{I}x$, and (79) as $(\exists x)\mathrm{N}x$ doesn't honor this entailment since $(\exists x)\mathrm{I}x$ doesn't imply $(\exists x)\mathrm{N}x$; indeed, if anything, the entailment goes the other way.

(78) is a generic, and indeed, a generic with complications beyond those of the usual generics. Understood purely as a generic, (78) allows exceptions just as,

80. Lions are tawny,

allows albino lions or ones of other colors. Indeed, a generic can be true in circumstances in which most of the entities described *don't* have the property in question. The complication talk of integers (and natural numbers, for that matter) introduces is that a variant of (80) is,

81. The lion is tawny.

These are typical generic-transformations, "The grasshopper is loud," "Grasshoppers are loud," "The dolphin is intelligent," "Dolphins are intelligent," are all acceptable. However,

82. The integer exists,
83. The natural number exists,

are odd. (They're acceptable, but sound grand—the sort of thing a philosopher might *proclaim*.) More natural are:

84. The integers exist,
85. The natural numbers exist.

Notice that typical generics *don't* allow a similar construction:

86. The lions exist,

and similar modifications of ordinary generics are unacceptable.[45]

---

[45] Collins (email May 31, 2019) notes that "definite generics … don't admit the plural with individual-level predicates," thus (86) is bad. He suggests, however, that there are contexts in which (82) and (83) are perfectly fine. Imagine one is presenting a standard construction of the numbers in ZF,

Regardless, we seem to have three possible readings of (78) "Integers exist." First there is

87. (All) integers exist,

second, there is the generic reading on which some but not all of those integers need exist. And the third is the already mentioned alternatives (84) "The integers exist," and (82) "The integer exists." Neither of the latter readings, to my ear, imply (87). A discovery that the integers exist, surely, is compatible with the simultaneous discovery that *not all* integers exist; similarly for the discovery that the integer exists. ("Sure, the integer exists, but not the even ones," is acceptable.)

How should regimentations of these readings of (78)—ones that eliminate ontological implicatures by incorporating them into the semantics of the translated expressions—go? (87) "(All) integers exist," is straightforward:

88. $(x)(\mathrm{I}x \rightarrow \mathrm{E}x)$.

Similarly, "(All) natural numbers exist" is rendered:

89. $(x)(\mathrm{N}x \rightarrow \mathrm{E}x)$.

(88) entails (89) given that $(x)(\mathrm{N}x \rightarrow \mathrm{I}x)$, which is true.

As Fine (2009, p. 166) notices, (87) is hard for the Quinean to *literally* render formally because the Quinean makes do only with the existential quantifier. But the Quinean has a response. This is that "All integers exist" is *genuinely distinct from* "An integer exists," or "There is an integer that exists" *only provided* something else holds that the Quinean rejects. "All integers exist" is presumably the denial of "There is an integer that doesn't exist"—$(\exists x)(\mathrm{I}x \,\&\, \neg\mathrm{E}x)$—which the Quinean treats as nonsensical. The Quinean, that is, denies that "Some integers exist and some don't," is *ever* cogent. Only if it *is* cogent does "All integers exist" express a different proposition from "there is an integer that exists."

What can force the Quinean to accept that "Some integers exist and some don't" is cogent? On the face of it, *nothing*. The English locution presupposes that there is a distinction between what there is and what exists, and the Quinean rejects this. We need an argument that the Quinean can't reject this, and it can't merely be that we can say in English, "Some integers exist and some don't." Not *everything* that we can say in English need be captured by conceptual engineering. It has to *matter* for other things we *need* to say that we can say this in English.

Fine (2009) suggests that what matters is that (78) "Integers exist," entails (79) "Natural numbers exist," but $(\exists x)\mathrm{I}x$ doesn't imply $(\exists x)\mathrm{N}x$. (78), as I noted earlier, doesn't entail (79) unless we have $(x)(\mathrm{N}x \rightarrow \mathrm{I}x)$. With it, (78) does entail (79). But so what? Is the regimentation, $(\exists x)\mathrm{I}x$, $(\exists x)\mathrm{N}x$, and $(x)(\mathrm{N}x \rightarrow \mathrm{I}x)$, of (78), (79) and "All natural numbers are integers," missing anything other than a capture of an internal implication between (78) and (79)—given $(x)(\mathrm{N}x \rightarrow \mathrm{I}x)$—the cogency of said

and has just finished defining the rationals set-theoretically. One can then say, "The irrational number exists." (For the record, this still sounds odd to me.)

implication turning on a distinction between "there is" and "exist"? If not, this consideration has no force.

What *does* matter, and what the Quinean *can't* handle, is that there are numerous sentences (I've given many in this paper) about "things" we take not to exist in any way whatsoever that the Quinean has to render true or false *in the same way*. For the Quinean, "There are more Greek gods than goddesses," and "There are more Greek goddesses than gods," *have to have the same truth values*. "Pegasus is a flying horse" and "Pegasus is a flying dog," must both be false, and "All flying horses are happy" and "All flying horses are unhappy" must both be true. *This* is the problem. The regimentation that solves *this* problem allows the sentences $(\exists x)\mathrm{F}x$ and $(\exists x)(\mathrm{Pegasus} = x)$ to be true, and allows, consequently, sentences like $(x)(\mathrm{F}x \rightarrow \mathrm{H}x)$ and F(Pegasus) to also be true, but disallows $(x)(\mathrm{F}x \rightarrow \neg\mathrm{H}x)$, and for that matter, $(\exists x)(\mathrm{F}x \,\&\, \mathrm{E}x)$ and E(Pegasus)—that is, it's a regimentation that enables distinguishable truths (and falsehoods) about what doesn't exist. As a *corollary* regimentations of "All integers exist" can be distinguished from regimentations of "There are integers that exist."

Last point: Generics, in any case, really aren't at issue here. Generics resist formalization because they still (to this day) resist a convincing linguistic analysis—but the generic readings of (78) allow some integers not to exist, and it's only this that the Quinean has a problem with because she's required to repudiate any interpretation of a proposition that turns on there being things some of which exist and some of which don't.

16.9 Our Concept of What Really Exists

Most of the time we (even we philosophers) unthinkingly treat *all* our concepts as represented in language by word-meanings. I've suggested in this paper that this isn't the case with our notion of *EXISTENCE*, where I'm here speaking of the concept central to metaphysics. Given the implicature view of ontological commitments and denials is right, our failure to see this is explicable. For, although the distinction between what is said and what is implicated is (generally) accessible to speaker-hearers, our tendency to hear many implicatures—generalized conversational ones, especially—as *asserted* runs interference. Thus, that ontological commitments and denials are uniformly implicatures isn't easily accessible because if we hear an implicature as asserted, we naturally think of it as arising from the meaning of the words we've used. The result is that it's never been considered that some of our concepts may arise not from the semantics of our words but purely from their implicatures: I suggest that the concept of EXIST is one of these. It doesn't correspond to the meaning of any word or phrase in the vernacular; rather, it's derived from (or expressed) only by *general conversational implicatures*. These make it natural to think "exist," for example, semantically expresses an ontologically-committing notion, EXIST. If my analysis is right, it explains a great deal about why we continue to fail to understand one of our most fundamental metaphysical concepts. One thing I am *not* claiming is that this fundamental metaphysical concept is undercut or de-legitimated by the discovery that it's rooted in a linguistic misapprehension.[46] On the contrary.

References

Asay, J. (2010). How to express ontological commitment in the vernacular. *Philosophia Mathematica (III), 18*, 293–310.

Azzouni, J. (2004). *Deflating existential consequence: A case for nominalism*. Oxford University Press.

Azzouni, J. (2007). Ontological commitment in the vernacular. *Noûs, 41*(2), 204–226.

Azzouni, J. (2010a). Ontology and the word 'exist': Uneasy relations. *Philosophia Mathematica (III), 18*, 74–101.

Azzouni, J. (2010b). *Talking about nothing: Numbers, hallucinations and fictions*. Oxford University Press.

Azzouni, J. (2012a). Referring to what is and to what isn't. In W. P. Kabasenche, M. O'Rourke, & M. H. Slater (Eds.), *Reference and referring* (pp. 253–269). The MIT Press.

Azzouni, J. (2012b). Responses to Gabriele Contessa, Erin Eaker, and Nikk Effingham. *Analysis, 72*(2), 366–379.

Azzouni, J. (2013a). Hobnobbing with the nonexistent. *Inquiry, 56*, 340–358.

Azzouni, J. (2013b). *Semantic perception: How the illusion of a common language arises and persists*. Oxford University Press.

Azzouni, J. (2017). *Ontology without borders*. Oxford University Press.

Balaguer, M. (2018). Fictionalism in the philosophy of mathematics. In E. N. Zalta (Ed.), *The Stanford encyclopedia of philosophy*. (Fall 2018 Ed.). https://plato.stanford.edu/archives/fall2018/entries/fictionalism-mathematics/

Carnap, R. (1937). *The logical syntax of language* (A. Smeaton, & C. von Zeppelin, Trans.). Harcourt Brace & Co.

Cartwright, R. (1954). Ontology and the theory of meaning. In *Philosophical essays (1987)* (pp. 1–12). MIT Press.

Chomsky, N. (2000). Explaining language use. In *New horizons in the study of language and mind* (pp. 19–45). Cambridge University Press.

Church, A. (1958). Ontological commitment. *Journal of Philosophy, 55*, 1008–1014.

Collins, J. (2017). The copredication argument. *Inquiry, 60*(7), 675–702.

Collins, John. (forthcoming). *Extending quantifier neutralism: On weak and strong determiners*.

Craig, W. L. (2017). *God and abstract objects: The coherence of theism: Aseity*. Springer.

Evans, G. (1982). *The varieties of reference*. Oxford University Press.

Fine, K. (2009). The question of ontology. In D. J. Chalmers, D. Manley, & R. Wasserman (Eds.), *Metametaphysics: New essays on the foundations of ontology* (pp. 157–177). Oxford University Press.

Grice, P. (1987). *Logic and conversation in studies in the way of words (1989)* (pp. 1–137). Harvard University Press.

Hofweber, T. (2007). Review of Jody Azzouni's deflating existential commitment. *The Philosophical Review, 116*(3), 465–467.

[46] E.g., I deny the claim that ontology, the study of EXISTENCE, isn't a real study because no word (in the vernacular) picks it out; I also deny Hofweber's (2007) charge that the resulting notion of EXISTENCE, on my story, is ineffable, or escapes characterization in some fashion. We can always coin a word that we stipulate to mean "EXIST." Philosophers do it all the time.

Hofweber, T. (2016). *Ontology and the ambitions of metaphysics*. Oxford University Press.
Moltmann, F. (2018). Existence predicates. *Synthese*. https://doi.org/10.1007/s11229-018-1847-z
Parsons, T. (1980). *Nonexistent objects*. Yale University Press.
Quine, W. V. (1953). On what there is. In *From a logical point of view (1980)* (pp. 1–19). Harvard University Press.
Routley, R. (1980). *Exploring Meinong's jungle and beyond: An investigation of noneism and the theory of items*. In interim (Ed.), *Philosophy department monograph* (Vol. 3). Research school of social sciences, Australian National University.
Schaffer, J. (2009). On what grounds what. In D. J. Chalmers, D. Manley, & R. Wasserman (Eds.), *Metametaphysics: New essays on the foundations of ontology* (pp. 347–383). Oxford University Press.
Sennet, A. (2016). Ambiguity. In E. N. Zalta (Ed.), *The Stanford encyclopedia of philosophy* (Winter 2016 ed.) https://plato.stanford.edu/archives/spr2016/entries/ambiguity/
Voltolini, A. (2018). Can one refer to and quantify over non-existent intentional objects. In P. Stalmaszczyk (Ed.), *Objects of inquiry in philosophy of language and linguistics* (pp. 97–115). Peter Lang.
Walton, K. (1990). *Mimesis as make-believe*. Harvard University Press.
Zalta, E. (1983). *Abstract objects: An introduction to axiomatic metaphysics*. Reidel.
Zwicky, A., & Sadock, J. (1975). Ambiguity tests and how to fail them. In Kimball (Ed.), *Syntax and semantics* (Vol. 4, pp. 1–36). Academic.

Chapter 17
The (Linguistic) Foundations of Arithmetic?

Robert Schwartzkopff

Abstract Tracing back to at least Frege, genuine higher-orderism sharply distinguishes between objects and categorically distinct higher-order entities and the expressions that denote them. Against the background of this view, Frege himself felt compelled to regard numbers as objects rather than higher-level entities—concepts, in Frege's terminology—because number words in arithmetical contexts appear to function as object-denoting expressions. In this paper, I argue that, from the perspective of natural language semantics, a case can be made that number words in such contexts function as concept-denoting rather than object-denoting expressions and that, consequently, numbers should be conceived of as concepts rather than objects. In developing this argument, I will develop a novel semantic analysis of expressions like 'the number of Mars' moons' as higher-order definite descriptions. Moreover, the view developed in this paper will shed new light on so-called easy arguments for the existence of numbers.

Keywords Frege · Copular clauses · Easy ontology · Higher-orderism · Nominalization · Numbers · Number words

---

Since the material contained therein partially dates back to 2014, this paper has been long in the making. During that time, many discussions with and comments from a great many people helped to vastly improve on the original idea. I wish to express my gratitude to all of them and, in particular, to Christian Folde, Stephan Krämer, Stefan Roski, Benjamin Schnieder, Nathan Wildman, and Richard Woodward as well as to the participants of the workshop *Higher-Order Metaphysics* (Oslo, June 2016), the workshop *Analyticity and Easy Being* (Oslo, May 2018), the workshop *Abstract Objects: Deflationary Approaches* (Santiago de Compostela, September 2021), the fourth FilMat conference *Philosophy of Mathematics: Foundations, Definitions, and Axioms* (Pavia, September 2022), as well as an anonymous referee for this volume. The opposite of gratitude is owed to the philosophy department at the University of Hamburg.

R. Schwartzkopff (✉)
Independent Scholar, Hamburg, Germany

X. de Donato-Rodríguez et al. (eds.), *Deflationist Conceptions of Abstract Objects*, Synthese Library 516, https://doi.org/10.1007/978-3-031-98717-5_17

17.1 Introduction

In late 1911 Wittgenstein was visiting Frege in Jena. As Wittgenstein was about to leave, he and Frege had an exchange that Wittgenstein recalled as follows:

> The last time I saw Frege, as we were waiting at the station for my train, I said to him "Don't you ever find *any* difficulty in your theory that numbers are objects?" He replied "Sometimes I *seem* to see a difficulty—but then again I *don't* see it".[1]

Which difficulty did Frege admit to sometimes seeming see? It is unlikely to have been the devastating difficulty for Logicism—Frege's theory of numbers—revealed by Russell's paradox. After all, having received Russell's fateful letter in late 1902, this difficulty had, at the time of Wittgenstein's visit, already been known to Frege for a good many years. Thus, Frege would have had no reason to only admit to seeming to see it. Perhaps, then, it was a difficulty concerning the linguistic justification of his conception of numbers as objects, a difficulty that Frege would only articulate in writing several years later.

Famously, Frege sharply distinguished between *objects* and categorically distinct higher-order entities called *concepts* (or, more generally, *functions*). Given this distinction, Frege felt compelled to regard numbers as objects because number words like 'two' in arithmetical contexts like 'Two is an even number' appear to function in the same way as proper names—i.e. as object-denoting expressions—such as, for instance, 'Sirius' in 'Sirius is a fixed star'. However, as evidenced by a diary entry from March 24th 1924, Frege had, by then, come to suspect that appearances might be deceptive:

> [O]ne comes to suspect that our way of using language is misleading, that number-words are not proper names of objects at all and words like 'number' and '[even] number' ... are not concept-words; and that consequently a sentence like '[Two] is [an even] number' simply does not express that an object is subsumed under a concept and so just cannot be construed like the sentence 'Sirius is a fixed star'. *But then how is it to be construed?*[2]

As I will argue in this paper, Frege's suspicion was warranted. In particular, I will argue that, from the perspective of natural language semantics, a case can be made that number words in arithmetical contexts function as concept-denoting rather than object-denoting expressions and that, consequently, numbers should be conceived of as concepts rather than objects. This view has two important consequences. First, it provides an answer to Frege's somewhat despairing question that concludes the quotation above. Second, it sheds new light on so-called easy arguments for the existence of numbers such as the following:[3] *Mars has (exactly) two moons. ∴ The number of Mars's moons is two. ∴ Something is the number of Mars's moons, viz. two ∴ Something is a number.*

The plan is as follows. Section 17.2 sketches the linguistic and metaphysical framework—a framework that traces back to Frege's distinction between objects and concepts and the expressions that denote them—that will inform the subsequent discussion. Section 17.3 develops a uniform analysis of number words as

---

[1] Reported by Geach in (Anscombe & Geach, 1961, p. 131).

[2] See (Frege, 1979, p. 263) emphasis added.

[3] See e.g. (Schiffer, 2003; Thomasson, 2015).

concept-denoting expressions that covers a wide array of number words' different uses and identifies the main challenges for the linguistic viability of this analysis. Section 17.4 is dedicated to meeting these challenges. Section 17.5 revisits the easy argument. Section 17.6 concludes.

17.2 Type-Theoretic Semantics and Higher-Orderism

The view to be defended in this paper will be developed within the kind of (extensional, functional) type-theoretic semantics widely used in contemporary semantic theorizing.[4] Type-theoretic semantics is closely related to Frege's own linguistic views. Indeed, it is sometimes said to 'execute the Fregean programme'.[5] Type-theoretic semantics comprises three main tenets (i) There are two basic semantic types: *e* (for entity or object) and *t* (for truth-value). (ii) Derivative types are licensed by the following rule: $\langle \rho,\sigma \rangle$ is a semantic type iff ρ and σ are types. (iii) Semantic composition proceeds (mainly) via the rule of *Functional Application*.

There are two truth-values, the True (T) and the False (F). Paradigm examples of type *t* expressions—i.e. of expressions that denote truth-values—are declarative sentences. Paradigm examples of objects are persons like Plato, paradigm examples of type *e* expressions—i.e. of object-denoting expressions—are proper names like 'Plato' when used 'referentially'. Entities of derivative types are functions. Functions whose values are truth-values are concepts. Concepts whose arguments are of type *e* are first-level concepts, concepts whose arguments are of type $\langle e,t \rangle$ are second-level concepts. Thus, both $\langle e,t \rangle$ and $\langle \langle e,t \rangle,t \rangle$ are types, entities of these types are (one-place) concepts, and entities of the former type are first-level whereas entities of the latter type are second-level concepts. Paradigm examples of type $\langle e,t \rangle$ expressions are expressions like 'is wise'. Paradigm examples of type $\langle \langle e,t \rangle,t \rangle$ expressions are expressions like 'some philosopher'. According to Functional Application, the denotation of a semantically complex expression α, with immediate constituents β and γ, is the value of the function detonated by, say, β when taking the denotation of γ as an argument. Later on we will have occasion to introduce further composition rules as well as so-called *semantic type-shifting* principles that will facilitate composition.[6]

Although similar to Frege's own view in many ways, standard type-theoretic semantics as practiced in contemporary linguistics departs from Frege in one crucial respect. Standard type-theoretic semantics treats higher-order entities—i.e. the denotations of derived functional types $\langle \rho,\sigma \rangle$—as *sets*. For instance, entities of type $\langle e,t \rangle$ are treated as certain sets of ordered pairs (and ordered pairs as sets). However, for Frege sets were themselves objects and therefore not categorically distinct from the latter. Thus, the functions of standard type-theoretic semantics are not genuinely higher-order entities.

---

[4] What follows presupposes some familiarity with type-theoretic semantics. For an introduction, see (Heim & Kratzer, 1998).

[5] Heim & Kratzer (1998, p. 13).

[6] For the latter see, in particular, (Partee, 1986).

Under the label of (genuine) *higher-orderism*, recent years have seen a renaissance of Frege's view that stresses the importance for ontological theorizing of taking Frege's categorical distinction between objects and higher-level entities seriously.[7] In a nutshell, higher-orderism has it that both denotation of and quantification over higher-level entities are *sui generis* kinds of denotation and quantification which are neither reducible to nor explicable in object-level terms.[8]

Methodologically, I will appeal to both of these approaches. Since the focus is on the semantic function of number words in natural language, I will freely avail myself of standard type-theoretic semantics and its set-theoretic framework. At the same time, since I intend my conclusions to be ontologically significant, I will regard this set-theoretic framework as a mere heuristic,[9] that—albeit technically convenient—can and should ultimately be dispensed with in favour of a genuinely higher-order framework.

17.3 Number Words: Many Uses, One Function

Number words in natural language can be used in a variety of different ways. Thus, consider the number word 'two' as it is used in the following sentences:[10]

(1)	a.	Mars's moons are *two*.	*predicative use*
	b.	Those are Mars's *two* moons.	*attributive use*
	c.	Mars has *two* moons.	*quantificational use*
	d.	The number of Mars's moons is *two*.	*specificational use*
	e.	*Two* is an even number.	*numeral use*

Do these different uses correspond to different semantic functions? Or can number words be understood semantically uniformly—i.e. as performing the same semantic function—in all of these different uses? I will assume that a difference in semantic function corresponds to a difference in semantic type.[11] If so, the first of these two questions receives an affirmative answer provided that the number word is of a different type (and has a correspondingly different denotation) in all of (1a–e). And the second question is to be answered affirmatively provided that the number word is of the same type (and has the same denotation) in all of them.

Recently, Snyder, Samuals, and Shapiro have defended a non-uniform, polymorphic account of the semantic function of 'two' in (1a–e).[12] According to this account, 'two' functions as a type *e* expression in (1d, e) but is of various higher-semantic types in (1a–c). Moreover, Snyder, et al. have it that (i) one of the different

[7] For a recent overview, see (Skiba, 2021).

[8] See (Skiba, 2021, p. 2). For a formal implementation, see (Krämer, 2014).

[9] In the sense of (Wright, 2007).

[10] The taxonomy below follows (Snyder, 2021, p. 42f). Snyder also discusses further uses of number words. Today's focus is on the uses in (1a-e), discussed in (Snyder et al., 2022).

[11] In doing so, I follow (Snyder et al., 2022). For dissent see (Hofweber, 2016, pp. 206–14).

[12] See (Snyder et al., 2022).

denotations that corresponds to these different types is lexical (roughly, basic) and that (ii) the other, non-lexical denotations derive from the lexical one via various semantic type-shifting operations.[13] If so, number words in arithmetical contexts like (1e) are expressions of type *e*—i.e. object-denoting expressions—and one should therefore countenance a conception of numbers as objects.

Today, in contrast, I will develop and defend a uniform account of 'two' according to which it functions as a type $\langle e,t \rangle$ expression (with the same denotation) in all of (1a–e). Since on this account, number words in arithmetical contexts like (1e) are first-level concept-denoting expressions, there would be no need to countenance a conception of numbers as objects and numbers should be conceived of as (first-level) concepts.

The five uses of number words on display in (1a–e) can be divided into two groups. First, the 'clearly' non-referential uses in (1a–c), i.e. those uses in which number words are typically taken to be of a higher semantic type rather than of the object type *e*. Second, the apparently referential uses in (1d, e), i.e. those uses that are typically taken to be type *e* uses. I will attend to these two groups in turn.

17.3.1 *The Clearly Non-Referential Uses*

Within linguistics, there is wide—albeit not universal—agreement that number words in their predicative, their attributive and their quantificational uses do not function referentially, i.e. not as expressions of type *e*.[14] This, of course, leaves open the question of which exact type (and denotation) number words have in these uses. In the remainder of this section, I will present analyses of (1a–c) according to which 'two' in all of these three uses uniformly functions as a type $\langle e,t \rangle$ expression (with the same denotation). Since these analyses only employ standard linguistic resources, their linguistic viability should not be matter of great controversy. However, taking the time to get into the details, has the benefit of introducing these resources before we turn to more controversial matters in Sect. 17.3.2.

17.3.1.1 The Predicative Use

We begin with the predicative use of number words on display in (1a) 'Mars's moons are two'. In (1a) 'Mars's moons' is a plural type *e* expression that denotes the plurality of moons that belong to Mars and 'two' is a plural type $\langle e,t \rangle$ expression that denotes what I will call the (first-level) *cardinality concept* of being two, a concept that subsumes exactly those type *e* pluralities that comprise exactly two atomic

---

[13] Snyder et al. (2022, p. 78) distinguish two variants of polymorphism that differ in which denotation is taken as basic. For today's purposes the details are irrelevant.

[14] See e.g. (Rothstein, 2017, pp. 9–25). for an overview. For dissent, see e.g. (Kennedy, 2013) who treats them as (a sub-type of) type *e* expressions. With (Heim & Kratzer, 1998, p. 61f.), I assume that 'are' in (1a,b) is a mere copula that can be treated as semantically vacuous.

parts.[15] Plural type e expressions are treated as denoting genuinely plural objects or pluralities understood as mereological sums and plural $\langle e,t\rangle$ expressions as denoting concepts that subsume pluralities so understood.[16]

Going forward, I will specify the denotation of predicatively used type $\langle e,t\rangle$ 'two' by way of (2), where $\sqsubseteq_{At}$ is the relation of atomic parthood, '$\exists_{=2}$' is an exact cardinality quantifier, and '$[\![a]\!]$' is read as 'the denotation of 'a''[17,18]

(2) $[\![\textbf{two}]\!] = \lambda x.\ \text{two}(x) = \lambda x.\ \exists_{=2}y\ (y \sqsubseteq_{At} x)$.[19]

The semantic composition of the proposed denotation of 'Mars's moons' proceeds as follows. First, the plural count noun 'moons' is treated as a plural type $\langle e,t\rangle$ expression with the following denotation (* is a closure operation under sum-formation):

(3) $[\![\textbf{moons}]\!] = \lambda x.\ {}^*\text{moon}(x)$.[20]

Second, assuming that 'Mars' is a singular type e expression with the denotation given in (4a), and treating the possessive morpheme '-'s' as having the denotation in (4b),[21] we can by Functional Application (FA)[22] derive a type $\langle\langle e,t\rangle,\langle e,t\rangle\rangle$ denotation for 'Mars's' given in (4c), which composes by FA with the type $\langle e,t\rangle$ denotation of 'moons' to yield the type $\langle e,t\rangle$ denotation of 'Mars's moons' given in (4d):[23]

---

[15] In the relevant sense, an F is atomic iff none of its proper parts is F, see e.g. (Krifka, 1989).

[16] See (Link, 1983). If concepts are understood set-theoretically, plural type $\langle e,t\rangle$ expressions thus denote characteristic functions of sets of sums. On Link's treatment, apparently plural reference and predication is, at bottom, singular reference to and singular predication of sums. There is also an alternative view according to which plural referring expressions and plural predicates plurally refer and apply to several non-plural objects at once. For a recent treatise on this issue, see (Florio & Linnebo, 2021).

[17] Variables 'x, y, ...' range unrestrictedly over the elements of the domain of objects D_e, which contains both atomic and plural objects. In general, for any type τ, D_τ is the domain of entities of type τ. Atomic parthood is defined as $a \sqsubseteq_{At} b \leftrightarrow a \sqsubseteq b \wedge \text{Atom}(a)$, where 'Atom' is a meta-language predicate that applies to atomic objects. The exact cardinality quantifier can be defined in the usual syncategorematic way.

[18] For a specification of the predicative denotation of 'two' along somewhat similar lines, see (Moltmann, 2013, p. 508). (Snyder et al., 2022, p. 78) opt for a different definition of $\lambda x.\ \text{two}(x)$—viz. $\lambda x.\ \#(x) = 2$—where '2' is a type e expression that denotes a number and # is a function that maps pluralities to numbers that represent how many atomic parts they have.

[19] For instance, if there are three atomic objects a, b, c, then $[\![\textbf{two}]\!]$ = the (characteristic function of the) set of sums with exactly two atomic parts = $\{a \sqcup b, a \sqcup c, b \sqcup c\}$.

[20] For instance, if there are three atomic moons a, b, c, then $[\![\textbf{moon}]\!] = \lambda x.\ \text{moon}(x)$ = the (characteristic function of the) set of atomic moons = $\{a, b, c\}$ and $[\![\textbf{moons}]\!] = \lambda x.\ {}^*\text{moon}(x)$ = the (characteristic function of the) set of pluralities of atomic moons = $\{a, b, c, a \sqcup b, a \sqcup c, b \sqcup c, a \sqcup b \sqcup c\}$.

[21] See (Barker, 1998).

[22] If α is a branching node, $\{\beta, \gamma\}$ is the set of α's daughters, and $[\![\beta]\!]$ is a function whose domain contains $[\![\gamma]\!]$, then $[\![\alpha]\!] = [\![\beta]\!]([\![\gamma]\!])$; see (Heim & Kratzer, 1998, p. 44).

[23] Variables 'X, Y, ...' range unrestrictedly over the elements $D_{\langle e,t\rangle}$, the domain of one-place first-level concepts.

(4) a. $[\![\textbf{Mars}]\!] = \text{Mars}.$
b. $[\![\textbf{–'s}]\!] = \lambda x.\lambda X.\lambda y.\ X(y) \wedge R(y,x).$
c. $[\![\textbf{Mars's}]\!] = \lambda X.\lambda y.\ X(y) \wedge R(y,\text{Mars}).$
d. $[\![\textbf{Mars's moons}]\!] = \lambda y.\ {*}\text{moon}(y) \wedge R(y, \text{Mars}).$

(N.B. That the variable 'R' is free indicates that its denotation is provided contextually. Thus, as a type $\langle e,t\rangle$ expression 'Mars's moons' denotes a unary concept that subsumes pluralities of moons that bear some contextually provided relation—in the case at hand, the relation of belonging to Mars.)

Third, the type-shifting operation IOTA,[24] defined in (5a), is applied to the type $\langle e,t\rangle$ denotation of 'Mars's moons' to derive the type e denotation, specified in (5b):

(5) a. $\text{IOTA} = \lambda \text{X}.\ \iota x[X(x)].$
b. $\text{IOTA}([\![\textbf{Mars's moons}]\!]) = \iota y[{*}\text{moon}(y) \wedge R(y, \text{Mars})].$

(N.B. 'ι' is a description operator that combines with a concept-denoting expression of type $\langle e,t\rangle$ to yield a type e expression that (i) denotes the unique (salient) atomic object subsumed under the denoted concept if that concept is singular or (ii) denotes the maximal (salient) plurality subsumed under the denoted concept if it is plural; provided there are some such, otherwise the ι-term will be empty.[25]).

Pulling these strings together, we can now represent the logical form of (1a) as follows:

(1) a.* $\text{two}\ (\iota y[\ {*}\text{moon}(y) \wedge R(y,\text{Mars})]).$

Given (1a*), (1a) is true iff the plurality denoted by 'Mars's moons' has exactly two atomic parts.

17.3.1.2 The Attributive Use

Let us now turn to the attributive use of number words on display in (1b) 'Those are Mars's two moons'. In (1b), the deictic expression 'those' is a plural type e expression and 'Mars's two moons' is a plural type $\langle e,t\rangle$ expression. Above we have seen how to obtain a type $\langle\langle e,t\rangle,\langle e,t\rangle\rangle$ denotation for 'Mars's'. We now face the problem of how to obtain a type $\langle e,t\rangle$ denotation for 'two moons' to compose it with. Since both 'two' and 'moons' are of type $\langle e,t\rangle$, the denotation of neither is in the domain of the denotation of the other. Thus, they cannot compose by FA. This difficulty can be resolved by appeal to a further composition rule—viz. *Predicate Modification*

[24] See (Partee, 1986, p. 121f.).

[25] See (Sharvy, 1980). For instance, when 'ι' is combined with a singular count noun like 'moon', '$\iota x[\text{moon}(x)]$' refers to the unique (salient) atomic moon (if there is one) and when combined with a plural count noun like 'moons', '$\iota x[{*}\text{moon}(x)]$' refers to the maximal (salient) sum of atomic moons (if there is one).

(PM)—that allows for the composition of two type $\langle e,t\rangle$ denotations.[26] The result of applying PM to the denotations of 'two' and 'moons', given in (6a), can then compose by FA with the denotation of 'Mars's', given in (4c), to yield the type $\langle e,t\rangle$ denotation for 'Mars's two moons', given in (6b):

(6) a. $[\![\textbf{two moons}]\!] = \lambda x.\ \text{two}(x) \wedge {}^*\text{moon}(x)$.[27]
b. $[\![\textbf{Mars's two moons}]\!] = \lambda x.\ \text{two}(x)) \wedge {}^*\text{moon}(x) \wedge R(y, \text{Mars})$.

The logical form of (1b) can be represented as follows:

(1) b*. two (those) ∧ *moon (those) ∧ R (those, Mars).

Given (1b*), (1b) is true iff the plurality denoted by 'those' consists of exactly two atomic parts, each of which is a moon belonging to Mars.

17.3.1.3 The Quantificational Use

Finally, let us attend to the quantificational use of number words on display in (1c) 'Mars has two moons'. In (1c), 'Mars' is a (singular) type e expression with the denotation in (4a), 'has' is a type $\langle e,\langle e,t\rangle\rangle$ expression that denotes the belonging relation in which moons stand to the planets they orbit, and 'two moons' is (part of) a quantificational type $\langle\langle e,t\rangle,t\rangle$ expression. Composition now faces two problems.

The first consists in compositionally deriving a suitable type $\langle\langle e,t\rangle,t\rangle$ denotation for 'two moons'. One way of doing so treats type $\langle e,t\rangle$ 'two moons' as the complement of an unpronounced determiner '∅' that is interpreted as the quantificational determiner 'some' of type $\langle\langle e,t\rangle,\langle\langle e,t\rangle,t\rangle\rangle$, as specified in (7a). The determiner's denotation then composes by FA with the denotation of 'two moons', given in (6a) to yield the type $\langle\langle e,t\rangle,t\rangle$ denotation given in (7b):[28]

(7) a. $[\![\textbf{some}]\!] = \lambda X.\ \lambda Y.\ \exists z(X(z) \wedge Y(z))$.
b. $[\![\varnothing\ \textbf{two moons}]\!] = \lambda Y.\ \exists z(\text{two}(z) \wedge {}^*\text{moon}(z) \wedge Y(z))$.[29]

---

[26] If α is a branching node, $\{\beta, \gamma\}$ is the set of α's daughters, and $[\![\beta]\!]$ and $[\![\gamma]\!]$ are in $D_{\langle e,t\rangle}$, then $[\![\alpha]\!] = \lambda x \in D_{\langle e,t\rangle}.\ [\![\beta]\!](x) = T \wedge [\![\gamma]\!](x) = T$; see (Heim & Kratzer, 1998, p. 65).

[27] Snyder et al. (2022, p. 79) resolve the difficulty of composing two type $\langle e,t\rangle$ expressions in a different, albeit functionally equivalent way, viz. by appeal to Landman's type-shifting operation ADJUNCT (2004, p. 12). Applying ADJUNCT to the type $\langle e,t\rangle$ denotation of 'two' yields a type $\langle\langle e,t\rangle,\langle e,t\rangle\rangle$ denotation—viz. $\lambda X.\ \lambda x.\ \text{two}(x) \wedge X(x)$—that can then compose by FA with the denotation of 'moons' to yield the same $\langle e,t\rangle$ denotation as the one given in (6a). *Prima facie*, neither the appeal to PM nor the appeal to ADJUNCT is superior to the other. Perhaps, it is even possible to argue that the appeal to PM is preferable; see (Pietroski, 2005).

[28] See (Ionin & Matushansky, 2006, p. 322) by way of (Krifka, 1999).

[29] Snyder et al. (2022, p. 79) resolve this compositional difficulty in a different, albeit functionally equivalent way, viz. by appeal to Partee's type-shifting operation A (1986, p. 126). When applied

The second problem consists in rendering (1c) as a whole interpretable. As a type $\langle e,\langle e,t\rangle\rangle$ expression 'has' 'expects' a type *e* argument, whereas its overt syntactic complement 'two moons' has been rendered as (part of) the type $\langle\langle e,t\rangle,t\rangle$ expression '∅ two moons'. Type mismatches such as these can be resolved via the mechanism of *Quantifier Raising* (QR). QR raises '∅ two moons' to sentence-initial position, leaving behind a numerically indexed trace 't_i'—effectively, a free variable—and, in order to bind this trace, inserting a (corresponding) numerical index 'i'—effectively, a lambda-operator—as the sister node of the 'open' sentence 'Mars has t_i'.[30] The resulting 'lambda-phrase' then composes by the composition rule of *Predicate Abstraction* (PA).[31] The resulting syntactico-semantic structure of (1c) can be represented as follows (arrows represent QR):

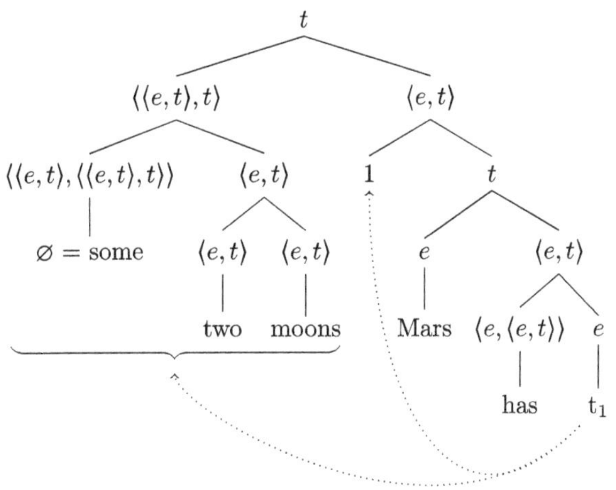

Fig. 17.1 Syntactic tree for (1c) "Mars has two moons" displaying its semantic composition

to the denotation of 'two moons' A yields the same type $\langle\langle e,t\rangle,t\rangle$ denotation as the one given in (7c). Neither the appeal to '∅' nor the appeal to A is *prima facie* superior to the other. Perhaps, it is even possible to argue that the appeal to '∅' is preferable; see (Ionin & Matushansky, 2006, p. 322).

[30] See (Heim & Kratzer, 1998, pp. 184–88).

[31] Let α be a branching node with daughters β and γ, where β dominates only a numerical index i. Then, for any variable assignment a, $[\![\alpha]\!]^{a} = \lambda v \in D_{\tau}.\ [\![\gamma]\!]^{a(v/i)}$ (where τ is determined by the type of 't_i'); see (Heim & Kratzer, 1998, p. 186).

Based on this structure, the logical form of (1c) can be represented as follows:

(1) c.* $\exists x(\text{two}(x) \wedge {}^*\text{moon}(x) \wedge \text{has}(x, \text{Mars}))$.

Given (1c*), (1c) is true iff there is at least one plurality having exactly two atomic parts, each of which is a moon belonging to Mars.[32]

This concludes the discussion of the clearly non-referential uses of number words. As we have seen, number words in all three of these uses can, using only accepted linguistics resources, be understood semantically uniformly as functioning as type $\langle e,t \rangle$ expressions that denote cardinality concepts.

17.3.2 The Apparently Referential Uses

Linguistically speaking, the claim that number words in their predicative, attributive, and quantificational uses can be understood as denoting type $\langle e,t \rangle$ cardinality concepts should be rather uncontroversial. Things are different, once we turn to their specificational and their numeral uses on display in respectively (1d) 'The number of Mars's moons is two' and (1e) 'Two is an even number'. With respect to these uses, both philosophical and linguistic orthodoxy has it that they should be understood as type *e* expressions. In this section I will present compositional analyses of (1d,e) according to which, appearances to the contrary notwithstanding, the number word 'two' functions as a type $\langle e,t \rangle$ expression that denotes a cardinality concept as well. The linguistic viability of these analyses will be defended Sect. 17.4.

17.3.2.1 Nominalization

There is no denying that number words in their specificational and their numeral uses appear to be type *e* expressions. Mainly, this appearance stems from their syntactic status. Number words in their predicative, attributive, and quantificational uses are non-nominal expressions, i.e. expressions that do not have the syntactic status of definite noun phrases. In contrast, number words in their specificational and their numeral uses are nominal expressions, i.e. expressions that do have the syntactic status of such phrases. At first glance, type *e* expressions seem—besides quantificational noun phrases of type $\langle \langle e,t \rangle, t \rangle$ like 'some philosopher'—to be the only expressions with that syntactic status. Since number words do not seem to be quantificational noun phrases, it therefore is tempting to treat them as type *e* expression.

This temptation can—and, I contend, should—be resisted. The phenomenon that (with a grain of salt) different uses of one and the same expression can differ in their syntactic status is known as *nominalization*. In the first instance, nominalization is a syntactic operation that changes an expression's syntactic type from non-nominal

[32] Note that (1c*) captures the lower-bounded (at least) reading of (1c) on which its truth is compatible with Mars having more than two moons.

to nominal. In what follows, I will mark the nominalization of a non-nominal expression by underscoring it. That is, where 'two' is non-nominal, 'two' is nominal.[33]

According to the orthodox view on nominalization, nominalizing an expression does not only change its syntactic type—from non-nominal to nominal—but is accompanied by a corresponding semantic type-shift—from the higher semantic type of the non-nominal expression to type *e*—with the result being that a nominalization denotes the *object correlate* of the higher-order entity denoted by the non-nominal expression whence it derives.[34]

In recent years, Hofweber has defended an alternative view on nominalization. According to this view, nominalizing an expression only effects a *merely syntactic type-shift* that occurs for broadly communicative and/or cognitive reasons.[35] On my preferred interpretation, Hofweber's view has it that nominalizing an expression only alters its syntactic type whilst leaving its semantic type unchanged.[36] If so, nominalizations are expressions of higher semantic types that are draped in the syntactic garments of type *e* expressions.[37]

17.3.2.2 The Specificational Use

Let us now turn to the specificational use of number words on display in (1d) 'The number of Mars's moons is two'. According to Frege's philosophically influential analysis,[38] sentences like (1d) are what linguists call (first-order) *equational sentences* in the following sense: (i) both the pre-copula *the-number-of* terms (e.g. 'the number of Mars's moons') and the post-copula (nominalized) number words (e.g. 'two') function as type *e* expressions, (ii) the copula 'is' functions as a type $\langle e,\langle e,t\rangle\rangle$ expression that denotes the identity relation between objects.

In contrast, Hofweber has argued that such sentences are not (first-order) equational sentences.[39] In particular, Hofweber has it that their post-copula nominalized number words are not type *e* expression but rather are of the same higher-semantic type as non-nominal number words in their quantificational use.[40] Alas, Hofweber

---

[33] The underscore notation is due to (Tennant, 1997, p. 310). Other examples are (i) colour adjectives: 'My socks are *green*' vs. 'The colour of my socks is *green*', (ii) other adjectives: 'Plato is *wise*' vs. 'Plato has *wisdom*', (iii) declarative sentences: '*Snow is white*' vs. 'It is true *that snow is white*'. As one can see, a nominalization is not always orthographically identical to the non-nominal expression whence it derives. Hence, my plea for a grain of salt.

[34] See (Chierchia, 1984) and (Partee, 1986, p. 121f).

[35] See (Hofweber, 2005a, b, 2007).

[36] Hofweber (2016, pp. 206–14) and Hofweber (2023, pp. 221–27) has disavowed this interpretation. However, it seems to me that his alternative interpretation is of questionable coherence.

[37] See (Hale & Linnebo, 2020) for discussion of a similar view.

[38] See (Frege, 1884, §57).

[39] See (Hofweber, 2005b, 2007).

[40] Hofweber takes the latter to be of type $\langle\langle e,t\rangle,\langle\langle e,t\rangle,t\rangle\rangle$, the type of (first-order) quantificational determiners like 'some'. As we have seen in §3.1.3, they can fruitfully be understood as expressions of type $\langle e,t\rangle$.

did not provide a compositional analysis of (1d) that would vindicate this claim, instead resting content with having argued that some such analysis must be correct.

It is to the great credit of, in particular, Moltmann and Felka that they have attempted to fill this lacuna.[41] According to Moltmann and Felka, sentences like (1d) are so-called *specificational sentences*.[42] Notably, both Moltmann and Felka make two assumptions. First, specificational sentences constitute a sentence type *sui generis* which, in particular, is not reducible to other types of *copular clauses* such as *predicational sentences* or *equational sentences*.[43] Second, specificational sentences should be analysed in accordance with the so-called *question-in-disguise* account, which allows them to spell out the post-copula number word in (1d) as the surface structure remainder of the partially elided sentence (1c)—i.e. as '~~Mars has~~ two ~~moons~~'—thereby vindicating Hofweber's central tenet.[44] However, as I have argued elsewhere neither Moltmann's nor Felka's specificational analysis of sentences like (1d) is entirely satisfactory.[45]

Taking a different tack, I would like to suggest that although Frege was correct to regard sentences like (1d) as equational sentences, he was mistaken about their order. In particular, I contend that such sentences are *higher-order equational sentences* in the following sense: (i) both the pre-copula *the-number-of* terms and the post-copula nominalized number words function as type $\langle e,t \rangle$ expressions that denote (first-level) cardinality concepts, (ii) the copula 'is' functions as a type $\langle \langle e,t \rangle, \langle \langle e,t \rangle, t \rangle \rangle$ expression that denotes the identity relation between first-level concepts.

This proposal faces two main challenges. First, we need a compositional semantics for *the-number-of* terms according to which they are *higher-order definite descriptions* that denote cardinality concepts. Second, a case has to be made that

[41] See (Moltmann, 2013; Felka, 2014, 2016) as well as (Knowles, 2015; Snyder, 2017).

[42] In treating (1d) as a specificational sentence, Moltmann and Felka follow (Higgins, 1973, p. 199) and (Romero, 2005). For arguments that such sentences are specificational sentences, see (Moltmann, 2013, p. 523) and (Felka, 2014, p. 269ff), respectively. For critique, see (Schwartzkopff, 2016, p. 2180f) and (Balcerak, 2013, p. 456f), respectively.

[43] For an overview on copular clauses, see (Mikkelsen, 2011). Paradigm examples of predicational sentences and equational sentences are 'The person Gottlob met was smart' and 'The person Gottlob met was the author of *On Certainty*', respectively.

[44] For details of the question-in-disguise account, see e.g. (Schlenker, 2003). According to Romero's *individual concept* account of specificational sentences, the post-copula number word in (1d) functions as a type e expression, the pre-copula *the-number-of* term as a type $\langle s,e \rangle$ expression (i.e. as denoting a function from worlds to objects), and the copula of specification as a type $\langle e, \langle \langle s,e \rangle, t \rangle \rangle$ expression. According to (Snyder et al., 2022, p. 79), the denotation of the type e nominalized number word is the object correlate of the cardinality concept denoted by the corresponding non-nominal type $\langle e,t \rangle$ number word.

[45] See (Schwartzkopff, 2016; Schwartzkopff, 2022).

this compositional semantics is linguistically viable. These challenges will be taken up in Sects. 17.4.1 and 17.4.2, respectively.[46]

17.3.2.3 The Numeral Use

In bringing this section to a close, let us attend to the numeral use of number words on display in (1e) 'Two is an even number'. Typically, (1e) is regarded as a *first-order predication* in which 'two' functions as a type *e* expression that denotes the number 2 (conceived of as an object) and 'is an even number' functions as a type $\langle e,t \rangle$ expression that denotes the first-level concept of being an even number.[47] In contrast, I would like to suggest that (1e) is a *higher-order predication* in which 'two' functions as a (nominalized) type $\langle e,t \rangle$ expression that denotes the first-level cardinality concept of being two and 'is an even number' functions as a type $\langle\langle e,t \rangle,t \rangle$ expression that denotes the second-level concept of being an even first-level cardinality concept.[48] If so, numbers—i.e. the denotations of number words in arithmetical contexts like (1e)—should be conceived of as cardinality concepts.[49]

The challenge for this proposal thus consists in making the case that the types and denotations it assigns to 'two' and 'is an even number' are linguistically viable. In order to meet this challenge, much would be gained if it was possible to identify a use of 'number' in which it functions as an expression of the proposed type and denotation. By identifying such a use, this challenge will be met in Sect. 17.4.2.

---

[46] The proposal to analyse sentences like (1d) as higher-order equational sentences is compatible with or at least congenial to treating such sentences as specificational sentences. First, there are reductive accounts of specificational sentences that reduce them to (potentially) higher-order equational sentences, see e.g. (Sharvit, 1999). Second, the proposed analysis of *the-number-of* terms as higher-order descriptions can fruitfully by employed to solve the difficulties that beset Moltmann's and Felka's specificational question-in-disguise analyses of such sentences; see (Schwartzkopff, 2022, p. 214ff). Third, this analysis of *the-number-of* terms could also be used to formulate a higher-order version of Romero's individual concept account of specificational sentences.

[47] On one of the variants of polymorphism discussed in (Snyder et al., 2022) the number 2 is the object correlate of the cardinality concept of being two. On the other, it is an object of a different kind.

[48] I take it that this understanding is a natural interpretation of Hofweber's contention (2016, p. 145) that "... the predicate 'is [an even] number' in [(1e)] is not predicated of the referent of ['two'], but forms a meaningful and true subject-predicate sentence nonetheless".

[49] By conceiving of numbers as cardinality concepts, my proposal resembles views such as those of (Hodes, 1984) and (Yi, 2018).

17.4 Linguistic Viability

Let us take stock. According to my view, number words in sentences like (1a–e) function semantically uniformly as expressions of type $\langle e,t\rangle$ that denote (first-level) cardinality concepts. In Sect. 17.3.1, we have seen that with respect to (1a–c) my view is linguistically viable. In Sect. 17.3.2, I have indicated how sentences like (1d,e) can be understood in accordance with my view and identified the main challenges of so understanding them. What remains to be seen is whether my view can rise to these challenges.

17.4.1 *The-Number-of Terms as Higher-Order Descriptions*

The first challenge consists in (i) giving a compositional semantics for *the-number-of* terms as higher-order descriptions and (ii) making the case that this semantics is linguistically viable. In this section, I will propose the sought-after semantics and begin to make the case for its viability, a case that will then be completed in Sect. 17.4.2.

17.4.1.1 Preparation

In preparation, it will be helpful to begin by considering the following ordinary first-order description that is structurally similar to *the-number-of* terms:

(8) the mother of Philip's children.

Suppose the relevant Philip is the late Duke of Edinburgh. Assuming a Strawsonian type *e* interpretation of descriptions[50], (8) then denotes Elizabeth, Queen of England.[51]

Compositionally, this denotation can be derived as follows. First, we assume that 'mother' is a relational noun of type $\langle e,\langle e,t\rangle\rangle$ that denotes the motherhood relation which mothers bear to their children (the superscripted '2' indicates that the expression is relational):

(9) $[\![\textbf{mother}^2]\!] = \lambda y.\lambda x.\ \text{mother}^2(x,y)$.

[50] See (Strawson, 1950).

[51] When I submitted this paper, Elizabeth was still alive. But even Her Majesty was not able to defeat father time, let alone the pace of academic publishing.

Second, we derive a type e denotation for 'Philip's children'[52]—the plurality of all of Philip's children—and assume that 'of' is semantically vacuous[53] so that:

(10) $[\![$**of Philip's children**$]\!] = \iota y[{}^*\text{child}(y) \wedge R(y,\text{Philip})]$.

Third, we compose the type e denotation of 'of Philip's children' with the type $\langle e,\langle e,t\rangle\rangle$ denotation of 'mother²' by FA. The result is the type $\langle e,t\rangle$ concept of standing in the motherhood relation to the plurality of all of Philip's children:

(11) $[\![$**mother of Philip's children**$]\!] =$
$\lambda x.\ \text{mother}^2(x,\iota x[{}^*\text{child}(x) \wedge R(x,\text{Philip})])$.

Fourth, we assume that 'the' denotes the first-order description function of type $\langle\langle e,t\rangle,e\rangle$:

(12) $[\![$**the**$]\!] = \lambda X.\ \iota x[X(x)]$.

Finally, we compose the denotations of 'the' and 'mother of Philip's children' by FA to obtain the following type e denotation for (8):

(13) $[\![$**the mother of Philip's children**$]\!] =$
$\iota x[\text{mother}^2(x, \iota x[{}^*\text{child}(x) \wedge R(x,\text{Philip})])]$.

Thus, as desired, (8) 'the mother of Philip's children' denotes the unique object that stands in the motherhood relation to the plurality of all of Philip's children. And this, of course, is none other than Elizabeth, Queen of England.

17.4.1.2 Execution

Let us now consider the *the-number-of* term:

(14) the number of Mars's moons.

According to my proposal, (14) is a higher-order description that denotes a cardinality concept, viz. the one that subsumes the plurality of all of Mars's moons. Compositionally, this denotation can be derived as follows.

First, we make the crucial assumption, to be defended later, that 'number' is a relational noun of type $\langle e,\langle\langle e,t\rangle,t\rangle\rangle$ that denotes the 'numbering relation' which only cardinality concepts bear to the pluralities that they subsume:

---

[52] The derivation runs parallel to the one for 'Mars's moons' in §3.1.1. The contextually provided relation R is the relation of having been fathered by Philip.

[53] See (Heim & Kratzer, 1998, p. 61f).

(15) $[\![\mathbf{number}^2]\!] = \lambda x.\lambda X.\ \mathrm{NUMBER}^2(X,x)$.

Second, we take the type *e* denotation of 'Mars's moons' derived in Sect. 17.3.1, assume that 'of' is semantically vacuous, and compose the type *e* denotation of 'of Mars's moons' with the denotation of 'number' by FA. The result is the second-level type $\langle\langle e,t\rangle,t\rangle$ concept of standing in the numbering relation to the plurality of all of Mars's moons:

(16) $[\![\textbf{number}^2\textbf{ of Mars's moons}]\!] =$
$\lambda X.\ \mathrm{NUMBER}^2(X,\iota y[{}^*\mathrm{moon}(y) \wedge R(y,\mathrm{Mars})])$.

Third, we assume that, on this occasion, 'the' denotes the higher-order description function of type $\langle\langle e,t\rangle,\langle\langle e,t\rangle,t\rangle\rangle$:[54]

(17) $[\![\mathbf{the}]\!] = \lambda\mathbf{X}.\ \iota Y[\mathbf{X}(Y)]$.

Finally, we compose the proposed denotations of 'the' and 'number of Mars's moons' by FA to obtain the following type $\langle e,t\rangle$ denotation for (14):

(18) $[\![\textbf{the number}^2\textbf{ of Mars's moons}]\!] =$
$\iota X[\mathrm{NUMBER}^2(X, \iota x[{}^*\mathrm{moon}(x) \wedge R(x,\mathrm{Mars})])]$.

According to (18), (14) 'the number of Mars's moons' is a higher-order description that denotes the unique type $\langle e,t\rangle$ concept that stands in the numbering relation to the plurality of all of Mars's moons. Since only cardinality concepts can stand in this relation, this denotation is a cardinality concept. Since as a matter of astronomical fact, Mars has (exactly) two moons, (14) denotes the cardinality concept of being two. Thus, a *the-number-of* term 'the number of *F*s' in general denotes the same cardinality concept as a type $\langle e,t\rangle$ number word '*n*' just in case there are (exactly) *n F*s.

Given my proposal, the logical form of (1d) therefore is the following ($\equiv$ is the higher-order identity relation between first-level concepts):

(1) d.* $\iota X[\mathrm{NUMBER}^2\,(X, \iota x[{}^*\mathrm{moon}(x) \wedge R(x,\mathrm{Mars})])] \equiv \underline{\lambda x.\mathrm{two}(x)}$.

Given (1d*), (1d) is true iff the first-level cardinality concept denoted by 'the number of Mars's moons' is identical to the first-level cardinality concept denoted by '<u>two</u>'.

The linguistic viability of this compositional semantics depends on the viability of three assumptions. First, that 'Mars's moons' denotes the type *e* plurality of all of Mars's moons. Second, that 'the' denotes the higher-order description function of

[54] Variables '**X**, **Y**, …' range unrestrictedly over the elements of $D_{\langle\langle e,t\rangle,t\rangle}$.

type $\langle\langle\langle e,t\rangle,t\rangle,\langle e,t\rangle\rangle$. Third and most crucially, that 'number' denotes the numbering relation between type *e* pluralities and type $\langle e,t\rangle$ concepts.

The first assumption can be substantiated as follows. First, 'Mars's moons' clearly *can* have this interpretation. In fact, in Sect. 17.3.1 we have already derived it using only generally accepted linguistic mechanism. Second, there is cross-linguistic evidence that, in the context of *the-number-of* term 'the number of Mars's moons', 'Mars's moons' *has* this interpretation. The German translation of this term is 'die Zahl der Marsmonde' (≈ 'the number the Martian moons'). Here 'Zahl' ('number') is followed by a plural definite description in the genitive case. Such descriptions are readily interpreted as (plural) type *e* expressions. Thus, the first assumption is linguistically viable.[55]

The second assumption may seem somewhat unorthodox. However, note that the denotations of both first-order and higher-order 'the'—specified in (12) and (17), respectively—instantiate the same generalised type $\langle\langle \mathrm{x},t\rangle, \mathrm{x}\rangle$, with respectively $\chi = e$ and $\chi = \langle e,t\rangle$: in both cases the description function takes a (singular) type $\langle \chi, t\rangle$ concept to the unique type χ entity that it subsumes. Thus, assigning these two different yet systematically related denotations to 'the' is but another example of expressions that are instances of generalized types such as, most prominently, *generalized conjunction*.[56] Thus, the second assumption is linguistically viable as well.

The case for the viability of the third assumption is somewhat involved and will have to await completion until the next section. Here, I lay the groundwork. According to the third assumption, 'number' in 'the number of Mars's moons' is a relational noun of type $\langle e,\langle\langle e,t\rangle,t\rangle\rangle$ that denotes the numbering relation between pluralities and the first-level (cardinality) concepts that subsume them. This relation can be defined by way of (19), where ≈ is the relation of *equinumerousity* between pluralities:

(19) $\lambda x.\lambda X.\ \mathrm{NUMBER}^2\ (X,x) =_{\mathrm{df.}} \lambda x.\lambda X.\ (\forall y((X(x) \wedge X(y)) \leftrightarrow x \approx y)).$

According to (19), a first-level concept X stands in the numbering relation to a plurality x iff for every plurality y, X subsumes both x and y iff x and y are equi-numerous.[57]

Is this interpretation of 'number' in *the number-of* terms linguistically viable? In order to answer this question affirmatively, much would be gained if this

[55] Although Moltmann's analysis of *the-number-of* terms (2013, p. 509f) differs from mine, her analysis would interpret 'Mars's moons' in the same way. Snyder's analysis of *the-number-of* terms (2017, p. 391f), on the other hand, would interpret 'Mars's moons' in 'the number of Mars's moons' as a type $\langle e,t\rangle$ expression.

[56] See (Partee, 1983).

[57] Equinumerousity between pluralities can be understood as one-to-one correspondence, which in turn can be defined in parallel to Frege's own definition (1884, pp. 83–86) of one-to-one correspondence between first-level concepts: $x \approx y \leftrightarrow \exists R \forall z_1((z_1 \sqsubseteq_{\mathrm{At}} x \rightarrow \exists! z_2(z_2 \sqsubseteq_{\mathrm{At}} y \wedge R(z_1,z_2))) \wedge (z_1 \sqsubseteq_{\mathrm{At}} y \rightarrow \exists! z_2(z_2 \sqsubseteq_{\mathrm{At}} x \wedge R(z_2,z_1))))$.

interpretation could be independently motivated. As we will see in the next section, there is indeed such a motivation, albeit an indirect one.

17.4.2 *Number as a Higher-Order Concept*

The second challenge for the viability of my proposal consisted in identifying a use of 'number' in which it functions as a type $\langle\langle e,t\rangle,t\rangle$ expression that denotes the second-level concept of being a first-level cardinality concept. This section is dedicated to meeting this challenge. As we will see presently, doing so also completes the case for the relational higher-order interpretation of 'number' that underlies my analysis of *the number-of* terms as higher-order descriptions.

17.4.2.1 Preparation

Once more, let us begin with some preparation. Consider the two occurrences of 'mother' in:

(20) a. the mother of Philip's children.
 b. Elizabeth is a mother.

As previously indicated, 'mother' in (20a) is a relational noun of type $\langle e,\langle e,t\rangle\rangle$ that denotes the distinctive motherhood relation that female parents bear to their children. However, in (20b) 'mother' functions as a non-relational noun of type $\langle e,t\rangle$, where it denotes the concept of being a mother (the superscripted '1' indicates that the expression is non-relational):

(21) $[\![\textbf{mother}^1]\!] = \lambda x.\ \text{mother}^1(x)$.

Clearly, these two uses of 'mother' are not unrelated. But what, exactly, is their relation? An answer to this question is suggested by the fact that the concept of being a mother is intrinsically relational. That is, being subsumed by that concept requires that there be some (plural) object that one stands in the motherhood relation to. By exploiting this fact, the non-relational denotation of 'mother' can be derived from its relational denotation by applying the type-shifting operation EX,[58] defined in (22a), to the denotation of relational 'mother', given in (8). Doing so yields the type $\langle e,t\rangle$ denotation for non-relational 'mother' specified in (22b):

(22) a. $\text{EX} = \lambda R.\lambda x.\ \exists y\ R(x,y)$.
 b. $\text{EX}([\![\textbf{mother}^2]\!]) = \lambda x.\ \exists y\ \text{mother}^2(x,y) = \lambda x.\ \text{mother}^1(x)$.

[58] See (Barker, 2011, p. 1116).

17.4.2.2 Execution

The relation between corresponding pairs of relational and non-relational nouns holds the key to completing the case for the viability of my proposal. To see why, consider the two occurrences of 'number' in (8) and (1e), now relabelled as (23a,b):

(23) a. the number of Mars's moons.
b. Two is an even number.

My proposal rests on two central tenets: (i) that 'number' in (23a) is a relation noun of type $\langle e,\langle\langle e,t\rangle,t\rangle\rangle$ that denotes the numbering relation between first-level cardinality concepts and pluralities, and (ii) that 'number' in (23b) is a non-relation noun of type $\langle\langle e,t\rangle,t\rangle$ that denotes the second-level concept of being a first-level cardinality concept:

(24) $[\![\textbf{number}^1]\!] = \lambda X.\ \text{NUMBER}^1(X)$.

Assuming this is correct, how—if at all—are these two uses of 'number' related? I contend that non-relational 'number' in (23b) stands to relational 'number' in (23a) as non-relational 'mother' in (20b) stands to relational 'mother' in (20a). That is, I contend that the second-level concept of being a first-level cardinality concept is intrinsically relational and that therefore a first-level concept being subsumed by this second-level concept requires that there be some (plural) object that it stands in the numbering relation to.[59] If so, the non-relational denotation of 'number' can be derived from its relational denotation by applying a version of the aforementioned typeshifting operation EX—viz. EX^+, defined in (25a)—that delivers the type $\langle\langle e,t\rangle,t\rangle$ denotation for non-relational 'number' specified in (25b):[60]

(25) a. $\text{EX}^+ = \lambda \mathbf{R}.\ \lambda X.\ \exists x\ \mathbf{R}(X,x)$.
b. $\text{EX}^+([\![\textbf{number}^2]\!]) = \lambda X.\ \exists x\ \text{NUMBER}^2(X,x) = \lambda X.\ \text{NUMBER}^1(X)$.[61]

---

[59] This contention is not uncontroversial. In fact, (Moltmann, 2013, p. 507) rejects it, at least implicitly. Evidently, I disagree.

[60] (25b) is reminiscent of Frege's definition of non-relational 'number', see e.g. (Zalta, 2022, §3.2): $[\![\textbf{number}]\!] = \lambda x.\ \exists F(x = Nx : F(x)))$, where '$Nx : \Psi(x)$' is the cardinality operator that neo-Fregeans (attempt to) implicitly define by way of stipulating Hume's Principle.

[61] In passing, let me flag a potential problem. According to (25b), numbers—conceived of as first-level cardinality concepts—are those first-level concepts that stand in the numbering relation to some (plural) object. Thus, my proposal would only treat the concept of being zero as a cardinality concept if one countenanced the existence of an 'empty plurality'; for a similar point see (Linnebo, 2018, p. 67). Doing so may seem *ad hoc*. However, as (Bylinina & Nouwen, 2018) have recently argued on independent linguistic grounds, even a satisfying account of type $\langle e,t\rangle$ 'zero' outside of arithmetical contexts requires that an empty plurality be accepted. Thus, treating the concept of being zero as a cardinality concept could by justified by appeal to their account

Building on this conjectured relation between relational and non-relational 'number', my argument for the third assumption regarding the type and denotation of (relational) 'number' in *the-number-of* terms now is the following. There is an attested use of 'number' in which it functions as a non-relational noun with the type and denotation that it would have if derived from the proposed interpretation of relational 'number' by way of EX^+. Thus, it stands to reason that there should be a use of relational 'number' whence this non-relational use derives. And this use, I contend, is precisely the use of 'number' in *the-number-of* terms. My argument for the existence of this use of non-relational 'number' is the following.

First, the inference from (1c)—relabelled as (26a)—to (26b) appears to be logically valid:

(26) a. Mars has two moons.
     b. Mars has an even number of moons.

Second, the validity of this inference is best accounted for by analysing (26a) and (26b) in such a way that their logical forms can be represented as follows:

(26) a*. $\exists x(\text{two}(x) \wedge {}^*\text{moon}(x) \wedge \text{has}\,(x,\text{Mars}))$.
     b*. $\exists X(\text{EVEN}(X) \wedge \text{NUMBER}^1(X) \wedge \exists x(X(x) \wedge {}^*\text{moon}(x) \wedge \text{has}\,(x,\text{Mars})))$.

Given (26a*), (26a) is true iff there is at least one plurality having exactly two atomic parts, each of which is a moon belonging to Mars. And given (26b*), (26b) is true iff there is at least one even first-level cardinality concept such that there is at least one plurality that this concept subsumes and each atomic part of which is a moon belonging to Mars. Thus, according to (26b*), (26b) is a restricted *higher-order quantification* over first-level concepts.

Third, I contend that analysing (26a) and (26b) as, respectively, (26a*) and (26b*) is linguistically viabile. If so, (26b) displays the very non-relational use of 'number' that my proposal predicts. In Sect. 17.3.1.3, we have already seen that analysing (26a) as (26a*) is viable. The remainder of this section is therefore dedicated to substantiating the viability of analysing (26b) as (26b*).

As a (higher-order) existential generalisation, (26b) is true iff the (infinite) disjunction of its instances is. That is, (26b) is true iff Mars has two moons or Mars has four moons or Mars has six moons ... or Mars has *n* moons, where '*n*' is an 'even' number word. Any viable compositional analysis needs to respect these truth-conditions. How can these truth-conditions be derived compositionally?

To start things off, note (i) that (26b) has been obtained from (26a) by replacing the number word 'two' with the complex phrase 'an even number (of)', and (ii) that the result of so replacing the number word is an interpretable sentence.[62] Thus, the question of how to analyse (26b) boils down to the question of how to interpret 'an even number (of)' in a way that (i) renders (26b) interpretable, and (ii) accords (26b) the correct truth-conditions.

[62] That 'a number of' can function as a replacement of a number word has also been observed by (Moltmann, 2013, p. 502).

In general, replacing an expression E of some type τ in a given sentence S with an expression E' will yield an interpretable sentence S' if E' is either itself of type τ or else if E' is of the corresponding 'quantificational type' $\langle\langle\tau,t\rangle,t\rangle$; at least if the composition of S' mirrors that of S. For instance, in 'Plato sits', 'Plato' is a type e expression and 'sits' is a type $\langle e,t\rangle$ expression. Thus, their denotations can compose by FA. Consequently, replacing 'Plato' with some expression E' will only yield an interpretable sentence if the denotation of E' can also compose with that of 'sits'. But this means that the candidates for the E'-role are either themselves type e expressions (e.g. 'Socrates') or else expressions of the corresponding quantificational type $\langle\langle e,t\rangle,t\rangle$ (e.g. 'some philosopher'). Since (26b) is the result of replacing the type $\langle e,t\rangle$ number word 'two' in (26a) with the phrase 'an even number (of)', this suggests that (26b) is interpretable if 'an even number (of)' is interpreted either as a type $\langle e,t\rangle$ expression or as an expression of type $\langle\langle\langle e,t\rangle,t\rangle,t\rangle$.

Interpreting 'an even number (of)' as a type $\langle e,t\rangle$ expression does not seem feasible. The phrase 'an even number (of)' is semantically complex, i.e. its denotation is derived from the denotations of its constituents. For instance, the denotation of 'even' makes a discernible contribution to that of the whole phrase, a contribution that is different from the contribution of the denotation of e.g. 'odd' in 'an odd number (of)'. Thus, one would be hard-pressed to come up with assignments of types and denotations to 'an' 'even', and 'number', that (i) can compose to yield a type $\langle e,t\rangle$ denotation for the whole phrase which (ii) would furnish (26b) with the correct truth-conditions.

Could 'an even number (of)' be interpreted as a type $\langle\langle\langle e,t\rangle,t\rangle,t\rangle$ expression? This interpretation faces three challenges. First, we need an assignment of types and denotations to its constituents 'an', 'even', and 'number' that allows for a compositional derivation of a type $\langle\langle\langle e,t\rangle,t\rangle,t\rangle$ denotation. Second, we need to show that composing this denotation with the 'rest' of (26b) yields a sentence with the correct truth-conditions.

As for the first challenge, I propose that 'an' is an expression of type $\langle\langle\langle e,t\rangle,t\rangle, \langle\langle\langle e,t\rangle,t\rangle,t\rangle,t\rangle\rangle$ and that 'even' and 'number' are type $\langle\langle e,t\rangle,t\rangle$ expressions. Moreover, I propose the following denotations:

(27) a. $[\![\mathbf{an}]\!] = \lambda\mathbf{X}.\lambda\mathbf{Y}.\ \exists Z(\mathbf{X}(Z) \wedge \mathbf{Y}(Z))$
b. $[\![\mathbf{even}]\!] = \lambda X.\ \mathrm{EVEN}(X)$.
c. $[\![\mathbf{number}^1]\!] = \lambda X.\ \mathrm{NUMBER}^1(X)$.

The proposed interpretation of 'an' is a higher-order version of its more familiar first-order interpretation in, for instance, 'An arithmetician wrote *Begriffsschrift*', where it denotes $\lambda X.\lambda Y.\ \exists z(X(z) \wedge Y(z))$. Thus, both these interpretations are but instances of a generalized denotation.[63] The denotation of 'number[1]' as been previously defined. The denotation of 'even' can be defined as follows. A plurality x is even just in case (i) x is finite and (ii) if x has any sub-pluralities, then x has disjoint

[63] That is, $[\![\mathbf{an}]\!]^{\text{gen.}} = \lambda V_1 \in D_{\langle\chi,t\rangle}.\ \lambda V_2 \in D_{\langle\chi,t\rangle}.\ \exists\, v \in D_\chi(V_1(v) \wedge V_2(v))$.

sub-pluralities y_1 and y_2 that jointly exhaust x such that y_1 and y_2 are equinumerous. A concept X, then, is even iff every plurality x that X subsumes is even:[64]

(28) a. $\lambda x.\ \text{even}(x) =_{\text{df.}} \lambda x.\ (\text{finite}(x) \wedge \forall y_1\ \forall y_2((y_1 \sqsubset x \wedge y_2 \sqsubset x \wedge \forall z(z \sqsubset x \leftrightarrow (z \sqsubset y_1 \leftrightarrow z \not\sqsubset y_2)) \rightarrow y_1 \approx y_2)))$.

b. $\lambda X.\ \text{EVEN}(X) =_{\text{df.}} \lambda X.\ \forall y(X(y) \rightarrow \text{even}(x))$.

As for the composition of the denotation of 'an even number (of)' from the denotations of its constituents, I propose that the denotations of 'even' and 'number' compose by a rule of *Higher-Order Predicate Modification*.[65] The resulting denotation of 'even number', specified in (29a), then composes with the denotation of 'an', given in (27a), by FA to yield the denotation for 'an even number' specified in (29b):

(29) a. $[\![\textbf{even number}^1]\!] = \lambda X.\ \text{EVEN}(X) \wedge \text{NUMBER}^1(X)$.

b. $[\![\textbf{an even number}^1]\!] = \lambda \mathbf{Y}.\ \exists X(\text{EVEN}(X) \wedge \text{NUMBER}^1(X) \wedge \mathbf{Y}(X))$.

Of course, as a type $\langle\langle\langle e,t\rangle,t\rangle,t\rangle$ expression, the denotation of 'an even number (of)' cannot compose with the type $\langle e,t\rangle$ denotation of 'moons' by any of our composition rules. Hence, 'an even number (of)' cannot not be interpreted *in situ*. However, we have already encountered a mechanism—viz. Quantifier Raising—to deal with this kind of type mismatch. Application of QR results in the following syntactico-semantic structure of (26b) which vindicates representing its logical form as (26b*):[66]

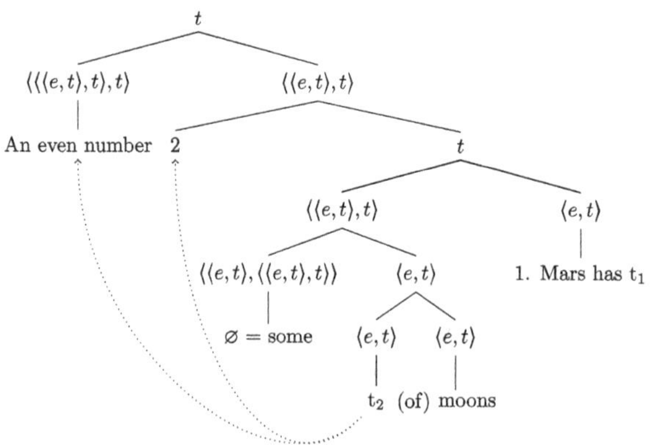

Fig. 17.2 Syntactic tree for (26b) "Mars has an even number of moons" displaying its semantic composition

[64] (Moltmann, 2013, p. 514) makes a similar suggestion.

[65] If α is a branching node, $\{\beta,\gamma\}$ is the set of α's daughters, and $[\![\beta]\!]$ and $[\![\gamma]\!]$ are in $D_{\langle\langle e,t\rangle,t\rangle}$, then $[\![\alpha]\!] = \lambda X.\ [\![\beta]\!](X) = \text{T} \wedge [\![\gamma]\!](X) = \text{T}$.

[66] Unlike in, for instance, 'the mother of Philip's children', 'of'-insertion is a late phenomenon that is needed to satisfy grammatical surface constraints but does not project a prepositional phrase; see (Chomsky, 1981).

What remains to be seen is whether the denotation of 'an even number' can compose with the denotation of the rest of (26b) in a way that delivers the correct truth-conditions. Unsurprisingly, it can. The rest of (26b) is the lambda-phrase given in (30a) which has the denotation specified in (30b):

(30) a. 2. [[∅[[t_2][moons]]][1. [[Mars][[has][t_1]]]]].
b. $\lambda X.\ \exists x(X(x) \wedge {}^*\text{moon}(x) \wedge \text{has}(\text{Mars}, x))$.

Composing (30b) with the denotation of 'an even number', given in (29b), thus yields a sentence that is true just in case the representation of (26b)'s logical form—viz. (26b*) '$\exists X(\text{EVEN}(X) \wedge \text{NUMBER}^1(X) \wedge \exists x(X(x) \wedge {}^*\text{moon}(x) \wedge \text{has}(x, \text{Mars})))$'—is true. Since the only values of 'X' that can render (26b*) true are the even first-level cardinality concepts of being two, being four, being six, ..., being n (for 'even' 'n'), the proposed interpretation of 'an even number' delivers the correct truth-conditions for (26b).

The case in favour of my proposal is now complete. Before moving on, we can finally give a rigorous answer the Frege's question of how else (1e) 'Two is an even number' is to be construed if not as expressing that an object is subsumed under a first-level concept. Given the results of this section, the logical form of (1e) is represented by (1e*), which construes (1e) as a higher-order predication, i.e. as expressing that a first-level concept is subsumed by a (complex) second-level concept:

(1) e.* $\text{EVEN}\ \underline{(\lambda x.\text{two}(x))} \wedge \text{NUMBER}^1\ \underline{(\lambda x.\text{two}(x))}$.

Given (1e*), (1e) is true iff the first-level concept of being two is an even cardinality concept.

17.5 The Easy Argument Revisited

According to my proposal, number words in their predicative, attributive, quantificational, specificational, and numeral use function semantically uniformly as type $\langle e,t \rangle$ expressions that denote first-level cardinality concepts. Thus far, we have seen that this proposal is linguistically viable. In bringing this paper to a close, let us revisit the easy argument for the existence of numbers (conceived of as (abstract) objects) in light of this proposal:

(31) Mars has exactly *two* moons.[67]
∴ (32) The number of Mars's moons is <u>*two*</u>.

[67] The premiss is (31) 'Mars has exactly two moons' instead of (1c) 'Mars has two moons' because (1c) is commonly taken to only possess a lower-bounded (at least) reading, whilst the additional communication of an upper-bounded (at most) reading is a matter of Gricean implicature; see (Horn, 1972). However, as pointed out by (Balcerak, 2013, p. 459), the truth of (32) 'The number of Mars's moons is two' requires that Mars have exactly—i.e. both at least and at most—two moons. If so, the truth of (1c) but not that of (31) would be compatible with the falsity of (32).

∴ (33) Something is the number of Mars's moons, viz. *two*.
∴ (34) Something is a number.

The argument starts from the uncontroversially true premiss (31), a premiss whose truth does, due to the absence of type *e* expressions for numbers, not seem to require the existence of numbers (conceived of as (abstract) objects). From there (32)—a statement that appears to contain a number-denoting type *e* expression and whose truth does therefore seem to require the existence of numbers—is inferred by way of an apparently trivial inference. Next, the position of the number-denoting type *e* expression in (32)—viz. that of the nominalized number word—is existentially generalized upon to yield (33). Finally, the conclusion (34)—according to which numbers exist—is inferred by appeal to a suitable definition of 'number'.

According to easy ontologist like Schiffer and Thomasson, the apparently trivial inference from (31) to (32) is merely analytic in nature. That is, it is backed by the allegedly analytic truth of the conditional 'If Mars has exactly two moons, then the number of Mars's moons is two', whose analyticity is said to flow from the rules that govern the use of the number-denoting type *e* expressions in (32). Moreover, easy ontologists contend that the ease with which such arguments establish an ontological conclusion, suggests taking an ontologically deflationary stance towards the kinds of objects—in the case at hand, numbers conceived of as (abstract) objects—whose existence they establish.

Easy arguments such as the above are not without distractors. One line of criticism rejects the validity of the inference from (31) to (32) on the grounds of their divergent existence requirements.[68] Another line of criticism accepts that the argument's inferences are valid but denies that its conclusion is ontologically significant in even a deflationary sense. This is because, so the criticism goes, (i) the truth of (32) requires the existence of numbers (conceived of as abstract objects) no more than does the truth of (31), and (ii) the inferences from (32) to (33) and from (33) to (34) are only valid if the existential quantifier is read quasi-substitutionally.[69]

According to the proposal defended today, both proponents and detractors of the easy argument are mistaken. *Pace* the latter, the argument is valid even if the quantifier is read non-substitutionally. This is because, given my proposal, the argument has the following logical form, which is valid in higher-order logic:[70]

---

[68] See e.g. (Yablo, 2000).

[69] See (Hofweber, 2005b, 2007).

[70] The uniqueness quantifier '$\exists!$' equips (31*) with the exactly-reading of (31). The inference from (33*) to (34*) requires some intermediary steps. First, (33*) entails:

(33a*) '$\exists Y\, \text{NUMBER}^2\left(\underline{Y}, \iota y\left[{*}\text{moon}(y) \wedge R(y,\text{Mars})\right]\right)$'.

Second, (33a*) entails:

(33b*) '$\exists Y \exists y\, \text{NUMBER}^2(\underline{Y}, y)$'.

Third, by appeal to the definition of 'NUMBER¹' in terms of 'NUMBER²', given in (25b), (33b*) entails (34*).

(31*) $\exists!x(\text{two}(x) \wedge {}^{*}\text{moon}(x) \wedge \text{has}(x, \text{Mars}))$.

∴ (32*) $\iota X[\text{NUMBER}^2\,(X, \iota y[{}^{*}\text{moon}(y) \wedge R(y, \text{Mars})])] \equiv \underline{\lambda x.\ \text{two}(x)}$.

∴ (33*) $\exists Y\,\iota X[\text{NUMBER}^2\,(X, \iota y[{}^{*}\text{moon}(y) \wedge R(y, \text{Mars})])] \equiv \underline{Y}$.

∴ (34*) $\exists Y\,\text{NUMBER}^1\,(\underline{Y})$.

Thus, the easy argument does establish an ontologically significant conclusion. However, *pace* the easy ontologists, the argument does not establish the existence of numbers conceived of as (abstract) objects but only the existence of numbers conceived of as higher-order entities, viz. first-level (cardinality) concepts. Of course, the ease with which the argument establishes their existence may still be taken to recommend a deflationary stance towards such higher-order entities. If so, easy ontologists might well not be altogether unhappy with the outcome.

17.6 Conclusion

Number words in natural language can be used in a variety of different ways. In some of these uses they function non-referentially, i.e. not as expressions of type *e* but as expressions of some higher-semantic type. However, in others uses—on display, in particular, in arithmetical contexts—number words appear to function referentially. If so, we should countenance a conception of numbers as (abstract) objects. In this paper, I argued that number words in all of the uses discussed today function semantically uniformly as type $\langle e,t \rangle$ expressions that denote first-level cardinality concepts. On this view, we should only countenance a conception of numbers as concepts. Of course, I do not expect my proposal to be uncontroversial. Which philosophical view ever is? I therefore had very much hoped to defend this proposal in my future work.[71] Alas, by now I have left academic philosophy. And I don't expect I shall return. In fact, I mean not to. Thus, let me answer the question of whether *I* ever find any difficulty in my view that numbers are concepts by echoing Frege's response to Wittgenstein: Sometimes I *seem* to see a difficulty—but then again I *don't* see it.

Funding This work was supported by the Deutsche Forschungsgemeinschaft [WO 1896/1-1 to Richard Woodward]; and the Research Council of Norway [250654 to Herman Cappelen and Øystein Linnebo].

[71] I imagine that, in particular, Moltmann and Snyder would have liked to have a word.

References

Anscombe, E., & Geach, P. (1961). *Three philosophers: Aristotle, Aquinas, Frege*. Cornell University Press.

Barker, C. (1998). Partitives, double genitives and anti-uniqueness. *Natural Language & Linguistic Theory, 16*(4), 679–717.

Barker, C. (2011). Possessives and relational nouns. In K. von Heusinger, C. Maienborn, & P. Portner (Eds.), *Semantics: An international handbook of natural language meaning* (Vol. 2, pp. 1009–1030). de Gruyter. https://www.degruyter.com/database/IBR/entry/ibr.ID_REZ-111732_ISBN9783110185232/html

Bylinina, L., & Nouwen, R. (2018). On "zero" and semantic plurality. *Glossa: A Journal of General Linguistics, 3*(1). https://doi.org/10.5334/gjgl.441

Chierchia, G. (1984). *Topics in the Syntax and Semantics of Infinitives and Gerunds* [Ph.D. thesis, Chierchia]. Topics in the syntax and semantics of infinitives and gerunds. University of Massachusetts.

Chomsky, N. (1981). *Lectures on government and binding*. Floris.

Felka, K. (2014). Number words and reference to numbers. *Philosophical Studies, 168*(1), 261–282. https://doi.org/10.1007/s11098-013-0129-3

Felka, K. (2016). *Talking about numbers: Easy arguments for mathematical realism*. Vittorio Klostermann.

Florio, S., & Linnebo, Ø. (2021). *The many and the one: A philosophical study of plural logic*. Oxford University Press.

Frege, G. (1884). *Die Grundlagen Der Arithmetik*. Verlag W. Koebner.

Frege, G. (1979). *Posthumous writings* (P. Long, R White, & R. Hargreaves, Trans.). Blackwell.

Hale, B., & Linnebo, Ø. (2020). Ontological categories and the problem of Expressibility. In *Essence and existence*. Oxford University Press.

Heim, I., & Kratzer, A. (1998). *Semantics in generative grammar*. Blackwell.

Higgins, F. R. (1973). *The pseudo-cleft construction in English* [PhD Thesis]. Massachusetts Institute of Technology. https://dspace.mit.edu/handle/1721.1/12988

Hodes, H. T. (1984). Logicism and the ontological commitments of arithmetic. *Journal of Philosophy, 81*(3), 123–149. https://doi.org/10.2307/2026440

Hofweber, T. (2005a). Number determiners, numbers, and arithmetic. *Philosophical Review, 114*(2), 179–225. https://doi.org/10.1215/00318108-114-2-179

Hofweber, T. (2005b). A puzzle about ontology. *Noûs, 39*(2), 256–283. https://doi.org/10.1111/j.0029-4624.2005.00501.x

Hofweber, T. (2007). Innocent statements and their metaphysically loaded counterparts. *Philosophers' Imprint, 7*, 1–33.

Hofweber, T. (2016). *Ontology and the ambitions of metaphysics*. Oxford University Press. https://doi.org/10.1093/acprof:oso/9780198769835.001.0001

Hofweber, T. (2023). Refocusing Frege's other puzzle: A response to Snyder, Samuels, and Shapiro. *Philosophia Mathematica, 31*(2), 216–235. https://doi.org/10.1093/philmat/nkad005

Horn, L. R. (1972). *On the semantic properties of logical operators* [Ph.D. thesis]. University of California.

Ionin, T., & Matushansky, O. (2006). The composition of complex cardinals. *Journal of Semantics, 23*(4), 315–360. https://doi.org/10.1093/jos/ffl006

Balcerak Jackson, B. (2013). Defusing easy arguments for numbers. *Linguistics and Philosophy, 36*(6), 447–461. https://doi.org/10.1007/s10988-013-9142-4

Kennedy, C. (2013). A scalar semantics for scalar readings of number words. In C. Cecchetto & I. Caponigro (Eds.), *From grammar to meaning: The spontaneous logicality of language* (pp. 172–200). Cambridge University Press. https://doi.org/10.1017/CBO9781139519328.010

Knowles, R. (2015). What "the number of planets is eight" means. *Philosophical Studies, 172*(10), 2757–2775. https://doi.org/10.1007/s11098-015-0443-z

Krämer, S. (2014). Semantic values in higher-order semantics. *Philosophical Studies, 168*(3), 709–724. https://doi.org/10.1007/s11098-013-0157-z

Krifka, Manfred. 1989. 'Nominal reference, temporal constitution and quantification in event semantics'. In Semantics and contextual expression, Renate Bartsch, Johan van Benthem, and P van Emde Boas, 75–116. De Gruyter. https://www.degruyter.com/document/doi/10.1515/9783110877335/html

Krifka, M. (1999). At least some determiners aren't determiners. In K. Turner (Ed.), *The semantics/pragmatics Interface from different points of view* (Vol. 1, pp. 257–291). Elsevier.

Landman, F. (2004). *Indefinites and the type of sets*. Blackwell.

Link, G. (1983). The logical analysis of plurals and mass terms: A lattice-theoretical approach. In R. Bäuerle, C. Schwarze, & A. von Stechow (Eds.), *Meaning, use, and interpretation of language* (pp. 303–323). de Gruyter. https://doi.org/10.1515/9783110852820

Linnebo, Ø. (2018). *Thin objects: An abstractionist account*. Oxford University Press.

Mikkelsen, Line. 2011. 'Copular Clauses'. In Semantics: An international handbook of natural language meaning, edited by Klaus von Heusinger, Claudia Maienborn, and Paul Portner, 2:1805–29. de Gruyter. https://www.degruyter.com/database/IBR/entry/ibr.ID_REZ-111732_ISBN9783110185232/html

Moltmann, F. (2013). Reference to numbers in natural language. *Philosophical Studies, 162*(3), 499–536. https://doi.org/10.1007/s11098-011-9779-1

Partee, B. (1983). Generalized conjunction and type ambiguity. In R. Bäuerle, C. Schwarze, & A. von Stechow (Eds.), *Meaning, use, and interpretation of language* (pp. 361–383). de Gruyter. https://doi.org/10.1515/9783110852820

Partee, B. (1986). Noun phrase interpretation and type-shifting principles. In J. Groenendijk, D. de Jongh, & M. Stokhof (Eds.), *Studies in discourse representation theory and the theory of generalized quantifiers* (pp. 115–143). Floris. https://doi.org/10.1515/9783112420027

Pietroski, Paul M. 2005. *Events and semantic architecture* (1. publ). Oxford Linguistics. Oxford University Press.

Romero, M. (2005). Concealed questions and specificational subjects. *Linguistics and Philosophy, 28*(6), 687–737.

Rothstein, S. (2017). *Semantics for counting and measuring. Key topics in semantics and pragmatics*. Cambridge University Press. https://doi.org/10.1017/9780511734830

Schiffer, S. (2003). *The things we mean*. Oxford University Press. https://doi.org/10.1093/0199257760.001.0001

Schlenker, P. (2003). Clausal equations (A note on the connectivity problem). *Natural Language & Linguistic Theory, 21*(1), 157–214.

Schwartzkopff, R. (2016). Number sentences and Specificational sentences: Reply to Moltmann. *Philosophical Studies, 173*(8), 2173–2192. https://doi.org/10.1007/s11098-015-0603-1

Schwartzkopff, R. (2022). How not to analyse number sentences. *Philosophia Mathematica, 30*(2), 200–222. https://doi.org/10.1093/philmat/nkac%20005

Sharvit, Y. (1999). Connectivity in Specificational sentences. *Natural Language Semantics, 7*(3), 299–339.

Sharvy, R. (1980). A more general theory of definite descriptions. *The Philosophical Review, 89*(4), 607–624. https://doi.org/10.2307/2184738

Skiba, L. (2021). Higher-order metaphysics. *Philosophy Compass, 16*(10). https://doi.org/10.1111/phc3.12756

Snyder, E. (2017). Numbers and cardinalities: What's really wrong with the easy argument for numbers? *Linguistics and Philosophy, 40*(4), 373–400. https://doi.org/10.1007/s10988-017-9215-x

Snyder, E. (2021). *Semantics and the ontology of number*. Cambridge University Press. https://doi.org/10.1017/9781108602259

Snyder, E., Samuels, R., & Shapiro, S. (2022). Resolving Frege's other puzzle. *Philosophia Mathematica, 30*(1), 59–87. https://doi.org/10.1093/philmat/nkab032

Strawson, P. F. (1950). On Referring. *Mind, 59*(235), 320–344. https://doi.org/10.1093/mind/LIX.235.320

Tennant, N. (1997). On the necessary existence of numbers. *Noûs, 31*(3), 307–336.
Thomasson, A. L. (2015). *Ontology made easy*. Oxford University Press. https://doi.org/10.1093/acprof:oso/9780199385119.001.0001
Wright, C. (2007). On quantifying into predicate position. In M. Leng, A. Paseau, & M. Potter (Eds.), *Mathematical knowledge* (pp. 150–174). Oxford University Press.
Yablo, S. (2000). Apriority and existence. In P. Boghossian & C. Peacocke (Eds.), *New essays on the a priori* (pp. 197–228). Oxford University Press. https://doi.org/10.1093/0199241279.001.0001
Yi, B.-u. (2018). Numerical cognition and mathematical knowledge: The plural property view. In S. Bangu (Ed.), *Naturalizing logico-mathematical knowledge* (pp. 52–85). Routledge.
Zalta, E. N. (2022). Frege's Theorem and Foundations for Arithmetic. In E. N. Zalta (Ed.), *The Stanford encyclopedia of philosophy*. Spring 2022. Stanford University: Metaphysics Research Lab. https://plato.stanford.edu/archives/spr2022/entries/frege-theorem/

Index

X. de Donato-Rodríguez et al. (eds.), *Deflationist Conceptions of Abstract Objects*, Synthese Library 516, https://doi.org/10.1007/978-3-031-98717-5

GPSR Compliance
The European Union's (EU) General Product Safety Regulation (GPSR) is a set of rules that requires consumer products to be safe and our obligations to ensure this.

If you have any concerns about our products, you can contact us on

ProductSafety@springernature.com

In case Publisher is established outside the EU, the EU authorized representative is:

Springer Nature Customer Service Center GmbH
Europaplatz 3
69115 Heidelberg, Germany

www.ingramcontent.com/pod-product-compliance
Lightning Source LLC
Chambersburg PA
CBHW070352091225
36514CB00003B/105
* 9 7 8 3 0 3 1 9 8 7 1 6 8 *